Normal Families and Normal Functions

This book centers on normal families of holomorphic and meromorphic functions and also normal functions. The authors treat one complex variable, several complex variables, and infinitely many complex variables (i.e., Hilbert space).

The theory of normal families is more than 100 years old. It has played a seminal role in the function theory of complex variables. It was used in the first rigorous proof of the Riemann mapping theorem. It is used to study automorphism groups of domains, geometric analysis, and partial differential equations.

The theory of normal families led to the idea, in 1957, of normal functions as developed by Lehto and Virtanen. This is the natural class of functions for treating the Lindelof principle. The latter is a key idea in the boundary behavior of holomorphic functions.

This book treats normal families, normal functions, the Lindelof principle, and other related ideas. Both the analytic and the geometric approaches to the subject area are offered. The authors include many incisive examples.

The book could be used as the text for a graduate research seminar. It would also be useful reading for established researchers and for budding complex analysts.

Peter V. Dovbush Dr. habil., Associate Professor, in Moldova State University, Institute of Mathematics and Computer Science. He received his Ph.D. in Lomonosov Moscow State University in 1983 and Doctor of Sciences in 2003. He has published over 50 scholarly articles.

Steven G. Krantz is a Professor of Mathematics at Washington University in St. Louis. He has previously taught at UCLA, Princeton University, and Penn State University. He received his Ph.D. from Princeton University in 1974. Dr. Krantz has directed 20 Ph.D. students and 8 Master's students.

He has published over 130 books and over 300 scholarly articles. Dr. Krantz is the holder of the Chauvenet Prize, the Beckenbach Book Award, and the Kemper Prize. He is a Fellow of the American Mathematical Society.

Normal Families and Normal Functions

Peter V. Dovbush and Steven G. Krantz

CRC Press
Taylor & Francis Group
Boca Raton London New York

CRC Press is an imprint of the
Taylor & Francis Group, an **informa** business
A CHAPMAN & HALL BOOK

Designed cover image from: *Visual Complex Functions* by Elias Wegert, Institute of Applied Analysis, TU Bergakademie Freiberg

First edition published 2024
by CRC Press
2385 Executive Center Drive, Suite 320, Boca Raton, FL 33431

and by CRC Press
4 Park Square, Milton Park, Abingdon, Oxon, OX14 4RN

CRC Press is an imprint of Taylor & Francis Group, LLC

Library of Congress Cataloging-in-Publication Data

Names: Dovbush, Peter V., author. | Krantz, Steven G. (Steven George), 1951- author.
Title: Normal families and normal functions / Peter V. Dovbush and Steven G Krantz.
Description: First edition. | Boca Raton : CRC Press, 2024. | Includes bibliographical references and index.
Identifiers: LCCN 2023040379 | ISBN 9781032666365 (hardback) | ISBN 9781032669878 (paperback) | ISBN 9781032669861 (ebook)
Subjects: LCSH: Holomorphic functions. | Functions, Meromorphic
Classification: LCC QA331 .D695 2024 | DDC 515/.98--dc23/eng/20231127
LC record available at https://lccn.loc.gov/2023040379

ISBN: 978-1-032-66636-5 (hbk)
ISBN: 978-1-032-66987-8 (pbk)
ISBN: 978-1-032-66986-1 (ebk)

DOI: 10.1201/9781032669861

Typeset in Latin Modern Roman
by KnowledgeWorks Global Ltd.

Publisher's note: This book has been prepared from camera-ready copy provided by the authors.

Contents

Preface

Normal families are a powerful tool in function theory, differential equations, geometry, and many other parts of mathematics.

The idea of normal family of holomorphic functions is usually associated with the name Montel. Although Montel was the first to formulate the phrase "normal family", the concept was in use well before Montel defined it. But the fact is that Montel's work was preceded by related ideas of Weierstrass, Stieltjes, Osgood, and others.

The idea of compactness had emerged as a fundamental concept in analysis during the nineteenth century: provided a set is bounded in $\mathbb{R}^n$, it is possible to define for any sequence $\{x_j\}$ of points of the set a subsequence $\{x_{j_k}\}$ which converges to a point of $\mathbb{R}^n$ (the Bolzano-Weierstrass theorem).

Riemann had sought to extend this extremely useful property to sets $\mathcal{E}$ of functions of real variables, but it soon appeared that boundedness of $\mathcal{E}$ was not sufficient.

Around 1880, Ascoli introduced the additional condition of equicontinuity of $\mathcal{E}$, which implies that $\mathcal{E}$ has again the Bolzano-Weierstrass property.

But at the beginning of the twentieth century, Ascoli's theorem had very few applications, and it was Montel who made it popular by showing how useful it could be for analytic functions of a complex variable.

His fundamental concept is what he called a normal family, which is a set $\mathcal{H}$ of functions defined in a domain $\Omega \subset \mathbb{C}$, taking their values in the Riemann sphere and meromorphic in Ω, and satisfying the following condition: from any sequence of functions of $\mathcal{H}$ it is possible to extract a subsequence that, on every compact subset of Ω, converges uniformly either to a holomorphic function or to the point ∞ of the Riemann sphere.

Most of Montel's mathematical papers are concerned with the theory of analytic functions of one complex variable, a very active field among French mathematicians between 1880 and 1940.

Montel's central observation is that if H consists of uniformly bounded holomorphic functions in Ω, then it is a normal family; this is a consequence of the Cauchy integral formula and of Ascoli's theorem. From this criterion follow many others; for instance, if the values of the functions of a set H belong to a domain Ω that can be mapped conformally onto a bounded domain, then H is a normal family. This is the case in particular when Ω is the complement of a set of two points in the complex plane $\mathbb{C}$.

Montel showed how the introduction of normal families may bring substantial simplifications in the proofs of many classical results of function theory such as the mapping theorem of Riemann and Hadamard's characterization of entire functions of finite order. An ingenious application is to the proof of Picard's theorem on essential singularities: suppose 0 is an essential singularity of a function f holomorphic in $\Omega := \{0 < |z| \leq 1\}$. Then Picard's theorem asserts that $f(z)$ takes on all finite complex values, with one possible exception, as z ranges through Ω. It can be proved by observing that if there are two values that f does not take in Ω, then the family of functions $f_n(z) = f(z/2^n)$ in the annulus $\Gamma : \frac{1}{2} \leq |z| \leq 1$ would be a normal family, and there would be either a subsequence (f_{m_k}) with $|f_{m_k}(z)| \leq M$ in Γ or a subsequence with $|f_{m_k}(z)| \geq 1/M$ in Γ, contradicting the assumption that 0 is an essential singularity of f.

Montel's works solidified this circle of ideas, and he provided the name "normal family". This led to work of Valiron that was later developed by Lohwater-Pommerenke and Zalcman. Montel also proved the very elegant theorem—characterizing normal families—that every mathematics student learns.

Today the theory of normal families has blossomed into a subject area all its own. There are a number of mathematicians (including these authors) who specialize in the study of normal families.

And, in the past 50 years, new ideas have grown out of Montel's concept. The idea of normal functions is one of the best and most fruitful of these. As we show in this text, normal functions (a natural generalization of bounded holomorphic functions) are the natural context in which to study normal families.

The genesis of the idea of normal function began with a paper of Kosaku Yosida in 1934. In 1939, Noshiro introduced the notion of normal function (although the notion of normal family existed at this time, he did not use that name; he says, after Yosida, that such function belongs to class (A)). This name was given to these functions much later in another pioneering paper in this vein—the 1957 paper of Lehto and Virtanen. In that article, Lehto and Virtanen showed that the notion of a normal meromorphic function is closely related to some of the most important problems of the boundary behavior of meromorphic functions. The 1957 paper of Lehto and Virtanen really formalized the concept of normal function and is the standard reference today.

What we try to do in this book is to put the study of normal families and normal functions into a natural, geometric context. Inspired by the ideas of H. H. Wu, we treat normal families and functions on complex manifolds. We describe the relationship of these ideas with invariant metrics.

We hope that this book will inspire other mathematicians to take up the gauntlet of normal families. We look forward to feedback from our readership.

Peter V. Dovbush
Steven G. Krantz

1

Introduction

The history of the concept of uniform convergence is a paradigm in the history of ideas in modern mathematics. One of the earliest results on the uniform convergence of a sequence of holomorphic functions is attributed to Weierstrass: if $f_1, f_2, \ldots$ are holomorphic in a plane region D and if $f_j \to f$ uniformly on D, then f is holomorphic in D. We refer the reader to [245] for a comprehensive survey of, and references to, Weierstrass's work.

The next major advance toward a theory of normal families was perhaps that of Stieltjes who, in 1894 in a paper on continued fractions [273], proved that if a sequence of holomorphic functions is uniformly bounded in a plane region D, and converges uniformly on some nonempty, open subset of D, then it converges, uniformly on each compact subset of D, to a function holomorphic in D. This may have been the first result in which convergence is obtained on a larger region than that covered by the given hypotheses.

Next, in 1901 Osgood showed that, for uniformly bounded sequences, it is sufficient to assume that convergence occurs on a dense subset of D [230] (see also [11]). In 1904 Porter [240] showed that it is sufficient for convergence to occur on a curve in D, and then later, Vitali [282, 283] and, independently, Porter [239], proved, again for uniformly bounded functions, that it is sufficient for the functions to converge on a sequence of points that converges to a point in D (see also [121, 174]). Note that these results imply that a uniformly bounded family of functions, each holomorphic in a plane region D, is (according to the terminology introduced later by Montel) a normal family.

Paul Montel created the theory of normal families of holomorphic (or meromorphic) functions over 115 years ago. Montel's work on normal families began in 1907, but the term "normal family" was first employed in 1911 [197]. Montel's theory of normal families was quite powerful. In the first two decades of the twentieth century Montel applied his theory of normal families to a variety of subjects within complex function theory. Montel's first mature works on normal families are the papers [200, written in 1912], [201, written in 1916] and [195, written in 1917]. The results from these papers are also presented in his book *Familles Normales*, first published in 1927. In both [200, written in 1912] and [201, written in 1916], he treated the Picard theory, and in [195, written in 1917], he proved the Riemann mapping theorem as well as several other theorems related to conformal mapping.

Although Montel was the first to use the phrase "normal family", the concept was in use well before Montel defined it. In an attempt to avoid

DOI: 10.1201/9781032669861-1

anachronistic use of the term, the phrase "Property N" will be used to indicate the use of the concept of normality prior to Montel's first use of the term. The nineteenth-century conception of Property N was a little different from the definition of normality given nowadays since the limit function usually was not allowed to be infinite. Principal among those who studied families with Property N in the nineteenth century were Cesare Ascoli (1843–1896) and Giulio Arzelà (1847–1912). Each of these mathematicians offered early proofs of what is nowadays a central theorem in the study of normal families, the so-called Ascoli-Arzelà Theorem, which is stated below for complex functions of a single variable.

Theorem 1.0.1 (Ascoli-Arzelà). *Let $\mathcal{F}$ be an equicontinuous family of complex functions defined on a domain $D \subset \mathbb{C}$ (i.e. for any given $\varepsilon > 0$ and any compact subset K in D, there exists a positive number δ, depending only on ε and K, such that $|g_j(z) - g(w)| < \varepsilon$ holds for all $z, w \in K$ and all j) with the property that for each $z \in D$ the set $\{f(z) : f \in \mathcal{F}\}$ is bounded. Then $\mathcal{F}$ is a normal family.*

It should be emphasized that Ascoli and Arzelà each proved this theorem only for real functions; in addition, both mathematicians were apparently working under the implicit assumption that the families of functions they considered were uniformly bounded.

At the beginning of the twentieth century Ascoli's theorem had very few applications, and it was Montel who made it popular by showing how useful it could be for analytic functions of a complex variable. Montel's first published version of the Ascoli-Arzelà theorem appeared in his *Comptes Rendus* note [194, written in 1904] where he claimed that uniform boundedness was a sufficient condition for Property N to hold. In [198, written in 1907] Montel also offered proofs of the Ascoli-Arzelà theorem for both real and complex functions, which he then used to prove what is often called Montel's theorem. Although he proved this theorem in [198, written in 1907], the following statement is from [193, written in 1910]:

Theorem 1.0.2 (Montel). *If a family of complex analytic functions $\mathcal{F}$ is uniformly bounded on a domain $D \subset \mathbb{C}$, then "from any infinite sequence of functions from the family, a new sequence can be extracted which converges uniformly in the domain D to an analytic limit function".*

This theorem was also established independently by Koebe [138].

The first step of Montel's proof of this theorem involved a very simple application of the Cauchy integral theorem which proves that uniformly bounded families of analytic functions are equicontinuous. His theorem follows immediately from the Ascoli-Arzelà Theorem.

Montel refers to Arzelà in the footnote on [193, p. 27], and, from then, essentially all accounts of normal families in complex analysis (see, for example, [46, 111, 179, 196, 245, 259, 260]) have followed this development from equicontinuity and convergence to normal families.

A more modern view is that a family is normal if it is relatively compact (that is, has compact closure) in some larger space of (usually continuous) functions endowed with the compact-open topology (see, for example, [56, 97, 126, 274]). From this perspective (for example, in several complex variables), the emphasis shifts from sequential compactness to compactness.

An overview of the early history of normal families can be found in the book of Daniel S. Alexander [7] or in the paper of Beardon and Minda [20]. This is a good place to mention that the short article [54, Dieudonné 1990] devoted to Montel by the *Dictionary of Scientific Biography* is a very efficient introduction to normal families.

Montel's treatise on the subject, *Leçons sur les Familles Normales de Fonctions Analytiques et leur Applications*, Recueillies et rédigées par J. Barbotte, Gauthier-Villars, Paris, 1927 (reissued in 1974 by Chelsea, New York), appeared 20 years after the subject began.

For 65 years after the publication of Montel's treatise, no book devoted to normal families appeared. And only in 1993, two books on normal families of one complex variable have appeared — the book by Joel L. Schiff [260] and one by Chi Tai Chuang [46].

The notion of a normal family plays an important role in the complex function theory of one complex variable. Montel used the normal family concept to prove the big Picard theorem, Schottky's theorem, and Landau's theorem. Normal families apply in the fields of entire and meromorphic functions, distribution of zeros of differential polynomials, ordinary differential equations, and functional equations. For instance, normal families are closely connected with a standard proof of the Riemann mapping theorem and with the circle of ideas surrounding classical theorems of Picard, Schottky, Landau, and Bloch.

One of the most fruitful developments of normal families in recent times is the well-known rescaling lemma of Zalcman [291]. For surveys of the numerous applications of this lemma, see [26, 27, 292].

A heuristic principle attributed to André Bloch says that a family of holomorphic functions which have a property P in common in a domain $\Omega \subset \mathbb{C}$ is [apt to be] a normal family in Ω if P cannot be possessed by nonconstant holomorphic functions in the whole plane $\mathbb{C}$ (that is, an entire function). [An example of such a P is "uniform boundedness".]

A rigorous formulation and proof of this result was given in 1975 by Zalcman [291], with extensions and counterexamples by Minda [189], Rubel [256], and Pang [231]. The important paper of Carathéodory [39] should also be mentioned in this context. Zalcman's work was inspired by the result of Lohwater and Pommerenke [176]. Their theorem deals with normal functions, but the proofs are almost identical, and Zalcman [291] gives major credit to Pommerenke for the idea. Pommerenke got the idea from reading Landau [155]. Landau in turn credits Valiron [280] with this idea. Of course what can be found in Valiron's paper is still far from the statement of the lemma in Zalcman's paper. We see that Zalcman's lemma is a beautiful example to show how mathematics is the work of many people, each one adding a new idea to the thoughts of the predecessors.

In 2017 appeared the book of Steinmetz [272]. In Section 3.3 the author gives the alternative proof of the famous Zalcman rescaling lemma. In the theory of normal families, Steinmetz played an important role by making use of the rescaling method, treated in detail in [272]. See also Chapter 4 of the recent book of Schiff [261] where an overview of the theory of normal families in one complex variables is given.

The concept of normal meromorphic function allows one to attribute to one function the properties related to some class of functions, namely with a normal family of meromorphic functions. The idea of such a transfer is simple; it was first introduced by Yosida [288] in 1934 and was discussed by Noshiro [226] in 1939. Although normal families had their beginning in 1911 with Paul Montel and normal functions are defined in terms of normal families, they did not have their formal beginning until 1957 when a large-scale study was undertaken by Lehto and Virtanen [164, 165]. They showed that the notion of a normal meromorphic function is closely related to some of the most important problems of the boundary behaviour of meromorphic functions. In [164], the authors extended the Lindelöf principle to meromorphic functions by requiring them to be normal.

In the years since these classic works, much has been written; the interested reader might consult the bibliography of [46, 260, 272] or that of A. J. Lohwater's paper [180].

In Chapter 9 of his treatise, Montel deals with the normal families of holomorphic functions of two complex variables. Since the publication of this book, many interesting results have been obtained in the theory of normal holomorphic functions (maps) and normal families of several complex variables but there had not appeared a single book on the subject. Therefore a book like the ones mentioned above has been waiting to be written for a long time.

We do not try to cover all developments in normal families since Montel's book, but concentrate mainly on our own results. This manuscript contains a number of our results concerning the theory of normal families and normal functions of several complex variables. We have somewhat revised and supplemented them with some of our later investigations.

In keeping with the usual practice in mathematics, we have attributed results to the author (s) who first proved them in one complex variable. But the form in which the results are presented in the manuscript is different from the original ones; at least all one-dimensional results are valid for meromorphic functions. The reader is advised to consult the relevant sources.

We have made some effort at the end of the book to give references to related work; any omissions of references are inadvertent and regretted by both of us. Typographical errors and mathematical slips are the pain and shame of any author. We would appreciate any such being called to our attention.

2

A Glimpse of Normal Families

Preamble: This short chapter is composed of very basic and very easy material, likely familiar to most readers in one form or other, but we record it here for later reference and to fix notation and terminology. Only very modest specific knowledge is required of the prospective reader. What is really expected of the reader is a level of mathematical maturity comparable to that fostered by books like Ahlfors [4, 5 Normal families, pp. 219–227], Greene and Krantz [91, Chapter 6], Marshall [185, Chapter 10.1], Gunning and Rossi [98, pp. 10–12], and Narasimhan [218, pp. 7–11], but we by no means require a mastery of the contents of these outstanding classics.

We will let $\mathbb{C}$ represent the field of complex numbers and $\mathbb{R}$ the subfield of real numbers.

Let $\mathbb{C}^n = \mathbb{C} \times \cdots \times \mathbb{C}$ denote the n-dimensional complex Euclidean space with product topology. The coordinates of $\mathbb{C}^n$ will be denoted by $z = (z_1, \ldots, z_n)$ with $z_j = x_j + iy_j, 1 \leq j \leq n$. Thus, $\mathbb{C}^n$ can be identified with $\mathbb{R}^{2n}$ in a natural manner, $z \mapsto (x_1, y_1, \ldots, x_n, y_n)$.

A *domain* is defined to be an open, connected set in $\mathbb{C}^n$.

Definition 2.0.1. Let $\{g_j\}$ be a sequence of complex-valued, continuous functions on a domain $\Omega \subset \mathbb{C}^n$ which converges to a limit function g. If, for any given $\varepsilon > 0$ and any compact subset K in Ω, there exists a positive integer j_0 depending only on ε and K, such that

$$|g_j(z) - g(z)| < \varepsilon$$

holds for all $z \in K$ and all $j > j_0$, then we say that $\{g_j\}$ converges uniformly on compact subsets of Ω.

Definition 2.0.2. A family, $\mathcal{F}$ of continuous functions on a domain $\Omega \subset \mathbb{C}^n$ is said to be *normal* on Ω provided every sequence $\{f_j\} \subseteq \mathcal{F}$ contains a subsequence which converges uniformly on compact subsets of Ω.

This definition does not require the limit functions of the convergent subsequences to be members $\mathcal{F}$.

In view of the Heine-Borel theorem, locally uniform convergence is the same as uniform convergence on each compact subset of Ω.

In what follows, $P(a, r)$ will denote the open polydisk centered at the point $a = (a_1, \ldots, a_n)$ and with polyradius $r = (r_1, \ldots, r_n)$. That is,

$$P(a, r) = \{z \in \mathbb{C}^n : |z_k - a_k| < r_k, k = 1, \ldots, n\} \subseteq \Omega$$

DOI: 10.1201/9781032669861-2

The corresponding closed polydisk will be denoted $\overline{P}(a,r)$. Note that $P(a,r)$ is just the Cartesian product of the open disks $D(a_\nu, r_\nu) \in \mathbb{C}$ and $\overline{D}(a,r)$ is the Cartesian product of the closed disks $\overline{D}(a_\nu, r_\nu)$. The set $\Gamma := \{z \in \mathbb{C}^n : |z_\nu - a_\nu| = r_\nu\}$ is usually called the distinguished boundary of $P(a,r)$.

The following lemma says that normality is a local property.

Lemma 2.0.3. *Suppose Ω is a domain in $\mathbb{C}^n$ and suppose that*

$$\Omega = \bigcup_{j=0}^{\infty} \overline{P}_j,$$

where $\overline{P}_j \subset \Omega$ are closed polydisks. A family of continuous functions $\mathcal{F}$ is normal on Ω if and only if, for each j, every sequence in $\mathcal{F}$ contains a subsequence which converges uniformly on $\overline{P}_j$.

Proof of Lemma 2.0.3. This lemma can be proved in a manner which is absolutely analogous to the case of one variable.

The only if part follows by definition. Suppose that, for each j, every sequence in $\mathcal{F}$ contains a subsequence which converges uniformly on $\overline{P}_j$. Suppose $\{f_n\} \subset \mathcal{F}$. Then there is a subsequence $\left\{f_n^{(1)}\right\} \subset \{f_n\}$ such that $\left\{f_n^{(1)}\right\}$ converges uniformly on $\overline{P}_1$. Likewise, there is a sequence $\left\{f_n^{(2)}\right\} \subset \left\{f_n^{(1)}\right\}$ which converges uniformly on $\overline{P}_2$, and indeed there is a sequence $\left\{f_n^{(k)}\right\} \subset \left\{f_n^{(k-1)}\right\}$ which converges uniformly on $\overline{P}_k$. Then the "diagonal" sequence $\left\{f_k^{(k)}\right\}$ converges uniformly on each $\overline{P}_j$, since it is a subsequence of $\left\{f_n^{(k)}\right\}$ for each k. A compact set $K \subset \Omega$ can be covered by finitely many polydisks, so the diagonal sequence converges uniformly on K. □

Remark 2.0.4. There is no loss of generality in assuming $\Omega = \cup_{j=0}^{\infty} \overline{P}_j$. To see this, we recall (see [271, Theorem 1 on p. 167]) that every nonvoid open set in $\mathbb{C}^n \approx R^{2n}$ is a union of a denumerable set of closed cubes $\{K_j\}$, $K_j \subset \Omega$, the sides of which are parallel to the coordinate planes and which are pairwise without common interior points. Since K_j is compact, there are finitely many points $a_j^1, \ldots, a_j^{N_j}$ contained in the interior of K_j, such that K_j is contained in a union of finitely many closed polydisk $\overline{P}_j^k(a_j^k, r_j^k) \subset \Omega$ with center at a_j^k and some polyradius r_j^k such that

$$K_j \subset \bigcup_{k=1}^{N_j} \overline{P}_j^k(a_j^k, r_j^k) \subset \Omega.$$

Hence

$$\Omega = \bigcup_{j=1}^{\infty} K_j \subseteq \bigcup_{j=1}^{\infty} \bigcup_{k=1}^{N_j} \overline{P}_j^k(a_j^k, r_j^k) \subseteq \Omega.$$

It follows that

$$\Omega = \bigcup_{j=1}^{\infty} \bigcup_{k=1}^{N_j} \overline{P}_j^k(a_j^k, r_j^k).$$

A countable union of finite sets is countable. After a suitable renumbering, we obtain that Ω may be written as a denumerable union of closed polydisks.

The space of continuous functions on an open set $\Omega \subset \mathbb{C}^n$ will be denoted by $\mathcal{C}(\Omega)$, the space of k times continuously differentiable functions on an open set Ω in a Euclidean space by $\mathcal{C}^k(\Omega)$, and the space of infinitely differentiable functions on Ω by $\mathcal{C}^\infty(\Omega)$.

For any $f \in \mathcal{C}(\Omega)$ and compact $K \subset \Omega$ put

$$\|f\|_K := \sup_{z \in K} |f|.$$

We can define a metric on the space $\mathcal{C}(\Omega)$ as follows. Let Ω be any proper open subset of $\mathbb{C}^n$. Now let $\{\overline{P}_j\}$ be a sequence of closed polydisks of Ω such that

$$\overline{P}_j \subset \Omega \text{ and } \Omega = \bigcup_{j=0}^{\infty} \overline{P}_j.$$

For any continuous functions $f, g \in \mathcal{C}(\Omega)$ define

$$d(f,g) := \sum_{j+1}^{\infty} \frac{1}{2^j} \frac{\|f-g\|_{\overline{P}_j}}{1 + \|f-g\|_{\overline{P}_j}}.$$

It is routine to confirm that d is a metric on $\mathcal{C}(\Omega)$:

1. If $d(f,g) = 0$, then $f = g$ on each $\overline{P}_j$ and hence on Ω,
2. We have $d(f,g) = d(g,f)$ for $f, g \in \mathcal{C}(\Omega)$,
3. We have $d(f,g) \leq d(f,h) + d(h,g)$ for $f, g, h \in \mathcal{C}(\Omega)$.

The following criterion is sometimes useful.

Theorem 2.0.5. *A sequence $\{f_j\}$ converges uniformly on compact subsets of Ω to $f \in \mathcal{C}(\Omega)$ if and only if $d(f_j, f) \to 0$.*

Proof. See the proof of Proposition 10.3 in [185] on page 157. □

We have shown that convergence with respect to the distance d is equivalent to convergence on compact sets. Compactness and sequential compactness are the same for metric spaces. A family is normal if and only if its closure is compact in this topology.

Definition 2.0.6. A function f is said to be *locally bounded* at a point $p \in \Omega$, if there is a neighborhood V of p such that $|f(z)|$ is bounded on V. A function $f(z)$ is said to be *locally bounded* if $f(z)$ is locally bounded at every point of Ω.

Definition 2.0.7. A family $\mathcal{F}$ of functions on Ω is called *locally bounded* or *bounded on compact sets* if, for every compact set K contained in Ω, there is a constant M (depending on K) such that $|f(z)| \le M$ for every z in K and every function f in the family $\mathcal{F}$.

In other words, $\mathcal{F}$ is uniformly bounded in a neighborhood of each point of Ω. In the literature this property is sometimes referred to as local uniform boundedness. Since any compact subset $K \subset \Omega$ can be covered by a finite number of such neighborhoods, it follows that a locally bounded family $\mathcal{F}$ is uniformly bounded on compact subsets of Ω.

Definition 2.0.8. A family of functions $\mathcal{F}$ defined on a set $S \subset \mathbb{C}^n$ is

1. *equicontinuous* at $w \in S$ if for each $\varepsilon > 0$ there exists a $\delta > 0$ so that if $z \in S$ and $|z - w| < \delta$, then $|f(z) - f(w)| < \varepsilon$ for all $f \in \mathcal{F}$;
2. *equicontinuous* on S if it is equicontinuous at each $w \in S$;
3. *uniformly equicontinuous* on S if for each $\varepsilon > 0$ there exists a $\delta > 0$ so that if $z, w \in S$ with $|z - w| < \delta$ then $|f(z) - f(w)| < \varepsilon$ for all $f \in \mathcal{F}$.

Next, the central result linking normal families and equicontinuity:

Theorem 2.0.9. *(Arzelà-Ascoli) A family $\mathcal{F}$ of continuous functions is normal on a domain $\Omega \subset \mathbb{C}^n$ if and only if*

1. *$\mathcal{F}$ is equicontinuous on Ω and*
2. *there is a point $z^0 \in \Omega$ so that the collection $\{f(z^0) : f \in \mathcal{F}\}$ is a bounded subset of $\mathbb{C}$.*

Proof. The proof of this theorem is the same as the proof of Arzelà-Ascoli's theorem in [185, Theorem 10.5, p. 158] for $n = 1$. □

Remark 2.0.10. The proof is actually valid for any metric space. For a proof see for example Lang [159, p. 57].

Definition 2.0.11. A complex-valued $\mathcal{C}^1$ function $f(z)$ defined on an open subset Ω of $\mathbb{C}^n$ is called *holomorphic*, denoted by $f \in \mathcal{O}(\Omega)$, if $f(z)$ is holomorphic in each variable z_j when the other variables are fixed. In other words, $f(z)$ satisfies

$$\frac{\partial f}{\partial \bar{z}_j} = 0$$

for each $j = 1, \ldots, n$, where

$$\frac{\partial}{\partial \bar{z}_j} = \frac{1}{2}\left(\frac{\partial}{\partial x_j} + i\frac{\partial}{\partial y_j}\right)$$

is the so-called Cauchy-Riemann operator.

A family $\mathcal{F}$ is said to be locally bounded in Ω if the functions in $\mathcal{F}$ are uniformly bounded in some neighborhood of each point of Ω. For families of holomorphic functions local boundedness is sufficient for normality, and indeed necessary, if the family is bounded at some point of the domain of definition. The proof essentially uses the Arzelà-Ascoli theorem that a family is normal if it is equicontinuous and pointwise bounded. An application of the Cauchy integral formula shows that if a family $\mathcal{F}$ of analytic functions is locally bounded, then so is $\mathcal{F}' = \{f' : f \in \mathcal{F}\}$, and equicontinuity follows.

Theorem 2.0.12. *The following are equivalent for a family $\mathcal{F}$ of holomorphic functions on a region $\Omega \subset \mathbb{C}^n$:*

(1) $\mathcal{F}$ is normal on Ω;

(2) $\mathcal{F}$ is locally bounded on Ω;

(3) $\mathcal{F}' = \{\nabla_z f : f \in \mathcal{F}\}$ is locally bounded on Ω and there is a $z_0 \in \Omega$ so that $\{f(z_0) : f \in \mathcal{F}\}$ is a bounded subset of $\mathbb{C}$. Here $\nabla_z f := \left(\frac{\partial f}{\partial z_1}(z), \ldots, \frac{\partial f}{\partial z_n}(z)\right)$.

Proof. The proof of this theorem is very similar to the proof for $n = 1$ [185, Theorem 10.7, p. 160]. The extra differences involved are only technicalities.

Suppose $\mathcal{F}$ is normal. By the proof of Theorem 2.0.9, for each $w \in \Omega$, $\{f(w) : f \in \mathcal{F}\}$ is bounded. By equicontinuity, each $w \in \Omega$ is contained in the interior of a closed polidisk $\overline{P}(w, r) \subset \Omega$ such that

$$|f(z) - f(w)| < 1 \text{ for all } z \in \overline{P}(w, r), f \in \mathcal{F}.$$

It follows $\mathcal{F}$ is locally bounded.

Now suppose $\mathcal{F}$ is locally bounded on Ω.

By hypothesis there exists a constant M such that $|f| \le M$ on a closed polydisk

$$P(z^1, r) = \left\{z \in \mathbb{C}^n : \left|z_k - z_k^1\right| \le r + \delta, k = 1, \ldots, n\right\} \subset \Omega$$

centered at z^1 with polyradius $r + \delta > 0$.

Then, by the Cauchy integral formula, for any $z \in P(z^1, r + \delta)$,

$$\frac{\partial f}{\partial z_\nu}(z) = \frac{1}{(2\pi i)^n} \int_\Gamma \frac{f(\xi)}{(z - \xi)(z_\nu - \xi_\nu)}\, d\xi, \tag{2.1}$$

where Γ is the distinguished boundary of $P(z^1, r + \delta)$, since $f \in H(\Omega)$. But since $|f|_\Gamma \le M$, from (2.1) we will have, for all $z \in P(z^1, r)$,

$$\left|\frac{\partial f}{\partial z_\nu}(z)\right| \le \frac{M}{\delta}\left(1 + \frac{r}{\delta}\right)^n;$$

from this it follows that $\mathcal{F}'$ is locally bounded at z^1, an arbitrary point of Ω, and **(3)** holds.

Finally, if **(3)** holds and $z \in \Omega$, then $|\nabla_z f| \leq L$ for z in a some polydisk $P(z^1, r) \subset \Omega$. The line segment from z to z^1 can be parameterized by $\gamma(t) = z^0 + t(z - z^1) \subset P(z^1, r)$, for $t \in [0,1]$ so that, if f is holomorphic in a neighborhood of γ, then

$$f'(\gamma(t)) = \sum_{\nu=1}^{n} \frac{\partial f}{\partial z_\nu}(\gamma(t))(z_\nu - z_\nu^1) = (\nabla_{\gamma(t)} f, \overline{z} - \overline{z}^1).$$

Integrating $f'(\gamma(t))$ along a line segment from 0 to 1, we have

$$f(z) - f(z^1) = \int_0^1 (\nabla_{\gamma(t)} f, \overline{z} - \overline{z}^1)\, dt.$$

By the Cauchy-Schwarz inequality,

$$\left|(\nabla_{\gamma(t)} f, \overline{z} - \overline{z}^1)\right| \leq \left|\nabla_{\gamma(t)} f\right| \left|z - z^1\right|.$$

Thus

$$\left|f(z) - f(z^1)\right| \leq \int_0^1 \left|\nabla_{\gamma(t)} f\right| \left|z - z^1\right| dt.$$

Since $|\nabla_z f| \leq L$ for $z \in P(z^1, r)$ we have $\left|f(z) - f(z^1)\right| \leq L\left|z - z^1\right|$ for all $f \in \mathcal{F}$. This confirms the equicontinuous $\mathcal{F}$ at z^1, an arbitrary point of Ω. By the Arzelà-Ascoli Theorem 2.0.9, $\mathcal{F}$ is normal. □

Holomorphic functions of several variables have compactness properties with respect to locally uniform convergence identical to the one variable case. A standard result in complex analysis says that the locally uniform limit of a sequence of holomorphic functions is again holomorphic.

The next two classical results of several complex variables find frequent application throughout the text.

Weierstrass proved that the set of holomorphic functions on a domain in $\mathbb{C}^n$ is closed under uniform convergence on compact sets.

Theorem 2.0.13 (Weierstrass). *Let $\{f_j\}$ be a sequence of holomorphic functions on a domain $\Omega \subset \mathbb{C}^n$ which converges uniformly on compact subsets of Ω to a function f. Then f is holomorphic in Ω, and for any $k = (k_1, \ldots, k_n)$ the sequence of derivatives $\frac{\partial^{|k|} f_j}{\partial z^k}$ converges uniformly on compact subsets of Ω to $\frac{\partial^{|k|} f}{\partial z^k}$.*

Proof. The proof is essentially the same as the classical proof of the Weierstrass theorem in one complex variable. For holomorphic functions of one variable, this is a standard result. Its proof is a simple application of Morera's theorem and Cauchy's formula. In the several variable case, we simply apply this result in each variable separately (with the other variables fixed) to conclude that the limit of a sequence of holomorphic functions, converging uniformly on compacta, is holomorphic in each variable separately. Such a limit is also obviously continuous, and so is holomorphic by Osgood's lemma (we could appeal to Hartog's theorem but we really only need the more elementary Osgood's lemma).

It is routine to confirm that a sequence of holomorphic functions $\{f_j\}$ converges uniformly on compact subsets of Ω if and only if $\{f_j\}$ converges uniformly in any polydisk which belongs to Ω. So, it is enough to prove the second part of the theorem for an open polydisk and the derivative with respect to a single variable z_ν. Let $a \in \Omega$, and choose $r > 0$ so that

$$P(a,r) = \{z \in \mathbb{C}^n : |z_k - a_k| < r, k = 1, \ldots, n\} \subset \Omega.$$

Let 2δ be the distance from $\overline{P}(a,r)$ to $\mathbb{C}^n \backslash \Omega$, so that $P(a,r) \subset \overline{P}(a, r+\delta) \subset \Omega$. Then, by the Cauchy integral formula, for any $z \in P(a, r+\delta)$,

$$\frac{\partial f_j}{\partial z_\nu} - \frac{\partial f}{\partial z_\nu} = \frac{1}{(2\pi i)^n} \int_\Gamma \frac{f_j(\xi) - f(\xi)}{(z-\xi)(z_\nu - \xi_\nu)}\, d\xi, \tag{2.2}$$

where Γ is the distinguished boundary of $P(a, r+\delta)$, since $f_j - f \in H(\Omega)$. But since $f_j \to f$ uniformly on Γ, it follows that for any $\varepsilon > 0$ there is a j_0 such that $|f_j - f|_\Gamma < \varepsilon$ for all $j > j_0$. Take K to be a compact subset of $P(a,r)$. From (2.2) we will have, for all $z \in P(a,r)$ and $j > j_0$,

$$\left| \frac{\partial f_j}{\partial z_\nu} - \frac{\partial f}{\partial z_\nu} \right| \leq \frac{\varepsilon}{\delta}\left(1 + \frac{r}{\delta}\right)^n;$$

from this it follows that $\frac{\partial f_j}{\partial z_\nu} \to \frac{\partial f}{\partial z_\nu}$ uniformly in $P(a,r)$. Since this can be proved for any $a \in \Omega$, $\frac{\partial f_j}{\partial z_\nu}$ converges uniformly in Ω.

What is true of the first derivatives is of course also true for higher derivatives. ☐

We will need the following theorem which is an n-dimensional version of Hurwitz's theorem. Observe that its proof uses nothing other than the maximum modulus theorem.

Theorem 2.0.14 (Hurwitz). *Let Ω be a domain in $\mathbb{C}^n$ and $\{f_j\}$ a sequence of nonvanishing holomorphic unctions f_j which converges uniformly on compact subsets to a holomorphic function f on Ω. Then f vanishes either everywhere or nowhere.*

Proof. (The proof comes from Fuks's book [80, Lemma 1, p. 274)].) Let $n = 1$. If the limit function is constant, our assertion is evident. If $f(z) \not\equiv$ const, then one can find for any point $z_0 \in \Omega$ a circle $\gamma_r = \{|z - z^0| = r\}$ on which $f(z) \neq 0$. Then there exists a number $\varepsilon > 0$ such that $|f(\zeta)| > \varepsilon$ and $|f_j(\zeta)| > \varepsilon/2$ when $\zeta \in \gamma_r$ and $j > j_0$. Here j_0 is some natural number chosen appropriately. Since $f_j(z) \neq 0$ for $z \in \Omega$, the maximum modulus theorem implies that for all z such that $|z - z^0| \leq r$ we have $|f_j(z)| > \varepsilon/2$. Hence it follows that $f(z^0) \neq 0$.

Let $n > 1$, $z^0 \in \Omega$ be an arbitrary point and the function $f(z) \not\equiv$ const in the domain $\Omega \subset \mathbb{C}^n$. Then one can choose a complex line $l_{z^0}(t) = \{z \in \mathbb{C}^n : z = z^0 + ta\}$ through z^0, where $|a| = 1$ and $t \in \mathbb{C}$, in such a way that the function $\phi(t) = f(l_{z^0}(t))$ is not constant in a neighborhood of the point $t = 0$. Set $\phi_j(t) = f_j(z^0 + at)$; then in some neighborhood of 0 all of the functions $\phi_j(t) \neq 0$ and $\lim_{j\to\infty} \phi_j(t) = \phi(t)$ is reached uniformly. Hence, as we have just seen, it follows that $\phi(0) = f(z^0) \neq 0$. The lemma is proved. □

For other proofs of these theorems, we refer the reader to [218, Proposition 5 (Weierstrass theorem), p. 7, and Corollary (Hurwitz's theorem), p. 80].

3

Normal Families in $\mathbb{C}^n$

Preamble: This chapter starts with the elegant definition of normal family in $\mathbb{C}^n$ from the Cima and Krantz article [47] and an extension of Marty's criterion of normality to holomorphic functions of several complex variables.

As an immediate consequence from Marty's theorem we obtain an extension of the famous Zalcman rescaling lemma to the several complex variables context.

Using Marty's criterion and Zalcman's lemma we obtain an extension of Montel's theorem.

After that we give the proof of the Riemann mapping theorem, which is one of the more famous results in the whole subject of mathematics. Together with its generalization to Riemann surfaces, the so-called uniformization theorem, it is without doubt a cornerstone of function theory.

As Royden [252] pointed out, although Marty's result is necessary and sufficient for the relative compactness of a family of holomorphic or meromorphic functions, it may not be easy to apply in certain instances. Zalcman's rescalling lemma can be used to prove the strengthening of the sufficiency part of Marty's normality criterion (Theorem 3.2.3 below). If $n = 1$ this Theorem coincides with Schwick's theorem [264], which is the generalization of a normality criterion of Royden [252].

By Marty's normality criterion, the normality of a family $\mathcal{F}$ of holomorphic functions on a domain $\Omega \subset \mathbb{C}^n$ is equivalent to the local boundedness of the corresponding family $\mathcal{F}^\sharp := \{f^\sharp(z) : f \in \mathcal{F}, z \in \Omega\}$ (this notation is defined below). It is rather surprising that there is a normality criterion based on a lower bound of the family $\mathcal{F}^\sharp$. We prove the extension of Grahl and Nevo's normality criterion [86] to a family of holomorphic functions of several complex variables.

For the sequences of holomorphic functions of one complex variables, examples due to A.F. Beardon and D. Minda [21], and K.R. Davidson [51] show that pointwise convergent sequences can have bad properties. We provide some of these results.

Zalcman's normality criterion and the one-dimensional Nevanlinna theorem are used to prove these results, which generalizes those obtained by Lappan [161].

DOI: 10.1201/9781032669861-3

We obtain an extension of Mandelbrojt's [184] boundedness criterion for a family $\mathcal{F}$ of zero-free holomorphic functions on a bounded domain in $\mathbb{C}^n$ to be normal.

The extension of the Zalcman-Pang lemma is also given.

3.1 Definitions and Preliminaries

In this chapter, we recall the basic properties of holomorphic functions of several complex variables using for references (mainly) the book by Narasimhan [218], Ohsawa [227], the book by Greene and Krantz [95].

For the most part, we assume only the background provided by an elementary graduate courses in real and complex analysis.

Let Ω be a domain in $\mathbb{C}$. By a domain Ω, we shall always mean an open connected subset of $\mathbb{C}$. Denote by $\mathcal{O}(\Omega)$ the set of all holomorphic functions in Ω.

We assume that the reader is familiar with the basic properties of holomorphic functions of a single variable as presented in standard texts (e.g. [95]) in the subject: A holomorphic function on Ω has a convergent power series expansion in a neighborhood of each point of Ω; a differentiable function is holomorphic if and only if it satisfies the Cauchy-Riemann equations; the space $\mathcal{O}(\Omega)$ is an algebra over the complex field under the operations of pointwise addition, multiplication, and scalar multiplication; if a sequence $\{f_j\}$ in $\mathcal{O}(\Omega)$ converges uniformly on each compact subset of Ω, then the limit function is also holomorphic on Ω; holomorphic functions satisfy the Cauchy integral theorem and formula, the identity theorem, and the maximum modulus theorem.

A function on an open set Ω in $\mathbb{C}^n$ is holomorphic if it is holomorphic in each variable separately (that this is equivalent to the existence of local multi-variable power series expansions of the function see, for example [227]).

Some of the properties of holomorphic functions, like the power series expansion, do extend from one variable to several variables. They differ, however, in many important aspects. It is therefore not correct to consider the theory of several complex variables as a straightforward generalization of that of one complex variable. For example, in one variable the zero set of a holomorphic function is a discrete set. The zero set of a holomorphic function in $\mathbb{C}^n, n \geq 2$, has a real $2n - 2$ dimension. In $\mathbb{C}$, it is trivial to construct a holomorphic function in a domain Ω which is singular at one boundary point $p \in \partial\Omega$. In contrast, in $\mathbb{C}^n$ when $n \geq 2$, it is not always possible to construct a holomorphic function on a given domain $\Omega \subset \mathbb{C}^n$ which is singular at one boundary point. This leads to the existence of a domain in several variables such that any holomorphic function defined on this domain can be extended holomorphically to a fixed larger set, a feature that does not exist in one variable.

We first fix our terminology.

The standard inner product on $\mathbb{C}^n$ is given by

$$\langle z, w \rangle = \sum_{\mu=1}^{n} z_\mu \overline{w}_\mu, \tag{3.1}$$

where $z = (z_1, \ldots, z_n)$ and $w = (w_1, \ldots, w_n)$, with $z_\mu, w_\mu \in \mathbb{C}$. This space is complete, and therefore it is a finite-dimensional Hilbert space.

$$|z| := \sqrt{\langle z, z \rangle} \quad (z \in \mathbb{C}^n)$$

denotes the associated norm.

Let $z \in \widehat{\mathbb{C}}^n$ and let $v \in \mathbb{C}^n$ be thought of as a tangent vector to $\widehat{\mathbb{C}}^n$ at z. Then the infinitesimal form of the Euclidean metric at z is defined by

$$de(z, v) = |v|.$$

The Euclidean length of a curve $\gamma : [a, b] \to \mathbb{C}$ is given by the integral

$$l(\gamma) := \int_\gamma |\gamma'(t)| dt.$$

The Euclidean distance between two points z and z' is defined to be the infimum of the Euclidean lengths of smooth paths joining the two points:

$$e(z, z') = \inf l(\gamma),$$

where the infimum is taken over smooth connecting paths γ. Taking the infimum over paths and keeping in mind that the shortest path between the two points in the Euclidean sense is the line segment joining them, we obtain the equality

$$e(z, z') = |z - z'|.$$

Theorem 3.1.1 (Cauchy-Schwarz). *If $z, w \in \mathbb{C}^n$, then*

$$|\langle z, w \rangle| \leq |z|\,|w|\,,$$

with equality holding if and only if z and w are linearly dependent.

Proof. One proof of the Cauchy-Schwarz inequality consists of the verification of one line, namely:

$$|z|^2 |w|^2 - |\langle z, w \rangle|^2 = \frac{1}{|z|^2} \left| |z|\, w - \langle w, z \rangle |z|^2 \right|^2$$

It might perhaps be more elegant to multiply through by $|z|^2$, so that the result should hold for $z = 0$ also, but the identity seems to be more appealing in the form given. This one line proves also that if the inequality degenerates to an equality, then w and z are linearly dependent. The converse is trivial: if w and z are linearly dependent, then one of them is a scalar multiple of the other, say $w = \alpha z$, and then both $|\langle z, w \rangle|^2$ and $\langle z, z \rangle \cdot \langle w, w \rangle$ are equal to $|\alpha|^2 \langle z, z \rangle^2$. □

By $B = B_n$ we denote the unit ball in $\mathbb{C}^n$. Thus B consists of all $z \in \mathbb{C}^n$ such that $|z| < 1$. To emphasize the special role played by the case $n = 1$, the open unit disk in $\mathbb{C}$ will be denoted by D in place of B_1.

We designate by $\mathcal{C}(\Omega)$ (respectively, $\mathcal{O}(\Omega)$) the family of functions continuous (respectively, holomorphic) in Ω. It is clear that $\mathcal{O}(\Omega) \subset \mathcal{C}(\Omega)$ and $\mathcal{C}(\Omega)$ is an algebra over $\mathbb{C}$ with respect to pointwise addition and multiplication, and that $\mathcal{C}(\Omega)$ is closed relative to uniform convergence on compact subsets of $\mathbb{C}^n$.

Given a domain $\Omega \subset \mathbb{C}^n$, we let $T_z(\Omega)$ denote the vector space of all tangent vectors at z of complex differentiable curves passing through z. The space $T_z(\Omega)$ is called the *tangent space* to Ω at z and $T(\Omega) := \bigcup_{z\in\Omega} T_z(\Omega)$ is called the *tangent bundle* of Ω. If $\Omega \subset \mathbb{C}^n$ then $T_z(\Omega)$ can be identified in a canonical way with $\{z\} \times \mathbb{C}^n$. We denote by $(z; v)$ the typical element of $T_z(\Omega)$.

Now let $\mathcal{C}^2(\{z\})$ denote the functions which are defined and twice continuously differentiable on a neighborhood of the point z in $\mathbb{C}^n$. We often call $\mathcal{C}^2(\{z\})$ the space of *germs* at z. For $\varphi \in \mathcal{C}^2(\{z\})$, we define the *Levi form* of φ at z, $L_z(\varphi, v)$, by

$$L_z(\varphi, v) := \sum_{k,l=1}^{n} \frac{\partial^2 \varphi}{\partial z_k \partial \overline{z}_l} v_k \overline{v}_l \qquad (v \in \mathbb{C}^n),$$

where $(z; v) = (z; (v_1, \dots, v_n)) \in T_z(\Omega)$.

For a holomorphic function f on Ω, set

$$f^\sharp(z) := \sup_{|v|=1} \sqrt{L_z(\log(1 + |f|^2), v)}. \tag{3.2}$$

It should be noted that the quantity $f^\sharp(z)$ is well defined since the Levi form $L_z(\log(1 + |f|^2), v)$ is nonnegative for all $z \in \Omega$.

In particular, for $n = 1$ the formula (3.2) takes the form

$$f^\sharp(z) := \frac{|f'(z)|}{1 + |f(z)|^2}$$

and $z^\sharp$ coincides with the spherical metric on $\widehat{\mathbb{C}}$.

The invariance of the Levi form.

The following property of the Levi form can be checked directly:

If $F : U \to \mathbb{C}^n$ is a holomorphic mapping and $\varphi \in C^2$ in a neighborhood of the point $F(z)$, then

$$L_z(\varphi \circ F, v) = L_{F(z)}(\varphi, F_* v) \qquad (v \in \mathbb{C}^n),$$

where $F_* = dF$ is the differential of the mapping F at z.

The last property shows that the Levi form is invariant relative to biholomorphic mappings.

The Riemann sphere is denoted by $\widehat{\mathbb{C}}$. Let $\widehat{\mathbb{C}} := \mathbb{C} \cup \{\infty\}$, where ∞ is a symbol not contained in $\mathbb{C}$. Introduce the following topology on $\widehat{\mathbb{C}}$. The open sets are the usual open sets $U \subset \mathbb{C}$ together with sets of the form $V \cup \{\infty\}$, where $V \subset \mathbb{C}$ is the complement of a compact set $K \subset \mathbb{C}$. With this topology, $\widehat{\mathbb{C}}$ is a compact Hausdorff topological space, homeomorphic to the 2-sphere S^2. Set

$$U_1 := \widehat{\mathbb{C}} \setminus \{\infty\} = \mathbb{C},$$
$$U_2 := \widehat{\mathbb{C}} \setminus \{0\} = \mathbb{C}^* \cup \{\infty\}.$$

Here and in what follows $\mathbb{C}^* = \mathbb{C} \setminus \{0\}$.

Define maps $\varphi_k : U_k \to \mathbb{C}$, $k = 1, 2$, as follows. The map φ_1 is the identity map and

$$\varphi_2 := \begin{cases} 1/z & \text{for } z \in \mathbb{C}^* \\ 0 & \text{for } z = \infty. \end{cases}$$

Clearly these maps are homeomorphisms and thus $\widehat{\mathbb{C}}$ is a two-dimensional manifold. Since U_1 and U_2 are connected and have nonempty intersection, $\widehat{\mathbb{C}}$ is also connected.

The complex structure on $\widehat{\mathbb{C}}$ is now defined by the atlas consisting of the charts $\varphi_k : U_k \to \mathbb{C}$, $k = 1, 2$. We must show that the two charts are holomorphically compatible. But $\varphi_1(U_1 \cap U_2) = \varphi_2(U_1 \cap U_2) = \mathbb{C}^*$ and

$$\varphi_2 \circ \varphi_1^{-1} : \mathbb{C}^* \to \mathbb{C}^*, \quad z \to 1/z,$$

is biholomorphic.

Further, let $z \in \widehat{\mathbb{C}}$ and let $v \in \mathbb{C}$ be thought of as a tangent vector to $\widehat{\mathbb{C}}$ at z. Then the *infinitesimal form of the spherical metric* at z is defined by

$$ds(z, v) = \frac{|v|}{1 + |z|^2}.$$

The *spherical length of a curve* $\gamma : [a, b] \to \widehat{\mathbb{C}}$ is given by the integral

$$\Lambda(\gamma) := \int_\gamma \frac{|dz|}{1 + |z|^2} = \int_a^b \frac{|\gamma'(t)|dt}{1 + |\gamma(t)|^2}.$$

which is easily seen to converge in case one (or both) of the end points lies at infinity. (Here, as is usual in the subject, a curve is defined to be a smooth function $\gamma : [a, b] \to \widehat{\mathbb{C}}$.) Indeed, the change of variable $z = w^{-1}$ gives $dz = -w^{-2}dw$, and this substitution does not alter the form of the integral; i.e., it becomes

$$\Lambda(\gamma) = \int_{\widetilde{\gamma}} \frac{|dw|}{1 + |w|^2}.$$

The spherical distance between two points z and z' is determined by the infimum of the spherical lengths of smooth paths joining the two points:

$$s(z, z') = \inf \Lambda(\gamma),$$

where, the infimum is taken over smooth connecting paths γ.

Then $s(w_1, w_2)$ is the Euclidean length of the shortest arc of the great circle on the Riemann sphere joining w_1 and w_1 and defines a metric on the sphere known as the *spherical metric.*

The spherical metric is equivalent to the Euclidean metric in the finite part of $\overline{\mathbb{C}}$ (but of course, they are not equal). The argument goes like this: Let z and z' be two points both lying in a disk E of radius r about the origin, and let γ be a path in E connecting the two; then the next two inequalities hold:

$$\frac{l(\gamma)}{1+r^2} < \int_\gamma \frac{2|dz|}{1+|z|^2} < l(\gamma)\,,$$

where $l(\gamma)$ denotes the Euclidean length of the path γ. Taking infimum over paths and keeping in mind that the shortest path between the two points in the Euclidean sense is the line segment joining them, we obtain the inequalities

$$\frac{|z-z'|}{1+r^2} < \rho\,(z, z') < |z-z'|\,.$$

By this we have established that the two metrics are equivalent in the finite part of the extended plane, but as the form of the integral giving the spherical metric is insensitive to the change of variables $w = z^{-1}$, the argument is even valid in neighborhoods about the point at infinity. Hence the two metrics are topologically equivalent.

If now $w = f(z)$ is a holomorphic function, then the spherical length of the image curve $f \circ \gamma$ is given by

$$l(f \circ \gamma) = \int_a^b \frac{|(f \circ \gamma)'(t)|dt}{1+|f \circ \gamma(t)|^2}.$$

Pseudometric spaces were introduced by Đurađ Kurepa [149] in 1934.

Definition 3.1.2. A *pseudometric space* (X, d) is a set X together with a nonnegative real-valued function $d : X \times X \longrightarrow \mathbb{R}_{\geq 0}$, called a pseudometric, such that for every $x, y, z \in X$,

1. $d(x, x) = 0$.
2. Symmetry: $d(x, y) = d(y, x)$
3. Subadditivity/Triangle inequality: $d(x, z) \leq d(x, y) + d(y, z)$.

Notice that a pseudometric space differs from a metric space in that, for the former, $d(x, y) = 0$ does not imply that $x = y$.

Definition 3.1.3. Let (X, ρ) and (Y, σ) be pseudometric spaces, let $f \in C(X, Y)$ and let $c > 0$. We will say that f is *Lipschitz* of order c with respect to metric ρ and σ if $\rho(f(a), f(b)) \leq c\sigma(a, b)$ for all $a, b \in X$ and that a family $\mathcal{F} \subset C(X, Y)$ is ρ-to-σ Lipshitz of order c with respect to ρ and σ if each $f \subset \mathcal{F}$ is Lipschitz of order c.

For our purposes it will often be the existence of c rather than its size that is important, in which case, we will shorten this terminology and say simply that $\mathcal{F}$ is Lipshitz.

If $\mathcal{F} \subset \mathcal{C}(X, Y)$ is Lipschitz of order 1 with respect to ρ and σ, we will say that F is *distance decreasing* with respect to ρ and σ.

Let X, Y be topological spaces. A family $\mathcal{F} \subset \mathcal{C}(X, Y)$ is said to be *evenly continuous* from $p \in X$ to $q \in Y$ if for each U open in Y about q, there exist V, W open in X, Y about p, q respectively such that

$$\{f \in F : f(p) \in W\} \subset \{f \in \mathcal{F} : f(V) \subset U\}.$$

If $\mathcal{F}$ is evenly continuous from each $p \in X$ to each $q \in Y$, we say that F is evenly continuous (from X to Y). We have the following Kelley-Morse topological version of the Ascoli-Arzelà theorem [126].

Theorem 3.1.4. *Let X be a regular locally compact space and let Y be a regular space. Then $\mathcal{F} \subset \mathcal{C}(X, Y)$ is relatively compact in $\mathcal{C}(X, Y)$ iff*

(a) F is evenly continuous, and

(b) $F(x) = \{f(x) : f \in \mathcal{F}\}$ is relatively compact in Y for each $x \in X$.

Let (X, ρ) and (Y, σ) be pseudometric spaces, let $f \in \mathcal{C}(X, Y)$, and let $c > 0$.

If Y is a topological space, Y^{∞} will represent the one-point compactification of Y if Y is not compact, $Y^{\infty} = Y$ if Y is compact.

The following theorem establishes an important property of distance decreasing families.

Theorem 3.1.5. *Let (Y, σ) be a locally compact metric space. Let X be a topological space and let ρ be a pseudometric on X which is continuous on $X \times X$. If $\mathcal{F} \subset \mathcal{C}(X, Y)$ is Lipschitz with respect to ρ and σ then $\mathcal{F}$ is relatively compact in $\mathcal{C}(X, Y^{\infty})$.*

Proof. Suppose $\mathcal{F}$ is Lipschitz of order c. We show that $\mathcal{F}$ is evenly continuous from X to Y^{∞}. If not, there exist $p \in X, q, s \in Y^{\infty}$ and nets $\{p_\alpha\}, \{f_\alpha\}$ in $X, \mathcal{F}$ respectively such that $p_\alpha \to p$, $s \neq q$, $f_\alpha(p_\alpha) \to s$, $f_\alpha(p) \to q$. If $q \in Y$, then for each α we have

$$\sigma(f_\alpha(p_\alpha), q) \leq \sigma(f_\alpha(p_\alpha), f_\alpha(p)) + \sigma(f_\alpha(p), q) \leq c\rho(p_\alpha, p) + \sigma(f_\alpha(p), q).$$

So $\sigma(f_\alpha(p_\alpha), q) \to 0$ and $q = s$, a contradiction. If $s \in Y$, then for each α we have

$$\sigma(f_\alpha(p), s) \leq c\rho(p, p_\alpha) + \sigma(f_\alpha(p_\alpha), s)$$

Hence $\sigma(f_\alpha(p), s) \to 0$ and $s = q$, a contradiction. The theorem is proved. □

3.2 Marty's Normality Criterion

L. Ahlfors said in [4, p. 225]:

> If a sequence tends to ∞, there is no great scattering of values, and it may well be argued that for the purposes of normal families such a sequence should be regarded as convergent. This is the classical point of view, and we shall restyle our definition to conform with traditional usage.

So the right way to give the definition of normal families is the following.

Definition 3.2.1. A family $\mathcal{F}$ of meromorphic functions from a domain $\Omega \subset \mathbb{C}^n$ to the Riemann sphere $\widehat{\mathbb{C}}$ is normal in Ω if every sequence of functions $\{f_j\} \subseteq \mathcal{F}$ contains a subsequence which converges uniformly on each compact subset of Ω.

This elegant definition of normal family is from the Cima and Krantz 1983 article [47, Definition 1.2, p. 306]. We think not of holomorphic functions but of meromorphic functions. Given that, one thinks of the functions as taking values in the Riemann sphere. If we do it that way, then we don't have to talk about compact divergence anymore. We can still say that a family is normal if there is a subsequence that converges uniformly on compact sets. Since the range is now the Riemann sphere, convergence to ∞ is a possibility.

For more information on this subject, see the book of Krantz [142, Chapter 2, Section 2.3, p.79].

A sequence which converges uniformly on compact subsets of a region Ω is sometimes said to converge normally on Ω or converge locally uniformly on Ω.

Definition 3.2.2. A family $\mathcal{F}$ is said to be *normal* at a point $p \in \Omega$ if it is normal in some neighborhood of p.

[Throughout this book, the term "neighborhood of p" refers to an open set that contains the point p.] In these circumstances, the convergence is then uniform on each compact subset of Ω.

It is routine to confirm that a family of holomorphic functions $\mathcal{F}$ is normal in a domain Ω if and only if $\mathcal{F}$ is normal at each point of Ω.

In terms of the spherical metric an extremily useful normality criterion for a family of meromorphic functions of one complex variable was given in 1931 by Frèdèric Ladislas Joseph Marty in his dissertation [187]. See Hayman [111, pp. 158–160], for references and a statement and for another proof of the result and Royden [252], Schiff [260] for an extension. The important paper of Carathéodory [38] should also be mentioned in this context. Marty's normality criterion implies many fundamental theorems in the one-dimensional theory.

We shall prove Marty's normality criterion for a family of holomorphic functions of several complex variables in terms of the spherical derivative.

Theorem 3.2.3 (Marty, see [68, Theorem 2.1]). *A family $\mathcal{F}$ of functions holomorphic on Ω is normal on $\Omega \subset \mathbb{C}^n$ if and only if for each compact subset $K \subset \Omega$ there exists a constant $M(K)$ such that at each point $z \in K$*

$$f^\sharp(z) \leq M(K) \tag{3.1}$$

for all $f \in \mathcal{F}$.

Proof. Of course it is enough to show that $\mathcal{F}$ is normal at every point $z_0 \in \Omega$. We appeal to the theorem of Arzelà-Ascoli (Theorem 2.0.9). As the Riemann sphere $\widehat{\mathbb{C}}$ is compact, the second condition of Arzelà and Ascoli is automatically fulfilled, and our task reduces to checking that a family is equicontinuous on compacts in Ω when it is spherically bounded on such compacta.

Let us show that (3.1) implies that the family $\mathcal{F}$ is equicontinuous. Fix an arbitrary $z_0 \in \Omega$ and choose $r > 0$ such that

$$K = \{z \in \mathbb{C}^n : |z - z_0| \leq r\} \subset \Omega.$$

Let $z(t) = z_0 + t(z_1 - z_0) \subset K$, for all $t \in [0, 1]$. It is elementary to see that

$$L_{z(t)}(\log(1 + |f|^2), v) = \frac{|Df(z)v|^2}{(1 + |f(z)|^2)^2} \geq 0, \tag{3.2}$$

where

$$Df(z)v := \sum_{k=1}^{n} \frac{\partial f}{\partial z_k}(z) v_k.$$

Hence

$$\frac{|df(z(t))/dt|^2}{(1 + |f(z(t))|^2)^2} = L_{z(t)}(\log(1 + |f|^2), z'(t)) \leq (f^\sharp(z(t))|z'(t)|)^2.$$

Then the spherical distance satisfies

$$s(f(z_0), f(z_1)) \leq \int_0^1 f^\sharp(z(t))|z'(t)|dt \leq M(K) \int_0^1 |z'(t)|dt \leq M(K)|z_1 - z_0|$$

for all $f \in \mathcal{F}$. Thus we see that the family $\mathcal{F}$ is equicontinuous on K—from the Euclidean metric to the spherical metric. The Ascoli-Arzelà theorem (Theorem 2.0.9) now implies that $\mathcal{F}$ is normal at z_0.

Conversely, assume that $\mathcal{F}$ is normal in Ω. If (3.1) fails for some compact set $K \subset \Omega$, then there exists a sequence $\{f_j\} \subset \mathcal{F}$ which converges locally uniformly in Ω to some $f \not\equiv \infty$ and such that the maximum of $f_j^\sharp$ over K tends to ∞. By Weierstrass's theorem (Theorem 2.0.13), $\partial f_j / \partial z_k$ converges uniformly to $\partial f / \partial z_k$ on K; therefore, there exists j_0 such that for $j > j_0$.

$$L_z(\log(1 + |f_j|^2), v) \leq L_z(\log(1 + |f|^2), v) + O(|v|^2) \text{ for all } z \in K, v \in \mathbb{C}^n.$$

It follows that

$$f_j^\sharp(z) \leq f^\sharp(z) + O(1) \text{ for all } j > j_0.$$

This is a contradiction, since the right-hand side is bounded on K.

If $f \equiv \infty$, then consider a domain Ω' containing K and satisfying $\overline{\Omega'} \subset \Omega$. Then there exists j_0 such that the functions $1/f_j$ are holomorphic in Ω' for $j > j_0$. Since $1/f_j \to 0$ uniformly in Ω', we see as above that $(1/(f_j)^\sharp)$ are bounded on K. Since the Levi form vanishes for any pluriharmonic function,

$$L_z(\log(1+|1/f_j|^2), v) = L_z(\log(1+|f|^2), v) - 2L_z(\log|f|, v) = L_z(\log(1+|f|^2), v). \quad (3.3)$$

Therefore

$$f_j^\sharp(z) = (1/f_j)^\sharp(z)$$

and we again obtain a contradiction. □

Remark 3.2.4. The proof of Marty's theorem shows that if holomorphic functions f_j converge uniformly to f in the spherical metric, then $f_j^\sharp$ converges to $f^\sharp$. The converse fails: consider a sequence of constants, or a sequence of rotations of a single function.

Specifically, Theorem 3.2.3 can be rephrased in this way:

Theorem 3.2.5 (see [62, Lemma (Marty's criterion)]). *A family $\mathcal{F}$ of functions holomorphic on Ω is normal on $\Omega \subset \mathbb{C}^n$ if and only if for each compact subset $K \subset \Omega$, there exists a constant $M(K)$ such that at each point $z \in K$ and all vectors $v \in \mathbb{C}^n$*

$$L_z(\log(1+|f|^2), v) \leq M(K)|v|^2 \quad (3.4)$$

for all $f \in \mathcal{F}$.

Remark 3.2.6. The inequality (3.4) is equivalent to the assertion that for each compact set $K \subset \Omega$, there exists a positive number $M(K)$ such that for all $z \in K$ and all $f \in \mathcal{F}$,

$$L_z(\log(1+|f|^2), v)/|v|^2 \leq M(K)$$

for all $v \in \mathbb{C}^n \setminus \{0\}$. When Ω is bounded domain in $\mathbb{C}^n$, this is equivalent to the existence of a positive constant $M'(K)$ such that

$$L_z(\log(1+|f|^2), v)/F_\Omega^K(z, v) \leq M'(K)$$

for all $z \in K$ and all $f \in \mathcal{F}$ because the Kobayashi and Euclidean metrics are bi-Lipschitz equivalent on compact subsets of Ω.

Remark 3.2.7. Condition (3.4) means that at each point $z \in K$ the Hermitian form

$$M(K)|v|^2 - L_z(\log(1+|f|^2), v)$$

is positive semi-definite.

Remark 3.2.8. Condition (3.4) implies that

$$s(f(z), f(w)) \leq M(K)|z - w| \text{ for all } f \in \mathcal{F} \text{ and all } z, w \in K.$$

The Marty criterion for a family $\mathcal{F}$ implies that the family $\mathcal{F}$ is s-to-e Lipschitz on compact subsets of Ω.

Loosely speaking, with the correct metrics, normality for families of holomorphic functions is equivalent to the existence of a uniform Lipschitz condition on compact subsets of the region.

As an immediate application, we have the following result.

Corollary 3.2.9. *Let $\{f_j\}$ be a sequence of holomorphic functions on $B(0, 1)$ which converges uniformly on $B(0, r)$, $r < 1$, to a function f, and let $\{z_j\}$ be a sequence converging to* 0. *Then the sequence*

$$g_j(z) := f(z + z_j)$$

also converges uniformly to f on $B(0, r_1)$, $r_1 < r$.

Proof. The hypothesis that $f_j \to f$ converges uniformly on $B(0, r)$, $r < 1$, implies the normality of the sequence $\{f_j\}$ on $B(0, 1)$. Therefore there exists $M > 0$ such that $f^\sharp(z) < M$ for $z \in B(0, (r + r_1)/2)$. Hence

$$\begin{aligned} s(f_j(z + z_j), f(z)) &< \\ &s(f_j(z + z_j), f_j(z)) + s(f_j(z), f(z)) < \\ &M|z_j| + s(f_j(z), f(z)). \end{aligned}$$

This shows that $f(z + z_j) \to f(z)$ uniformly on $B(0, r_1)$. □

Remark 3.2.10. Theorem 3.4 is similar to the one obtained by Timoney [276, Theorem 3.1, p. 252], but the article of the latter was submitted for consideration on October 17, 1979, while the article [62] was submitted on July 18, 1979.

Theorem 3.2.5 has been the impetus for a large number of investigations. Several authors, including Cima and Krantz [47], Hahn [105], Josep and Kwack [124] and Funahashi [81], to name only a few, proved Marty's criterion for holomorphic mappings and extended to higher dimensions some classical theorems.

Marty's normality criterion should not be taken lightly, for it has many important applications. Using Theorem 3.2.5 in Chapter 4, we will find a way to extend the definition of normal holomorphic function in the unit disk to arbitrary domains in $\mathbb{C}^n$, $n > 1$.

3.3 Zalcman's Rescaling Lemma

Marty's normality criterion has a host of applications. Here we generalize Zalcman's rescalling lemma ([291, Lemma, p. 814]) to the theory of holomorphic functions in $\mathbb{C}^n$. The provided proof is fairly short and elementary; it uses only Marty's normality criterion (Theorem 3.2.3). The proof of Zalcman's rescaling lemma in $\mathbb{C}^n$ roughly follows the proof given in ([272, Theorem 4.1, p. 113]) for the one-dimensional case. Professor R. Steinmetz informed me, via email, that the proof in his book does not belong to him but is essentially the proof he learned from Professor L. Zalcman.

By an *entire function* we shall mean an element of $\mathcal{O}(\mathbb{C}^n)$.

Theorem 3.3.1 (see [68, Theorem 3.1]). *A family $\mathcal{F}$ of functions holomorphic on $\Omega \subset \mathbb{C}^n$ is not normal at some point $z_0 \in \Omega$ if and only if there exist sequences $f_j \in \mathcal{F}$, $z_j \to z_0$, $\rho_j = 1/f_j^\sharp(z_j) \to 0$, such that the sequence*

$$g_j(z) := f_j(z_j + \rho_j z)$$

converges locally uniformly in $\mathbb{C}^n$ to a nonconstant entire function g satisfying $g^\sharp(z) \le g^\sharp(0) = 1$.

It is remarkable that non-normality can be described in terms of a convergent sequence. If $\{f_j\}$ were convergent, then the functions g_j would converge to a constant on compact subsets of $\mathbb{C}^n$, since the radii ρ_j tend to 0. Zalcman's lemma (Theorem 3.3.1) says that arbitrarily small balls centered at z_j can be found where f_j is close to a non-trivial entire (holomorphic in $\mathbb{C}^n$) function, after rescaling.

Proof of Theorem 3.3.1. The proof will be based on Marty's normality criterion (Theorem 3.2.3). Assume $\mathcal{F}$ is not normal at a point $z_0 \in \Omega$. To simplify matters we assume that $z_0 = 0$ and all functions under consideration are holomorphic on the unit ball $\{z \in \mathbb{C}^n : |z| < 1\}$. By Marty's normality criterion (Theorem 3.2.3), $\mathcal{F}$ contains functions f_j, $j \in \mathbb{N}$, satisfying $\sup_{|z|<1/(2j)} f_j^\sharp(z) > 2j^3$. Hence there exists a ξ_j with $|\xi_j| < 1/j$ such that

$$\max_{|z|\le 1/j} (1 - j|z|) f_j^\sharp(z) = (1 - j|\xi_j|) f_j^\sharp(\xi_j) \ge j^3.$$

Put $r_j = 1/f_j^\sharp(\xi_j)$. Then from the above inequality we have $(1-j|\xi_j|)/j^2 \ge j \cdot r_j$. Set $h_j(z) = f_j(\xi_j + r_j z)$. We claim that appropriately chosen subsequences $z_k = \xi_{j_k}$, $\rho_k = r_{j_k}$, and $g_k = h_{j_k}$ will do. First of all, h_j is defined on $|z| < (1 - |\xi_j|)/r_j = (1 - |\xi_j|) f_j^\sharp(\xi_j)$, hence on $|z| < j^3$. By the invariance of the Levi form under biholomorphic mappings, we have

$$L_z(\log(1 + |h_j|^2), v) = L_{\xi_j + r_j z}(\log(1 + |f_j|^2), r_j v)$$

and hence

$$h_j^\sharp(z) = r_j f_j^\sharp(\xi_j + r_j z) = \frac{f_j^\sharp(\xi_j + r_j z)}{f_j^\sharp(\xi_j)}.$$

Recall that $|z| < j$ and $j \cdot r_j \leq (1 - j|\xi_j|)/j^2$. It follows that for all $|z| < j$,

$$|\xi_j + r_j z| \leq |\xi_j| + r_j|z| < |\xi_j| + \frac{1}{j^2}(1 - j|\xi_j|) <$$
$$\left(1 - \frac{1}{j}\right)|\xi_j| + \frac{1}{j^2} < \frac{1}{j} - \frac{1}{j^2} + \frac{1}{j^2} = \frac{1}{j}.$$

Therefore

$$(1 - j|\xi_j + r_j z|)f_j^\sharp(\xi_j + r_j z) \leq (1 - j|\xi_j|)f_j^\sharp(\xi_j)$$

and

$$h_j^\sharp(z) = \frac{f_j^\sharp(\xi_j + r_j z)}{f_j^\sharp(\xi_j)} \leq \frac{1 - j|\xi_j|}{1 - j|\xi_j + r_j z|} \leq$$
$$\frac{1 - j|\xi_j|}{1 - j|\xi_j| - j \cdot r_j|z|} \leq \frac{1}{1 - 1/j} \quad (|z| < j)$$

where we again have used $j \cdot r_j \leq \frac{1}{j^2}(1 - j|\xi_j|)$. For every $m \in \mathbb{N}$ the sequence $\{h_j\}_{j>m}$ is normal in $|z| < m$ by Marty's normality criterion (Theorem 3.2.3). The well-known Cantor diagonalization process yields a subsequence $\{g_k = h_{j_k}\}$ which converges uniformly on every ball $|z| < R$. The limit function g satisfies $g^\sharp(z) \leq \limsup_{j\to\infty} h_j^\sharp(z) \leq 1 = g^\sharp(0)$. Clearly, g is nonconstant because $g^\sharp(0) = 1$.

Conversely, suppose that there exist sequences $f_j \in \mathcal{F}$, $z_j \to z_0$, $\rho_j = 1/f_j^\sharp(z_j) \to 0$, such that the sequence

$$g_j(z) = f_j(z_j + \rho_j z)$$

converges locally uniformly in $\mathbb{C}^n$ to a nonconstant entire function g satisfying $g^\sharp(z) \leq g^\sharp(0) = 1$, but $\mathcal{F}$ is normal. Without loss of generality we may suppose that $z_0 = 0$. By Marty's normality criterion (Theorem 3.2.3), there exists a constant $M > 0$ such that

$$\max_{|z|<1/j_0} f_j^\sharp(z) < M.$$

Since $z_j \to 0$, $\rho_j = 1/f_j^\sharp(z_j) \to 0$, we see that for $|z| < 1/j_0$, there exists j_1 such that for all $j > j_1$

$$|z_j + \rho_j z| \leq |z_j| + \rho_j|z| \leq |z_j| + \rho_j/j_0 < 1/j_0.$$

Therefore

$$g_j^\sharp(z) = f_j^\sharp(z_j + \rho_j z)/f_j^\sharp(z_j) = f_j^\sharp(z_j + \rho_j z)\rho_j \leq M\rho_j \to 0.$$

But then $g^\sharp(0) = 0$, a contradiction, since $g^\sharp(0) = 1$. This contradiction concludes the proof of the theorem. □

Remark 3.3.2. In [6] Aladro and Krantz gave a theorem of a normality criterion for families of holomorphic mappings $\{f_j\}$ from a domain Ω of $\mathbb{C}^n$ into a complete Hermitian complex space M. Unfortunately, they failed to deal with the compactly divergent case of the obtained mapping $g_j(\xi) = f_j(p_j + \rho_j\xi)$ $(\rho_j \searrow 0)$. Do Duc Thai, Pham Nguyen Thu Trang, and P. D. Huong [275] fixes the statement of the theorem and completes the proof. They give a correct generalization of Zalcman's lemma to complex spaces ([275, Theorem 2.5 and Corollaries 2.8, 2.13, and Proposition 2.11]).

Zalcman in his 1993 article "Normal families revisited", Complex Analysis and Related Topics (J. J. O. O. ' Wiegerinck, ed.), Univ. ' of Amsterdam, Amsterdam, 1993, pp. 149–164., on page 156, made the following comment about the use of Marty's theorem: "The proof of the lemma is one of the few really effective uses of Marty's theorem of which I am aware. It is of such a general character that it is rather easily adapted to other situations. Appropriate versions thus hold for families of quasiregular or quasimeromorphic functions in space [191] and certain holomorphic mappings in $\mathbb{C}^n$ [6], leading to versions of Theorem 1 for these classes. One almost immediate consequence of the (proof of the) Lemma is Brody's theorem. A compact complex manifold is hyperbolic if and only if it contains no complex lines".

Zalcman's lemma is of such a general character that it is rather easily adapted to other situations (see for example [191, Theorem 1, p. 35] and [279]). Higher-dimensional analogue of Zalcman's lemma have also proven to be quite useful [15, 186].

3.4 Pointwise Limits of Holomorphic Functions

While most elementary texts on real analysis contain an example to show that a pointwise limit of differentiable functions need not be differentiable, hardly any texts on complex analysis contain an example to show that the pointwise limit of analytic functions need not be analytic.

Beardon and Minda [51] provide a simple example to show that the pointwise limit of a sequence of analytic functions need not be analytic.

From a modern viewpoint, it is easy to see from Montel's theory of normal families why an example of a nonanalytic pointwise limit of analytic functions requires a little work. Montel proved that the family of functions f that are analytic in a region Ω, and that map Ω into the complement of three given points w_1, w_2, and w_3 in the extended plane, is normal in Ω. It follows from this that if a sequence f_n of analytic functions is pointwise convergent in Ω, and if each f_n fails to take any of the values w_j there, then the convergence is locally uniform, and the limit is analytic. Thus, in any example f_n of the type we are seeking, $f_n(\Omega)$ must cover all but two points of the complex sphere.

However, Beardon and Minda [51] take a different route, and they did not use Montel's theorem at all in that paper.

The example of a nonanalytic limit given in the article of Beardon and Minda [51] needs only Cauchy's integral formula; it is a slight modification of Exercise 11 in [259, p. 321], and a more detailed discussion (though not on this aspect) can be found in [71, 5, pp. 160–162], where it is shown to be closely related to the so-called Mittag-Leffler function.

Example 3.4.1. We shall construct an entire function E that has a radial limit 0 at ∞ in each direction except along the positive real axis $\mathbb{R}^+$, where $\mathrm{Re}[E(x)] \to +\infty$ as $x \to +\infty$. Given this, we let $F(z) = \exp(-E(z))$, so that F is an entire function with radial limit 1 at ∞ in all directions except along $\mathbb{R}^+$, where it has radial limit 0. It follows that the sequence of entire functions F_n defined by $F_n(z) = F(nz)$ converges pointwise on $\mathbb{C}$ to a function that is 1 on the complement of $[0, +\infty)$ and 0 on $(0, +\infty)$.

For each positive a, let γ_a be the boundary curve of the half-strip H_a given in $\mathbb{R}^2$ by $(a, +\infty) \times (-\pi, \pi)$, and let E_a be the exterior of H_a. Notice that if $a < b$ then $E_a \subset E_b$, and that $H_a \cap E_b$ is the open rectangle $(a, b) \times (-\pi, \pi)$.

Now let $f(z) = \exp(\exp(z))$. As $|f(x \pm i\pi)| = 1/\exp(\exp x)$, $|f|$ is integrable on γ_a with integral $\|f\|_{1,a}$ with respect to $|dz|$, say, and this ensures that the function

$$I_a(z) = \frac{1}{2\pi i} \int_{\gamma_a} \frac{f(w)}{w - z} dw$$

exists and is analytic in the complement of γ_a. Obviously

$$|I_a(z)| \leq \frac{\|f\|_{1,a}}{2\pi \operatorname{dist}(z, \gamma_a)},$$

and this shows (i) that I_a has radial limit 0 at ∞ in all directions except along $\mathbb{R}^+$, and (ii) that I_a is bounded on the real segment $[a+1, +\infty)$. In fact, $I_a(x) \to 0$ as $x \to +\infty$, but we do not need this.

Next, suppose that $0 < a < b$ and that $z \notin \gamma_a \cup \gamma_b$. Then

$$I_b(z) - I_a(z) = \frac{1}{2\pi i} \int_{\partial(H_a \cap E_b)} \frac{f(w)}{w - z} = f(z)\chi_{a,b}(z),$$

where $\chi_{a,b}$ is the characteristic function of the rectangle $H_a \cap E_b$. This shows that $I_b(z) = I_a(z)$ on E_a, and as $\cup_a E_a = \mathbb{C}$; this means that we can define an entire function E by $E(z) = I_a(z)$ on E_a, for each a. In particular, E has radial limit 0 at ∞ in every direction except possibly along $\mathbb{R}^+$. If $x \in H_a \cap \mathbb{R}^+$ then we choose any b with $b > x$; then $x \in E_b$ and

$$E(x) = I_b(x) = I_a(x) + f(x) = O(1) + \exp(\exp x)$$

as $x \to +\infty$ so that $\mathrm{Re}[E(x)] \to +\infty$ as $x \to +\infty$. This completes the example.

We remark that (as stated in [71, 5, p. 327]) the function F in this example also provides an interesting example relating to Liouville's theorem, for the function $[1 - F(z)][1 - F(-z)]$ has radial limit zero in all directions, but it is not identically zero.

There are also some positive results.

Proposition 3.4.2. *Let* $\mathcal{F}$ *be a pointwise bounded family of holomorphic functions on* $\Omega \subset \mathbb{C}^n$. *There is a (maximal) dense open set* $\Omega_0 \subset \Omega$ *such that* $\mathcal{F}$ *restricted to* Ω_0 *is locally bounded.*

Proof. For each z in Ω, let $\phi(z)$ denote the least upper bound for

$$\{|f(z)| : f \in \mathcal{F}\}.$$

Because each f in $\mathcal{F}$ continuous, it is easy to see that $K_j = \{z : \phi(z) \leq j\}$ is (relatively) closed in Ω. Since $\phi(z)$ is finite for each z, Ω is the union of the K_j. The Baire category theorem shows that, for every ball B in Ω, $K_j \cap B$ has interior for large j. Thus, the union Ω_0 of the interiors $\operatorname{int} K_j$ is a dense open subset of Ω. Obviously $\mathcal{F}$ bounded on each $\operatorname{int} K_j$, and hence $\mathcal{F}$ is locally bounded on a dense set Ω_0. Conversely, if $\mathcal{F}$ bounded by j on an open set Ω, then B is contained in $\operatorname{int} K_j$ and thus is in Ω_0. □

Corollary 3.4.3 (Osgood)**.** *Suppose that* $f_1, f_2, \ldots$ *are holomorphic in a domain* $\Omega \subset \mathbb{C}^n$, *and that* $f_j \to f$ *pointwise in* Ω. *Then there is a dense open subset* Ω_0 *of* Ω *on which the convergence is locally uniform, and in which* f *is holomorphic.*

Proof. A sequence of functions converging pointwise is pointwise bounded. By Proposition 3.4.2, $\{f_j\}$ is locally bounded on Ω_0. By Montel's Theorem 2.0.12, a subsequence converges uniformly on compact subsets to f on Ω_0, and hence f is holomorphic on Ω_0. □

The set where f is holomorphic does not necessarily coincide with the set on which convergence is locally uniform (see [51, Example 1, 2]). No proof is included, for all existing proofs rely on mathematical machinery that we do not wish to introduce in the present text.

Proposition 3.4.4. *If* $\{f_j\}$ *is a sequence of bounded holomorphic functions on* $\Omega \subset \mathbb{C}^n$ *converging pointwise to* f, *then* f *is holomorphic and the convergence is uniform on compact subsets.*

Proof. The family $\mathcal{F} = \{f_j\}$ is bounded and thus locally bounded. By Montel's Theorem 2.0.12, every subsequence of $\{f_j\}$ has a sub-subsequence which converges uniformly on compact subsets. This limit is f perforce, so f is holomorphic. If the whole sequence does not converge to f uniformly on compact subsets, there would be a compact set K in Ω, an $\varepsilon > 0$, a subsequence $\{f_{j_k}\}$, and points z_j in K such that

$$|f_{j_k}(z_k) - f(z_k)| \geq \varepsilon$$

for $k \geq 1$. Thus no subsequence of $\{f_{j_k}\}$ could converge uniformly to f on K, which is a contradiction. □

Theorem 3.4.5 is an extension of Proposition 3.4.4.

Theorem 3.4.5. *If $\{f_j\}$ is a sequence of holomorphic functions on $\Omega \subset \mathbb{C}^n$ whose ranges omit 0 and 1 there, and converging pointwise to f, then f is holomorphic and the convergence is uniform on compact subset.*

Proof. Because the family $\mathcal{F}$ manifestly omits two values in Ω, Montel's theorem in $\mathbb{C}^n$ empowers us to extract from the sequence $\{f_j\} \subseteq \mathcal{F}$ a subsequence $\{f_{j_k}\} \subseteq \mathcal{F}$ that converges normally in Ω to a function f. This gives the conclusion by repeating verbatim the argument at the end of the proof of Proposition 3.4.4. □

As an illustration of the direct use of normality rather than equicontinuity, we consider Vitali's theorem: this exploits analytic continuation and so enables us derive information about the limit f of a sequence $\{f_j\}$ throughout a domain Ω simply from a knowledge of f on only a small part of Ω.

Theorem 3.4.6 (Vitali's Theorem). *Suppose that the family $\mathcal{F} = \{f_j\}$ of holomorphic functions is normal in a domain Ω, and that f_j converges pointwise to some function f on some nonempty open subset W of Ω. Then f extends to a function F which is holomorphic on Ω, and $f_j \to f$ locally uniformly on Ω.*

Proof. As $\mathcal{F}$ is normal in Ω, there is some subsequence of $\{f_j\}$ which converges locally uniformly in Ω to some function F. By Theorem 2.0.12 F is holomorphic in Ω, and $F = f$ on W.

Now suppose that if $\{f_j\}$ fails to converge locally uniformly on Ω to F. Then there is a compact subset K of Ω, a positive ε, and a subsequence, say $\{g_j\}$, of $\{f_j\}$ such that for all j,

$$\sup_K s(g_j(z), F(z)) \geq \varepsilon.$$

However, by normality, there is some subsequence, say $\{h_j\}$, of $\{g_j\}$ which converges locally uniformly to some function h on Ω. Clearly, $h = f = F$ on W, and as h is holomorphic in Ω, we must have $h = F$ throughout Ω. It follows that

$$\sup_K s(h_j(z), F(z)) \to 0.$$

and as $\{h_j\}$ is a subsequence of $\{g_j\}$, this violates the preceding inequality (which holds for all j). The proof is complete. □

3.5 Montel's Normality Criteria

Marty's normality criterion provides less than complete information, principally because condition (3.1) is generally very difficult to verify in those situations in which it is not already evident that the family $\mathcal{F}$ is normal. Accordingly, there has been a continuing search for other conditions which imply normality. One of the earliest such results in the classical theory of normal families of one complex variable, and certainly the most important, was found by Montel himself [200].

The first, and simpler, version of the Montel theorem states that a family of holomorphic functions defined on an open subset of the complex plane $\mathbb{C}$ is normal if and only if it is locally uniformly bounded.

The first version of Montel's theorem is a direct consequence of Marty's theorem and the Cauchy integral formula.

For the formulation of the second (strongest) version of Montel's theorem, we need the following definition.

Definition 3.5.1. A complex quantity $a \in \mathbb{C}$ is an *omit value* of a family $\mathcal{F}$ of holomorphic functions defined on a domain Ω if

$$a \notin \bigcup_{f \in \mathcal{F}} f[\Omega].$$

The stronger version of Montel's theorem (occasionally referred to as the fundamental normality test) states that a family of holomorphic functions, all of which omit the same two values $a, b \in \mathbb{C}$ is normal.

The second version of Montel's theorem can be deduced from the first by using the fact that there exists a holomorphic universal covering from the unit disk to the twice punctured plane $\mathbb{C} \setminus \{a, b\}$. (Such a covering is given by the elliptic modular function.)

We will prove the extension of Montel's normality criterion to the context of several complex variables.

Theorem 3.5.2 (Montel, see [66, Theorem 2.1])**.** *Let $\mathcal{F}$ be a family of holomorphic functions on an open set $\Omega \subseteq \mathbb{C}^n$ that omit two fixed complex values. Then each sequence of functions in $\mathcal{F}$ has a subsequence which converges uniformly on compact subsets.*

We shall give two proofs of Montel's normality criterion in $\mathbb{C}^n$ based on Zalcman's rescalling lemma (Theorem 3.3.1).

First proof of Montel's normality criterion 3.5.2. : Composing the functions of $\mathcal{F}$ with a linear fractional transformation, we may assume that the omitted values are 0 and 1. Suppose $\mathcal{F}$ is not normal on Ω. Then, by Zalcman Lemma (Theorem 3.3.1), there exist $f_j \in \mathcal{F}$, $z_j \in \Omega$ and $\rho_j \to 0+$ such that $f_j(z_j + \rho_j\zeta) = g_j(\zeta) \to g(\zeta)$ uniformly on compact subsets of $\mathbb{C}^n$, where g is a nonconstant entire function. By Hurwitz's theorem (Theorem 2.0.14), g does not take on the values 0 and 1, since no f_j does. Let $b \in \mathbb{C}^n$. The function

$$g_b(\lambda) := g(\lambda \cdot b)\,,$$

is an entire function on $\mathbb{C}$, satisfying

$$g_b(0) = g(0), \quad g_b(1) = g(b)$$

and

$$g_b(\mathbb{C}) \subset g(\mathbb{C}^n) \subseteq \mathbb{C} \setminus \{0, 1\}.$$

But then, by the one-dimensional version of Picard's little theorem, g_b is constant, hence $g(0) = g(b)$ for all $b \in \mathbb{C}^n$, a contradiction. □

We also give a simple proof of the theorem of Montel, based on the idea of A. Ros [249, Theorem 2, p. 240] (see [291, p. 218]).

Second proof of Montel's normality criterion 3.5.2. : Since normality is a local notion, we may suppose that $\Omega = B$, the unit ball. So let $\mathcal{F}$ be as in the statement of Montel's normality criterion (Theorem 3.5.2) and suppose that $\mathcal{F}$ is not normal. Composing with a linear fractional transformation, we may assume that the omitted values are 0 and 1. This implies that if $f \in \mathcal{F}$ and $m \in \mathbb{N}$, then there exists a function g holomorphic in Ω such that $g^{2^m} = f$. Let $\mathcal{F}_m$ be the family of all such functions g. Note that

$$\left|\frac{1}{2^m}\right|^2 \Big(\frac{|f|^{-1} + |f|}{|f|^{-1/2^m} + |f|^{1/2^m}}\Big)^2 L_z(\log(1 + |f|^2, v)) = L_z(\log(1 + |g|^2, v))$$

for all $(z, v) \in \Omega \times \mathbb{C}^n$. This implies that

$$g^\sharp(z) = \frac{1}{2^m} \frac{|f|^{-1} + |f|}{|f|^{-1/2^m} + |f|^{1/2^m}} f^\sharp(z) \geq \frac{1}{2^m} f^\sharp(z) \quad (z \in \Omega),$$

where we have used the basic inequality $a^{-1} + a \geq a^{-t} + a^t$ valid for $a > 0$ and $0 < t < 1$. By Marty's normality criterion (Theorem 3.2.3), the family $\{f^\sharp : f \in \mathcal{F}\}$ is not locally bounded. We deduce that for fixed $m \in \mathbb{N}$, the family $\{g^\sharp : g \in \mathcal{F}_m\}$ is not locally bounded. Using Marty's normality criterion (Theorem 3.2.3) again, we find that $\mathcal{F}_m$ is not normal, for all $m \in \mathbb{N}$. Note that if $g \in \mathcal{F}_m$, then g omits the values $e^{2\pi ik/2m}$ for $k, m \in \mathbb{Z}$. From the Zalcman rescalling lemma (Theorem 3.3.1), we thus deduce that there exists an entire function g_m omitting the values $e^{2\pi ik/2m}$ and satisfying $g_m(z) \leq g_m^\sharp(0) = 1$. The g_m thus form a normal family and we have $g_{m_j} \to G$ for some subsequence $\{g_{m_j}\}$ and some nonconstant entire function G. By Hurwitz's theorem (Theorem 2.0.14), G omits the values $e^{2\pi ik/2m}$ for all $k, m \in N$. Since $G(\mathbb{C}^n)$ is open this implies that $|G(z)| \neq 1$ for all $z \in \mathbb{C}^n$. Thus either $|G(z)| < 1$ for all $z \in \mathbb{C}^n$ or $|G(z)| > 1$ for all $z \in \mathbb{C}^n$. In the first case G is bounded and thus constant by Liouville's theorem. In the second case $1/G$ is bounded. Again $1/G$ is constant and thus G is constant. Thus we get a contradiction in both cases. □

We shall now show that the following result, in which the values omitted are allowed to vary with the function, as long as they do not approach one another too closely, is also true.

Theorem 3.5.3 (Montel-Carathéodory's theorem). *Let* $\mathcal{F}$ *be a family of functions holomorphic on* $\Omega \subset \mathbb{C}^n$. *Suppose that for some* $\varepsilon > 0$, *there exist for each* $f \in \mathcal{F}$ *distinct points* $a_f, b_f \in \mathbb{C}$ *such that for all* $z \in \Omega$, $f(z) \neq a_f, b_f$ *and*

$$s(a_f, \infty)s(a_f, b_f)s(\infty, b_f) > \varepsilon.$$

Then $\mathcal{F}$ *is normal on* Ω.

Proof. Otherwise, there exists some ball B in Ω, which we may assume to be unit ball, on which $\mathcal{F}$ fails to be normal. Then, by Zalcman's Rescalling Lemma 3.3.1, there exist $f_j \in \mathcal{F}$, $z_j \in B$ and $\rho_j \to 0+$ such that $f_j(z_j + \rho_j\zeta) = g_j(\zeta) \to g(\zeta)$ uniformly on compact subsets of C^n, where g is a nonconstant entire function. Taking successive subsequences and renumbering, we can assume that $a_{f_j} \to a$ and $b_{f_j} \to b$, where a and b are distinct points in $\mathbb{C}$. Since $g_j(\zeta) - a_{f_j} \neq 0$ and g is nonconstant, it follows from the Hurwitz Theorem (Theorem 2.0.14) that $g(\zeta) \neq a$. Similarly, $g(\zeta) \neq b$. Again, we may assume that the omitted values are 0 and 1. The reasoning used at the end of Theorem 3.5.2 now shows that g is a constant, a contradiction. □

Montel's theorem remains valid if the omitted values are replaced by omitted functions, so long as the omitted functions never take on the same value at points of Ω.

Theorem 3.5.4 (Fatou's theorem). *Let* $a(z)$ *and* $b(z)$ *be functions holomorphic on* $\Omega \subset \mathbb{C}^n$ *such that* $a(z) \neq b(z)$ *for each* $z \in \Omega$. *Let* $\mathcal{F}$ *be a family of functions holomorphic on* Ω *such that, for each* $z \in \Omega$,

$$f(z) \neq a(z) \quad , \quad f(z) \neq b(z)$$

for all $f \in \mathcal{F}$. *Then* $\mathcal{F}$ *is normal on* Ω.

Proof. Consider the family of functions

$$\mathcal{G} = \left\{ \frac{f(z) - a(z)}{f(z) - b(z)} \text{ for all } f \in \mathcal{F} \right\}.$$

Then each $g \in \mathcal{G}$ is holomorphic on Ω; and if $g \in \mathcal{G}$, then $g(z) \neq 0, 1$ for $z \in \Omega$. Thus $\mathcal{G}$ is normal on Ω by Montel's Theorem 3.5.2. But then, as is easily seen, $\mathcal{F}$ is normal on Ω as well. □

3.6 Application of Montel's Theorem

In this section we give an application of the theorem of Montel (Theorem 3.5.2) and, us the graph of a function, we give another description of the "Conjecture of Julia". We also prove that, if U is a domain in the complex plane, then we get the converse of the theorem of Hurwitz.

Recall that $\Omega \subset \mathbb{C}^n$ is called a domain of normality if (1) there exists a family $\mathcal{F}$, holomorphic in the domain Ω, which which forms in this domain a normal family; (2) there is no domain $\widetilde{\Omega}$ properly containing Ω possessing the above property for the $\mathcal{F}$.

In such a case the domain Ω is said to be the *domain of normality* for the sequence $\mathcal{F}$.

The *domain of holomorphy* of a family $\mathcal{F}$ is the intersection of the domains of holomorphy for the functions belonging to the family $\mathcal{F}$ (thus, in the family $\mathcal{F}$, we can certainly find functions which have different functional elements at different points of the domain Ω with identical coordinates).

Now, if we consider the normality of a family of holomorphic functions in a domain $\Omega \subset \mathbb{C}^n$ in the strict sense, that is, if every sequence of the family has a subsequence which converges uniformly to a holomorphic function on every compact set in Ω, then the domain of normality is a domain of holomorphy.

This is known as the "Conjecture of Julia": namely, whether the normality domain is the domain of holomorphy. This question was answered affirmatively by H. Cartan and P. Thullen (their results can be found in [80, Chapter 1]).

Using the word "graph" of a holomorphic function, this is described in other words as follows:

Let $\mathcal{F}$ be a family of holomorphic functions in a domain Ω in $\mathbb{C}^n$. Suppose that the set $\mathcal{F}(p) = \{f(p); f \in \mathcal{F}\}$ is bounded for every point $p \in \Omega$. If $\Omega \times \mathbb{C}$ is a domain of normality of the family Γ_F of graphs of $\mathcal{F}$, then Ω is a domain of holomorphy.

We shall give a proof of the above in what follows. For this purpose the following definitions are necessary:

Definition 3.6.1. Let $\{E_j\}$ be a sequence of subsets of a metric space X. We say that it converges geometrically to a set $E \subset X$ (and write $E_j \to E$) if

(a) E coincides with the limit set of the sequence $\{E_j\}$, i.e. consists of all points of the form $\lim_{j_v \to \infty} x_{jv}, x_{j_v} \in E_{j_v}$ (in particular, E is closed in X);

(b) for any compact set $K \subset E$ and any $\epsilon > 0$ there is an index $j(\epsilon, K)$ such that K belongs to the ϵ-neighborhood of E_j in X for all $j > j(\epsilon, K)$.

A nonempty analytic set A on a domain Ω is called *principal* if there is a function $f \in \mathcal{O}(\Omega)$, not identically vanishing on any component of Ω, such that $A = \{z \in \Omega : f(z) = 0\}$. An analytic set is called *locally principal* if it is principal in a neighborhood of each of its points.

Definition 3.6.2. Let $\{A_\nu\}$ be a sequence of principal analytic sets in Ω. This sequence is said to *converge analytically* to an analytic set A if and only if, given a point $p \in \Omega$, there exists a neighborhood U of p and holomorphic functions $\{f_\nu\}, f$ in U, such that:

(i) f is not identically zero.

(ii) $A_\nu \cap U = \{q \in U; f_\nu(q) = 0\}, A \cap U = \{q \in U; f(q) = 0\}$.

(iii) The sequence $\{f_\nu\}$ converges uniformly to f.

From this definition, the following remarks are easily seen.

1. If A is not empty, then A is a principal analytic set in Ω.
2. If a sequence $\{p_\nu\}$ with $p_\nu \in A_\nu$ has accumulation points, then they belong to A.
3. The sequence $\{A_\nu\}$ converges geometrically to A.

Definition 3.6.3. A family $\mathcal{F}$ of holomorphic functions in Ω is said to be *bounded* if and only if the set $F = \{f(p), f \in \mathcal{F}\}$ is bounded in the complex plane $\mathbb{C}$.

Definition 3.6.4. An analytic set A in $\Omega \times \mathbb{C}(w)$ is said to be *fine* in w at p if and only if the set $A(p) = \{(p, w) \in \mathbb{C}^{n+1}\} \cap A$ has no finite accumulating point, where $\mathbb{C}(w)$ is a complex plane with coordinate w.

From this definition, as is easily seen, A is not fine in w at p if and only if A includes the complex plane $\{(z, w) \in \mathbb{C}^{n+1}; z = p\}$.

For simplicity, in the sequel, we consider the case of two complex variables $z, \zeta \in \mathbb{C}$.

Lemma 3.6.5. *Let $\{f_j\}$ be a sequence of holomorphic functions in $\Omega \subset \mathbb{C}^2$ satisfying the following conditions:*

(i) *$\{f_j\}$ is bounded at each point of Ω.*

(ii) *The sequence $\{\Gamma_j\} := \{(p, f_j(p)) \in \Omega \times \mathbb{C}; p \in \Omega\}$ of graphs of $\{f_j\}$ converges analytically to an analytic set A in $\Omega \times \mathbb{C}(w)$.*

Then $\{f_j\}$ converges uniformly to a holomorphic function on every compact set in Ω.

Proof. This argument belongs to Professor H. Fujimoto. Let

$$E = \{p \in \Omega : A \text{ is not fine in } w \text{ at } p\}.$$

Then E is *a proper analytic set in* Ω.

In fact, A is a proper analytic set in $\Omega \times \mathbb{C}(w)$ and E does not coincide with $\mathbb{C}$. Take a point $p \in \Omega$ and a polydisk $P \subset \Omega$ with center at p. Since $P \times \mathbb{C}$ is a domain in which we can solve second Cousin problem, there exists a holomorphic function $\varphi(z, \zeta, w)$ in $P \times \mathbb{C}$ such that

$$A \cap (P \times C) = \{(z, \zeta, w) \in P \times \mathbb{C}; \varphi(z, \zeta, w) = 0\}.$$

Then it is easily seen that

$$\begin{aligned} E \cap P &= \{(z, \zeta) \in P; \varphi(z, \zeta, c) = 0 \text{ for all complex numbers } c\} \\ &= \bigcap_{c \in \mathbb{C}} \{(z, \zeta) \in P; \varphi(z, \zeta, c) = 0\}. \end{aligned}$$

Thus E is a proper analytic set in Ω.

Now, we show that the sequence $\{f_j\}$ is uniformly bounded on every compact set in Ω. To prove this, we have only to show that for any point $p \in \Omega$ there exists a neighborhood U of p such that $\{f_j\}$is uniformly bounded on U. Let $p_0 \in \Omega \setminus E$. Since $\{f_j(p_0)\}$ is bounded, there exists a subsequence of $\{f_j(p_0)\}$ which converges to a complex number q_0. Since $(p_0, q_0) \in A$, there exists a neighborhood $U := U_1 \times U_2$ of (p_0, q_0) with $U_1 \subset \Omega \setminus E$ and holomorphic functions $\{\psi_j\}, \psi$ on U such that

$$\Gamma_j \cap U = \{(x, y, w) \in U; \psi_j(x, y, w) = 0\},$$

$$A \cap U = \{(x, y, w) \in U; \psi(x, y, w) = 0\}$$

and that the sequence $\{\psi_j\}$ converges uniformly to ψ on U. Since the point $p_0 \notin E$, $\psi(p_0, w)$ is not identically zero. Therefore there exists a small positive number r such that $\psi(p_0, w) \neq 0$ if $|w - q_0| = r$. Thus if we choose U_1 sufficiently small, we have $\psi \neq 0$ on $\overline{U}_1 \times \Theta$, where $\Theta = \{w \in \mathbb{C}; |w - q_0| = r\}$. Put

$$m = \min\left\{|\psi(p, w)|; (p, w) \in \overline{U}_1 \times \Theta\right\}.$$

Since $\{\psi_j\}$ converges uniformly to ψ there exists a j_0 such that $|\psi_j - \psi| < m/2$ on $\overline{U}_1 \times \Theta$ for all $j \geq j_0$. Then for any point $p \in U_1$, by the theorem of Hurwitz, the equation $\psi_j(p, w) = 0$ has at least one root w_j in the disk $\{w \in \mathbb{C}; |w - q_0| < r\}$ for all $j \geq j_0$. Thus $\{f_j\}$ is uniformly bounded on U_1. Let $p_0 \in E$. Since E is a proper analytic set, by a linear change of coordinate if necessary, we can choose a polydisk $U = U_1 \times U_2$ of p_0 such that $\left(\partial U_1 \times \partial U_2\right) \cap E = \emptyset$. Then $\{f_j\}$ is uniformly bounded on $\partial U_1 \times \partial U_2$ and by the maximal principle $\{f_j\}$ is uniformly bounded on $U_1 \times U_2$.

Now by the theorem of Montel (Theorem 3.5.2), any sequence $\{f_j\}$ has a convergent subsequence. To prove that $\{f_j\}$ converges uniformly on every compact set in Ω, we have only to show that the limit function is independent of the choice of the convergent sequence. Let $\left\{f_j^{(1)}\right\}$ and $\left\{f_j^{(2)}\right\}$ be two sequences of $\{f_j\}$ which converge to $f^{(1)}$ and $f^{(2)}$ respectively on every compact set in Ω. Then it is easily seen that

$$A = \text{graph of } f^{(1)} = \text{ graph of } f^{(2)},$$

so that $f^{(1)} = f^{(2)}$. This complete the proof of Lemma 3.6.5. □

Definition 3.6.6. A family of principal analytic sets in Ω is said to be *analytically normal* in Ω if and only if, given a sequence in the family, there exists a subsequence which converges analytically to an analytic set in Ω.

Definition 3.6.7. Let Γ (resp. $\mathcal{F}$) be a family of principal analytic sets (resp. holomorphic functions) on $D \subset \mathbb{C}^2$. Then D is called a *domain of normality* of Γ (resp. $\mathcal{F}$) if and only if Γ (resp. $\mathcal{F}$) is normal in D and for any schlicht domain $\tilde{D}$ in $\mathbb{C}^2$ such that $D \subsetneq \tilde{D}$, Γ (resp. $\mathcal{F}$) is no longer normal in $\tilde{D}$.

For a domain of normality of a family of holomorphic functions, the so-called conjecture of G. Julia is well known (see for example [80, Chapter 1]). In connection with this conjecture, we have the following.

Theorem 3.6.8. *Let $\mathcal{F}$ be a family of holomorphic functions in $\Omega \subset \mathbb{C}^2$. Suppose that $\mathcal{F}(p)$ is bounded at each point p of Ω. If $\Omega \times \mathbb{C}$ is a domain of normality of a family $\Gamma_{\mathcal{F}}$ of graphs of $\mathcal{F}$, then Ω is a domain of holomorphy.*

Proof. Suppose that Ω is not a domain of holomorphy. Then since Ω is not holomorphycally convex, there exists a compact set K in Ω such that

$$\widetilde{K} = \left\{ (z, \zeta) \in \Omega; \ \ |f(z, \zeta)| \leqq \sup_K |f| \ \text{ for all holomorphic functions in } \Omega \right\}$$

is not compact in Ω. Put $\rho = d(K, \partial\Omega)$ and take a point $p_0 = (z_0, \zeta_0) \in \tilde{K}$ such that $d(p_0, \partial\Omega) < \frac{\rho}{2}$. It is well known that any holomorphic function in Ω is also holomorphic in the polydisk

$$\Delta = \left\{ (z, \zeta) \in \mathbb{C}^2; |z - z_0| < \rho, |\zeta - \zeta_0| < \rho \right\}.$$

Put

$$\Delta' = \left\{ (z, \zeta) \in \mathbb{C}^2; |z - z_0| < \frac{2}{3}\rho, |\zeta - \zeta_0| < \frac{2}{3}\rho \right\}.$$

Since $d(p_0, \partial D) < \frac{\rho}{2}$, $\Delta' \setminus D \neq \phi$.

We shall show that

$\Gamma_{\mathcal{F}}$ is normal in $(D \cup \Delta') \times \mathbb{C}$.

In fact, take a sequence $\{\Gamma_\nu\}$ from $\Gamma_{\mathcal{F}}$. Since $\Gamma_{\mathcal{F}}$ is normal in $\Omega \times \mathbb{C}$, there exists a subsequence $\{\Gamma_{\nu}\}$ which converges analytically to an analytic set A in $\Omega \times \mathbb{C}$. Let f_{ν_j} be the function which represents the graph Γ_{ν_j}. Then by Lemma 3.6.5 $\{f_{\nu_j}\}$ converges uniformly to a holomorphic function η on every compact set in Ω. Expand f_{ν_j} into the convergent Taylor series in Δ':

$$f_{\nu_j}(z, \zeta) = \sum a_{k_1,k_2}^{\nu_j} (z - z_0)^{k_1} (\zeta - \zeta_0)^{k_2}$$

where

$$a_{k_1,k_2}^{(\nu_j)} = \frac{1}{k_1! k_2!} \frac{\partial^{k_1+k_2} f_{\nu_j}(p_0)}{\partial z^{k_1} \partial \zeta^{k_2}}.$$

Put $K' = \overline{K^{(r)}}$, where $r = \frac{3}{4}\rho$. Since K' is compact, it holds that $\sup_{\nu_j} \sup_{K'} \left| f_{\nu_j} \right| = M$. Then by the estimates of Cauchy, we have

$$\left| a_{k_1,k_2}^{(\nu_j)} \right| \leqq \frac{1}{k_1! k_2!} \sup_{p \in K} \left| \frac{\partial^{k_1+k_2} f_{\nu_j}(p)}{\partial z^{k_1} \partial \zeta^{k_2}} \right| \leqq \frac{M}{\left(\frac{3}{4}\rho\right)^{k_1+k_2}}.$$

Therefore $\{f_{\nu_j}\}$ is uniformly bounded on Δ'. Thus there exists a subsequence of $\{f_{\nu_j}\}$ which converges uniformly to a holomorphic function in Δ'. For simplicity, we may assume that $\{f_{\nu j}\}$ converges uniformly to $\widetilde{\eta}$ in Δ'.

Since $\widetilde{\eta} = \eta$ in $\Delta' \cap \Omega$, the holomorphic function ξ such that $\xi = \eta$ in Ω and $\xi = \widetilde{\eta}$ in Δ' is determined. Then $\{\Gamma_{\nu j}\}$ converges analytically to the analytic set $B = \{(p, \xi(p)); p \in \Delta' \cup D\}$ in $(\Delta' \cup \Omega) \times \mathbb{C}$. Since $\Omega \times \mathbb{C}$ is a domain of holomorphy of $\Gamma_{\mathcal{F}}$, this is a contradiction. □

Remark for Lemma 3.6.5.

Suppose in Lemma 3.6.5 we omit the condition of the boundedness of $\{f_j\}$ at each point and consider under the condition only of an analytic convergence of graphs. Put

$$\Omega = \{p \in D \setminus E; \{f_j(p)\} \text{ is bounded }\}.$$

It is easily seen that the boundedness at p is equivalent to the condition that $A(p) \neq \phi$. Moreover, since A is fine in w at $p \in D \setminus E$, it is easy to see that $\Omega = \pi(A) \setminus E$, where π is the projection given by $\pi(p, w) = p$. Therefore Ω is open. Thus, if Ω is not empty, then there exists a holomorphic function η in Ω such that $\{f_j\}$ converges uniformly to η on every compact set in Ω. If $\Omega = \phi$, then $[(D \setminus E) \times \mathbb{C}] \cap A = \phi$. Since $\{E_j\}$ converges analytically to A, for any compact set K_1 in $D \setminus E$ and for any positive number a, there exists a $j_\circ$ such that $\mathbb{S}_j \cap K = \phi$ for all $j \geqq j_0$, where $K = K_1 \times \{w \in C; |w| \leqq a\}$. Thus if $p \in K_1$ then $(p, f_j(p)) \notin K$, i.e., $|f_j(p)| > a$ for all $j \geqq j_0$. This means that $\{f_j\}$ is compactly divergent in $D \setminus E$.

Let D be a domain of holomorphy and let $\Omega \neq \phi$; then it is easily seen that Ω is a holomorphic open set; i.e., any connected component of Ω is a domain of holomorphy.

Lemma 3.6.9. *Let Ω be a domain in $\mathbb{C}^2$ and let $\{f_j\}, f$ be holomorphic functions in Ω. Let $\Lambda = \{a, b, c\}$ be a set consisting of three different complex numbers. Suppose that for any $\alpha \in \Lambda$, the sequence $\{A_{j,\alpha}\}$ of analytic sets in D given by*

$$A_{\nu,\alpha} = \{(x, y) \in \Omega; f_j(x, y) = \alpha\}$$

converges geometrically to an analytic set $A_\alpha = \{(x, y) \in \Omega; f(x, y) = \alpha\}$. Then $\{f_\nu\}$ is normal in Ω.

Proof. Take a point $p \in \Omega$. If $f(p) \notin \Lambda$, then there exists a connected open neighborhood V of p such that $V \subset\subset \Omega$ and $f(\overline{V}) \cap \Lambda = \emptyset$. Therefore, $\Lambda \setminus f(\overline{V}) = \Lambda$. If $f(p) \in \Lambda$, we can choose a connected open neighborhood U of p such that $U \subset\subset \Omega$ and $f(\overline{U}) \cap [\Lambda \setminus \{f(p)\}] = \emptyset$. In any case we can choose a connected open neighborhood U of p such that $U \subset\subset \Omega$ and $\Lambda \setminus f(\overline{U})$ contains at least two complex numbers. Let $\{a, b\} \subset \Lambda \setminus f(\overline{U})$, then $A_a \cap \overline{U} = \emptyset$ and $A_b \cap \overline{U} = \emptyset$. Since $\{A_{j,a}\}$, $\{A_{j,b}\}$ converge geometrically to A_a, A_b respectively, there exists a j_0 such that $A_{j,a} \cap \overline{U} = \emptyset$, $A_{j,b} \cap \overline{U} = \emptyset$ for all $j \geq j_0$. That is, the sequence $\{f_j\}_{j \geq j_0}$ does not take two different values a, b in U. Thus, by the theorem of Montel (Theorem 3.5.2), it is normal in U. Since p is an arbitrary point of Ω, $\{f_j\}$ is normal at each point of Ω. Then by the diagonal method $\{f_j\}$ is normal in Ω. □

Definition 3.6.10. A set ω is called a *set of uniqueness* if and only if any holomorphic function in Ω which is zero on ω is identically zero.

Theorem 3.6.11 (see [284, Theorem 3, p. 645]). *Let $\{f_j\}$, f be holomorphic functions on Ω and let f be not constant. Let ω be a set of uniqueness. Let $A_{j,a}, A_a$ be the same as in Lemma 3.6.9. Suppose that for any $a \in f(\omega)$ the sequence $\{A_{j,a}\}$ of analytic sets in Ω converges geometrically to an analytic set A_a, then $\{f_j\}$ converges uniformly to f on every compact set in Ω.*

Proof. First, the set $f(\omega)$ contains infinitely many complex numbers. In fact, suppose that $f(\omega)$ contains only finitely many complex numbers. Let $f(\omega) = \{a_1, a_2, \ldots, a_m\}$. Put $S_i = \{p \in \Omega; f(p) = a_i\}$. Then the analytic set $S = \cup S_i$ is given by

$$S = \left\{p \in \Omega; g(p) = \prod_{i=1}^{i=m} [f(p) - a_i] = 0\right\}.$$

Since f is not constant, g is also not constant. But since $S \supset \omega$ and since $g = 0$ on S, g is identically zero, this is a contradiction. Thus, by Lemma 3.6.9, $\{f_\nu\}$ is normal in Ω.

To prove that the sequence $\{f_\nu\}$ converges uniformly to f on every compact set in Ω, we have only to show that any sequence of $\{f_\nu\}$ contains a subsequence which converges uniformly to f on every compact set in Ω. Take a subsequence $\{f_{\nu_j}\}$ of $\{f_\nu\}$. We may assume that $\{f_{\nu_j}\}$ is either compactly uniformly convergent or compactly divergent. Since $\{A_{\nu,a}\}$ converges geometrically to A_a and since A_a is not empty, the case of compact divergence does not occur. Let $\{f_{\nu j}\}$ converge uniformly to a holomorphic function h on every compact set in Ω. Let $p_0 \in \omega$ and let $a = f(p_0)$. Then, since $p_0 \in A_a$, there exists a sequence $\{p_{\nu_j}\}$ such that $p_{\nu_j} \in A_{\nu_j,a}$ and $p_{\nu_j} \to p_0$. Let $K = \{p_0, p_{\nu_j}; j = 1, 2, \ldots\}$. Since K is compact, for any positive number ε there exists a j_0 such that

$$\sup_{p \in K} \{|f_{\nu_j}(p) - h(p)|\} < \varepsilon, |h(p_{\nu_j}) - h(p_0)| < \varepsilon$$

for all $j \geq j_0$. Then we have

$$|f_{\nu j}(p_{\nu j}) - h(p_0)| \leq |f_{\nu_j}(p_{\nu_j}) - h(p_{\nu j})| + |h(p_{\nu_j}) - h(p_0)| < 2\varepsilon.$$

Since $p_{\nu_j} \in A_{\nu_j,a}$ and since $f_{\nu_j}(p_{\nu_j}) = a = f(p_0)$, we have $f(p_0) = h(p_0)$. That is, $f = h$ on ω, so that $f = h$ on Ω. □

Now we consider the case of just one complex variable and we get the converse of the theorem of Hurwitz.

In Theorem 3.6.11, we cannot omit the condition that f is not constant. But since Lemma 3.6.9 holds in the case that f is constant, by the same method we can prove the following:

Theorem 3.6.12. *Let U be an open set in the complex plane $\mathbb{C}$ such that $U \cap f(\Omega) \neq \emptyset$. If $\{A_{\nu,a}\}$ converges geometrically to A_a for all $a \in U$, then $\{f_\nu\}$ converges uniformly to f on every compact set in Ω.*

If a sequence $\{f_j\}$ of holomorphic functions in a domain $\Omega \subset \mathbb{C}$ converges to a nonconstant holomorphic function f on every compact set in Ω, then for any point $p \in \Omega$ and for any complex number a, there exists a positive number r which satisfies the following property:

> For any r with $0 < r < r_0$, there exists a j_0 such that the number of points of f_j in the disk $\{z \in \mathbb{C}; |z-p| < r\}$ are equal for all $j \geq j_0$, counted according to multiplicities.

In fact, if $f(p) \neq a$, then there exists a neighborhood U of p such that $U \Subset \Omega$ and $\min = \{|f(z) - a|\,;\, z \in \overline{U}\} = \delta > 0$. Since $\{f_j\}$ converges uniformly to f on $\overline{U}$, there exists a j such that

$$\max\{|f_j(z) - f(z)|\,;\, z \in \overline{U}\} < \frac{\delta}{2}$$

for all $j \geq j_0$. Then

$$|f_j(z) - a| \geq |f(z) - a| - |f_j(z) - f(z)| > \frac{\delta}{2}$$

for all $j \geq j_0$ and $z \in \overline{U}$. That is $f_j(z) \neq a$ in $\overline{U}$ for all $j \geq j_0$. If $f(p) = a$, then this is just the theorem of Hurwitz.

Since Theorem 3.6.11 holds for a domain in the complex plane, as the converse of the above, we have

Corollary 3.6.13 (Corollary of Theorem 3.6.11)**.** *Let $\{f_j\}$, f be holomorphic functions in a domain Ω in $\mathbb{C}$ and let f be not constant. Suppose that for any point $p \in \Omega$ and for any complex number a, there exists a positive number r_0 satisfying the following property:*

> *For any r with $0 < r < r_0$, there exists a j_0 such that the number of a-points of f_j and f are equal in the disk $\{z \in \mathbb{C}; |z-p| < r\}$ for all $j \geq j_0$, counted according to multiplicities.*

Then $\{f_j\}$ converges uniformly to f on every compact set in Ω.

Proof. Because of Theorem 3.6.12 we have only to show that for every point $p \in \Omega$ the sequence $\{A_j\}$ of analytic sets given by $A_j = \{z \in \Omega; f_j(z) = f(p)\}$ converges geometrically to an analytic set $A = \{z \in \Omega; f(z) = f(p)\}$.

It is easily seen that the condition **(a)** in Definition 3.6.1 is trivially fulfilled.

Let K be a compact set in Ω such that $A \cap K = \emptyset$. For any point $q \in K$, there exists a positive number δ such that $f(z) - f(p) \neq 0$ in the $d = \{z \in \Omega; |z - q| < \delta\}$. Evidently we may assume that $\delta < r_0$. Then there exists a $j_0(q)$ such that $f_j(z) - f(p) \neq 0$ in the disk d for all $j \geq j_0(q)$. Since K is compact we can choose finitely many such disks d_ν with center at q_ν and radius δ_ν such that $K \subset \cup d_\nu$. Put $j_0 = \max j_0(q_\nu)$. Then since $f_j(z) - f(p) \neq 0$ in K for all $j \geq j_0$, we have that $A_j \cap K = \emptyset$ for all $j \geq j_0$. Thus the condition **(b)** in Definition 3.6.1 proved. The theorem now follows from Theorem 3.6.12. □

3.7 Riemann's Theorem

The Riemann Mapping Theorem: Statement and Idea of Proof

Let us consider the question of conformal equivalence of planar domains. A fundamental question here is "Which planar domains are conformally equivalent to the unit disk D"?

We begin by examining the Cayley transform

$$z \to \frac{z - i}{z + i} .$$

Under this mapping the point ∞ is sent to the point 1, the point 1 is sent to the point $(1 - i)/(1 + i) = -i$, and the point -1 is sent to $(-1 - i)/(-1+i) = i$. Thus the image under the Cayley transform (a linear fractional transformation) of three points on $\mathbb{R} \cup \{\infty\}$ contains three points on the unit circle. Since three points determine a (generalized) circle, and since linear fractional transformations send generalized circles to generalized circles, we conclude that the Cayley transform sends the real line to the unit circle. Now the Cayley transform is one-to-one and onto from $\mathbb{C} \cup \{\infty\}$ to $\mathbb{C} \cup \{\infty\}$. By continuity, it either sends the upper half plane to the (open) unit disk or to the complement of the closed unit disk. The image of i is 0, so in fact, the Cayley transform sends the upper half plane to the unit disk.

The argument just given, while wordy, is conceptually much simpler and quicker than the sort of calculations that we have given previously.

The conformal equivalence of open sets in $\mathbb{C}$ is generally not very easy to compute or to verify, even in the special case of sets bounded by (generalized) circular arcs. For instance, it is the case (as will be verified soon) that the square $\{z = x + iy : |x| < 1, |y| < 1\}$ is conformally equivalent to the unit disk $D = \{z \in \mathbb{C} : |z| < 1\}$. But it is quite difficult to write down explicitly the conformal equivalence. Even when a biholomorphic mapping is known (from other considerations) to exist, it is generally quite difficult to find it. Thus there is a priori interest in demonstrating the existence of the conformal equivalence of certain open sets by abstract, nonconstructive methods. In this section we will be concerned with the question of when an open set $U \subseteq \mathbb{C}$ is conformally equivalent to the unit disk.

Several restrictions must obviously be imposed on U. For instance, U obviously cannot be $\mathbb{C}$ because every holomorphic function $f : U \to D$ would then be constant (Liouville's theorem). Also, U must be *topologically equivalent* (i.e., homeomorphic) to the unit disk U since we are in fact demanding much more in asking that U be conformally equivalent. For instance, U could not be conformally equivalent to $\{z : 1 < |z| < 2\}$ since it is not topologically (i.e., homeomorphically) equivalent. [Although at the moment we have no precise proof that the two are not homeomorphic, this assertion is at least intuitively plausible.]

Surprisingly, the two restrictions just indicated (that U not be $\mathbb{C}$ and that U be topologically equivalent to D) are not only necessary but they are sufficient to guarantee that U be conformally equivalent to the disk U.

To give this notion a more precise formulation, we first define formally what we want to mean by topological (homeomorphic) equivalence:

Definition 3.7.1. Two open sets U and V in $\mathbb{C}$ are *homeomorphic* if there is a one-to-one, onto, continuous function $f : U \to V$ with $f^{-1} : V \to U$ also continuous. Such a function f is called a *homeomorphism* from U to V.

[It is a fact that the inverse of a continuous, one-to-one, onto function from one open subset of $\mathbb{R}^2$ to another is *automatically* continuous. Thus the condition in the definition of homeomorphism that f^{-1} be continuous is actually redundant. But this "automatic continuity" for the inverse function is rather difficult to prove and plays no role in our later work. Thus we shall skip the proof; the interested reader may consult a topology text.]

Theorem 3.7.2 (The Riemann mapping theorem). *If U is an open subset of $\mathbb{C}$, $U \neq \mathbb{C}$ and if U is homeomorphic to D, then U is conformally equivalent to D.*

Riemann enunciated this theorem in 1851. Unfortunately Riemann's proof was flawed. He constructed his conformal mapping by way of the so-called Dirichlet principle, and the version of that principle that he needed was not fully understood at the time. Experts have said that it is not possible to make Riemann's argument rigorous.

It was many years before a rigorous proof of the Riemann mapping theorem was finally produced—long after Riemann's death. The first rigorous proof was given by William Fogg Osgood in 1900. In modern times, Greene and Kim have found a way to make Riemann's original argument correct and rigorous (see [92]). In fact, they won a prize for that paper.

The proof of the Riemann mapping theorem consists of two parts. First, one shows that if a domain U is homeomorphic to the disk, then each holomorphic function on U has an antiderivative (this property is called "holomorphic simple connectivity"). The argument to prove this assertion is essentially topological; and indeed the conclusion will be deduced from a topological condition called simple connectivity that appears to be weaker than homeomorphism to the disk, though it is in fact equivalent to it for open sets in $\mathbb{C}$. The second step uses the first to construct a one-to-one holomorphic mapping from U into the unit disk. We then construct a mapping from U to D that is *both* one to-one and onto by solving an "extremal problem"; that is, we shall find a function that maximizes some specific quantity (details will be provided momentarily). That is where normal families comes in.

The second part of the proof of the Riemann mapping theorem will occupy us for the rest of this section (Theorem 3.7.7—the analytic form of the Riemann mapping theorem). The first, essentially topological, part of the proof will be given separately. Before presenting the (second part of the) proof, let us discuss the significance of the Riemann mapping theorem and of some of the issues that it raises.

First, it is important to realize that homeomorphism is, in general, quite far from implying biholomorphic equivalence:

Example 3.7.3. The entire plane $\mathbb{C}$ is topologically equivalent to the unit disk U. An equivalence is given by

$$f(z) = \frac{z}{1+|z|}.$$

However, as previously noted, $\mathbb{C}$ is *not* conformally equivalent to U.

It is a fact that two annuli $\{z : r_1 < |z| < r_2\}$ and $\{z : s_1 < |z| < s_2\}$ are conformally equivalent if and only if $r_2/r_1 = s_2/s_1$. Yet two such annuli are always homeomorphic (exercise). Thus we see that the Riemann mapping theorem is startling by comparison.

The idea of the proof of the Riemann mapping theorem comes from the Schwarz lemma. Namely, one can think of this lemma in the following form: Consider a holomorphic (not necessarily one-to-one and onto) function $f : D \to D$ with $f(0) = 0$. Then f is conformal (i.e., one-to-one and onto) if and only if $|f'(0)|$ is as large as possible, that is, if and only if

$$|f'(0)| = \sup\{|h'(0)| : h : D \to D, h(0) = 0, h \text{ holomorphic}\}.$$

The special attention paid to the point 0 here is inessential. In fact if $f : D \to D$ is a holomorphic function such that

$$|f'(P)| = \sup\{|h'(P)| : \quad h : D \to D,\ h(P) = P,\ h \text{ holomorphic}\},$$

then f must be conformal. (Exercise: Prove this, and find an analogous statement to cover the case when f maps P to Q and $Q \neq P$.)

If U satisfies the hypotheses of the Riemann mapping theorem, then one might look for a conformal mapping of U to D by choosing a point $P \in U$ and then looking for a holomorphic function $f : U \to D$ such that $f(P) = 0$ and $|f'(P)|$ is maximal. It turns out to be technically simpler to consider only functions f that are assumed in advance to be one-to-one, and we shall take advantage of the resulting simplification.

Several questions arise about our plan of attack. First, do there exist any one-to-one holomorphic functions $f : U \to D$? Assuming that there are such functions, is the set of values $\{|f'(P)|\}$ bounded? Also, is the least upper bound attained? That is, is there a holomorphic function f from U to D such that $f(P) = 0$ and $|f'(P)|$ is as large as possible?

The latter question is of particular historical interest because Riemann in fact neglected in his proof to verify that a similar sort of maximum was assumed by some particular function. Thus his theorem was in doubt for some time until a suitable version of what has become known as "Dirichlet's principle" gave a method to fill the gap. The modern approach to the Riemann mapping theorem involves a convergence idea for holomorphic functions called "normal families", which we develop in the next section.

Note that the issue of the boundedness of the derivatives of f at P is easily dispatched. For let $f : U \to D, f(P) = 0$. Choose $r > 0$ such that $\overline{D}(P, r) \subseteq U$. Then, of course, $|f(z)| \leq 1$ for all z; hence the Cauchy estimates imply that $|f'(P)| \leq 1/r$. Thus we have an a priori upper bound on the size of derivatives of f at P. In the next section we turn our attention to establishing the existence of an f whose derivative $f'(P)$ has modulus that *actually equals* the least upper bound of the absolute values of such derivatives.

Let us conclude this section by settling the derivative-maximizing holomorphic function issue that was raised earlier.

Proposition 3.7.4. *Let $U \subseteq \mathbb{C}$ be any open set. Fix a point $P \in U$. Let $\mathcal{F}$ be the family of holomorphic functions from U into the unit disk D that takes P to 0. Then there is a holomorphic function $f_0 : U \to D$ that is the normal limit of a sequence $\{f_j\}, f_j \in \mathcal{F}$, such that*

$$|f_0'(P)| \geq |f'(P)|$$

for all $f \in \mathcal{F}$.

Proof. We have already noted that the Cauchy estimates give a finite upper bound for $|f'(P)|$ for all $f \in \mathcal{F}$. Let

$$\lambda = \sup \{|f'(P)| : f \in \mathcal{F}\}.$$

By the definition of supremum, there is a sequence $\{f_j\} \subseteq \mathcal{F}$ such that $\left|f_j'(P)\right| \to \lambda$. But the sequence $\{f_j\}$ is bounded by 1 since all elements of $\mathcal{F}$ take values in the unit disk. Therefore Montel's theorem applies and there is a subsequence $\{f_{j_k}\}$ that converges uniformly on compact sets to a limit function f_0.

By the Cauchy estimates, the sequence $\left\{f_{j_k}'(P)\right\}$ converges to $f_0'(P)$. Therefore $|f_0'(P)| = \lambda$ as desired.

It remains to observe that, a priori, f_0 is known only to map U into the closed unit disk $\overline{D}$. But the maximum modulus theorem implies that if $f_0(U) \cap \{z : |z| = 1\} \neq \emptyset$, then f_0 is a constant (of modulus 1). Since $f_0(P) = 0$, the function f_0 certainly cannot be a constant of modulus 1. Thus $f_0(U) \subset D$. $\square$

Holomorphically Simply Connected Domains

The derivative-maximizing holomorphic function from an open set U to the unit disk (as provided by Proposition 3.7.4) turns out to have some interesting special properties when U resembles the disk in a certain sense. The type of resemblance that we wish to consider is called "holomorphic simple connectivity". This terminology is a temporary one that we use for convenience; it is not used universally (in particular, it is not found in most other texts on complex function theory).

The concept of "holomorphically simply connected" (h.s.c.) arises naturally in our present context. However, it turns out to be implied by a more easily verified topological condition that is called simple connectivity. Simple connectivity is in fact a universally used mathematical concept; it is a fact that that simple connectivity implies holomorphic simple connectivity. Recall the definition of holomorphic simple connectivity:

Definition 3.7.5. A connected open set $U \subseteq \mathbb{C}$ is *holomorphically simply connected* if, for each holomorphic function $f : U \to \mathbb{C}$, there is a holomorphic antiderivative F, that is, a function satisfying $F'(z) = f(z)$.

Example 3.7.6. As established in [95, Chapter 4], open disks and open rectangles are holomorphically simply connected.

If $U_1 \subseteq U_2 \subseteq \ldots$ are holomorphically simply connected sets, then their union $U = \bigcup U_j$ is also holomorphically simply connected. In particular, the plane $\mathbb{C}$ is holomorphically simply connected.

Theorem 3.7.7 (Riemann mapping theorem: analytic form). *If U is a holomorphically simply connected open set in $\mathbb{C}$ and $U \neq \mathbb{C}$, then U is conformally equivalent to the unit disk.*

The proof of this result involves some intricate constructions. We present them in the next section. The rest of the present section is devoted to the consideration of some preliminary results that are needed for the proof. These results also have considerable intrinsic interest.

Lemma 3.7.8 (holomorphic logarithm lemma). *Let U be a holomorphically simply connected open set. If $f : U \to \mathbb{C}$ is holomorphic and nowhere zero on U, then there exists a holomorphic function h on U such that*

$$e^h \equiv f \text{ on } U.$$

Proof. The function $z \to f'(z)/f(z)$ is holomorphic on U, because f is nowhere zero on U. Since U is holomorphically simply connected, there is a function $h : U \to \mathbb{C}$ such that $h'(z) = f'(z)/f(z)$ on U. Fix a point $z_0 \in U$. Adding a constant to h if necessary, we may suppose that

$$e^{h(z_0)} = f(z_0).$$

After this normalization, we can now demonstrate that $e^h \equiv f$ on U.

It is enough to show that $g(z) \equiv f(z)e^{-h(z)}$ satisfies $g' \equiv 0$ on U. For then g is necessarily a constant. Since we have arranged that $g(z_0) = 1$, we conclude that $g(z) \equiv 1$. This is equivalent to what we want to prove. Now

$$\begin{aligned}
\frac{\partial g}{\partial z}(z) &= \frac{\partial}{\partial z}\left(f(z)e^{-h(z)}\right) \\
&= \frac{\partial f}{\partial z}(z)e^{-h(z)} + f(z)\left(-\frac{\partial h}{\partial z}(z)e^{-h(z)}\right) \\
&= e^{-h(z)}f(z)\left(\frac{\partial f(z)/\partial z}{f(z)} - \frac{\partial h}{\partial z}(z)\right) = 0
\end{aligned}$$

by the way that we constructed h. □

Corollary 3.7.9. *If U is holomorphically simply connected and $f : U \to \mathbb{C} \setminus \{0\}$ is holomorphic, then there is a function $g : U \to \mathbb{C} \setminus \{0\}$ such that*

$$f(z) = [g(z)]^2$$

for all $z \in U$.

Proof. Choose h as in the lemma. Set $g(z) = e^{h(z)/2}$. □

It is important to realize that the possibility of taking logarithms, or of taking roots, depends critically on the holomorphic simple connectivity of U. There are connected open sets (such as the annulus $\{z : 1 < |z| < 2\}$) for which these processes are not possible even for the function $f(z) \equiv z$.

The Proof of the Analytic Form of the Riemann Mapping Theorem

Let U be a holomorphically simply connected open set in $\mathbb{C}$ that is not equal to all of $\mathbb{C}$. In the sequel we call such set a *proper domain*. Fix a point $P \in U$ and set

$$\begin{aligned}
\mathcal{F} = \{f : f &\text{ is holomorphic on } U, f : U \to D, \\
&f \text{ is one-to-one}, f(P) = 0\}.
\end{aligned}$$

We shall prove the following three assertions:

(1) $\mathcal{F}$ is nonempty.

(2) There is a function $f_0 \in \mathcal{F}$ such that

$$|f_0'(P)| = \sup_{h \in \mathcal{F}} |h'(P)|\,.$$

(3) If g is any element of $\mathcal{F}$ such that $|g'(P)| = \sup_{h\in\mathcal{F}} |h'(P)|$, then g maps U onto the unit disk D.

The proof of assertion (**1**) is by direct construction. Statement (**2**) is almost the same as Proposition 3.7.4 (however there is now the extra element that the derivative-maximizing map must be shown to be one-to-one). Statement (**3**) is the least obvious and will require some work: If the conclusion of (**3**) is assumed to be false, then we are able to construct an element $\widehat{g} \in \mathcal{F}$ such that $|\widehat{g}'(P)| > |g'(P)|$. Now we turn to the proofs.

Proof of (1). If U is bounded, then this assertion is easy: If we simply let

$$a = 1/(2 \sup \{|z| : z \in U\})$$

and $b = -aP$, then the function $f(z) = az + b$ is in $\mathcal{F}$.

If U is unbounded, we must work a bit harder. Since $U \neq \mathbb{C}$, there is a point $Q \notin U$. The function $\phi(z) = z - Q$ is nonvanishing on U, and U is holomorphically simply connected. Therefore there exists a holomorphic function h such that $h^2 = \phi$. Notice that h must be one-to-one since ϕ is. Also there cannot be two distinct points $z_1, z_2 \in U$ such that $h(z_1) = -h(z_2)$ [otherwise $\phi(z_1) = \phi(z_2)$]. Now h is a nonconstant holomorphic function; hence an open mapping. Thus the image of h contains a disk $D(b, r)$. But then the image of h must be disjoint from the disk $D(-b, r)$. We may therefore define the holomorphic function

$$f(z) = \frac{r}{2(h(z) + b)}.$$

Since $|h(z) - (-b)| \geq r$ for $z \in U$, it follows that f maps U to U. Since h is one-to-one, so is f. Composing f with a suitable automorphism of U (a Möbius transformation), we obtain a function that is not only one-to-one and holomorphic with image in the disk, but also maps P to 0. Thus $f \in \mathcal{F}$. □

Proof of (2). Proposition 3.7.4 will yield the desired conclusion once it has been established that the limit derivative-maximizing function is itself one-to-one. Suppose for notation that the $f_j \in \mathcal{F}$ converge normally to f_0, with

$$|f_0'(P)| = \sup_{f \in \mathcal{F}} |f'(P)|.$$

We want to show that f_0 is one-to-one into U. The argument principle, specifically Hurwitz's theorem, will now yield this conclusion:

Fix a point $b \in U$. Consider the holomorphic functions $g_j(z) \equiv f_j(z) - f_j(b)$ on the open set $U \setminus \{b\}$. Each f_j is one-to-one: hence each g_j is nowhere vanishing on $U \setminus \{b\}$. Hurwitz's theorem guarantees that either the limit function $f_0(z) - f_0(b)$ is identically zero or is nowhere vanishing. But, for a function $h \in \mathcal{F}$, it must hold that $h'(P) \neq 0$ because if $h'(P)$ were equal to zero, then that h would not be one-to-one. Since $\mathcal{F}$ is nonempty, it follows that $\sup_{h \in \mathcal{F}} |h'(P)| > 0$. Thus the function f_0, which satisfies $|f_0'(P)| = \sup_{h \in \mathcal{F}} |h'(P)|$, cannot have $f_0'(P) = 0$ and f_0 cannot be constant. The only possible conclusion is that $f_0(z) - f_0(b)$ is nowhere zero on

$U \setminus \{b\}$. Since this statement holds for each $b \in U$, we conclude that f_0 is one-to-one. □

Proof of (3). Let $g \in \mathcal{F}$ and suppose that there is a point $R \in U$ such that the image of g does not contain R. Set

$$\phi(z) = \frac{g(z) - R}{1 - g(z)\bar{R}}.$$

Here we have composed g with a transformation that preserves the disk and is one-to-one. Note that ϕ is nonvanishing.

The holomorphic simple connectivity of U guarantees the existence of a holomorphic function $\psi : U \to \mathbb{C}$ such that $\psi^2 = \phi$. Now ψ is still one-to-one and has range contained in the unit disk. However, it cannot be in $\mathcal{F}$ since it is nonvanishing. We repair this by composing with another Möbius transformation: Define

$$\rho(z) = \frac{\psi(z) - \psi(P)}{1 - \psi(z)\overline{\psi(P)}}$$

Then $\rho(P) = 0$, ρ maps U into the disk, and ρ is one-to-one. Therefore $\rho \in \mathcal{F}$. Now we will calculate the derivative of ρ at P and show that it is actually larger in modulus than the derivative of g at P.

We have

$$\begin{aligned}\rho'(P) &= \frac{\left(1 - |\psi(P)|^2\right) \cdot \psi'(P) - (\psi(P) - \psi(P))\left(-\psi'(P)\overline{\psi(P)}\right)}{\left(1 - |\psi(P)|^2\right)^2} \\ &= \frac{1}{1 - |\psi(P)|^2} \cdot \psi'(P).\end{aligned}$$

Also

$$2\psi(P) \cdot \psi'(P) = \phi'(P) = \frac{(1 - g(P)\bar{R})g'(P) - (g(P) - R) \cdot (-g'(P)\bar{R})}{(1 - g(P)\bar{R})^2}.$$

But $g(P) = 0$; hence

$$2\psi(P) \cdot \psi'(P) = \left(1 - |R|^2\right) g'(P).$$

We conclude that

$$\begin{aligned}\rho'(P) &= \frac{1}{1 - |\psi(P)|^2} \cdot \frac{1 - |R|^2}{2\psi(P)} g'(P) \\ &= \frac{1}{1 - |\phi(P)|} \cdot \frac{1 - |R|^2}{2\psi(P)} g'(P) \\ &= \frac{1}{1 - |R|} \cdot \frac{1 - |R|^2}{2\psi(P)} g'(P) \\ &= \frac{1 + |R|}{2\psi(P)} g'(P).\end{aligned}$$

However, $1 + |R| > 1$ (since $R \neq 0$) and $|\psi(P)| = \sqrt{|R|}$. It follows, since $(1 + |R|)/(2\sqrt{|R|}) > 1$, that

$$|\rho'(P)| > |g'(P)| .$$

Thus, if the mapping g of statement (3) at the beginning of the section were not onto, then it could not have property (**2**)—of maximizing the absolute value of the derivative at P.

We have completed the proofs of each of the three assertions and hence of the analytic form of the Riemann mapping theorem. □

The proof of statement (**3**) may have seemed unmotivated. Let us have another look at it. Let

$$\mu(z) = \frac{z - R}{1 - \bar{R}z},$$

$$\tau(z) = \frac{z - \psi(P)}{1 - \overline{\psi(P)}z},$$

and

$$S(z) = z^2$$

Then, by our construction, with $h = \mu^{-1} \circ S \circ \tau^{-1}$:

$$\begin{aligned} g &= \mu^{-1} \circ S \circ \tau^{-1} \circ \rho \\ &\equiv h \circ \rho. \end{aligned}$$

Now the chain rule tells us that

$$|g'(P)| = |h'(0)| \cdot |\rho'(P)| .$$

Since $h(0) = 0$ and since h is not a conformal equivalence of the disk to itself, the Schwarz lemma will then tell us that $|h'(0)|$ must be less than 1.

But this says that $|g'(P)| < |\rho'(P)|$, giving us the required contradiction.

The Riemann mapping theorem has no analogue in higher dimensions: A theorem of Liouville (see e.g. [219],[83]) states that every conformal map of a domain in $\mathbb{R}^n$ with $n > 2$ is a Möbius transformation.

One sometimes adds this misleading comment that there is no Riemann mapping theorem in several variables. We next describe two positive mapping theorems. It is not hard to see that the only plane domains $\Omega \neq \mathbb{C}$ for which the automorphism group is transitive are those biholomorphic to the disk. Bun Wong [286] showed that the corresponding statement is true for strictly pseudoconvex domains in $\mathbb{C}^n$, $n > 1$. Later, Rosay [250] generalized this result to the following statement.

Theorem 3.7.10. *A bounded domain in* $\mathbb{C}^n$ *with* C^2 *boundary whose automorphism group is transitive is biholomorphic to the unit ball in* $\mathbb{C}^n$*; i.e. a bounded homogeneous domain with* C^2 *boundary is biholomorphic to the ball.*

It should be remarked that bounded homogeneous domains are always (without any assumptions on the boundary) domains of holomorphy, i.e. pseudoconvex. There are, of course, many such domains without smooth boundary.

The next theorem, due to B. Fridman [77], may claim to be an approximate Riemann mapping theorem.

Theorem 3.7.11. *Let Ω be a bounded pseudoconvex domain in $\mathbb{C}^n$ diffeomorphic to the unit ball B^n. Given $\varepsilon > 0$, there exist domains $\Omega_1 \subset \Omega, \Omega_2 \subset B^n$ whose boundaries are within ε of those of Ω and B^n respectively such that Ω_1 and Ω_2 are biholomorphically equivalent.*

Let us mention also a result of Bedford and Pinchuk [22] which confirms, at least partially, a conjecture of Greene and Krantz [94].

Theorem 3.7.12. *A (weakly) pseudoconvex domain in $\mathbb{C}^2$ with real analytic boundary whose automorphism group is non-compact is biholomorphic to a domain of the form*

$$\left\{(z, w) \in \mathbb{C}^2 : |z|^2 + |w|^{2p} < 1\right\}$$

for $p > 0$ an integer.

For more on this subject see Bell and Narasimhan [23].

3.8 Julia's Theorem

Let Ω be a domain in $\mathbb{C}^n$ containing the origin 0. Let $\mathcal{F}$ be a family of holomorphic functions in Ω. Consider the set

$$Z_0 := \{z = (z_1, z_2, \ldots, z_n) : z_j = 0 \quad (j = 1, \ldots, n-1), \quad 0 < |z_n| < r\},$$

in Ω and assume that $\mathcal{F}$ is a normal family at each point of Z_0; i.e., for any $p \in Z_0$ there exists a connected neighborhood V of p in Ω such that $\mathcal{F}$ is normal on V.

Under this assumption we have the following theorem.

Theorem 3.8.1. *(Julia [125]). Suppose $\mathcal{F}$ is not normal at the origin 0. Then, given any r' with $0 < r' < r$, there exists $\rho > 0$ such that for any $z' = (z'_1, \ldots, z'_{n-1}, 0) \in \mathbb{C}^n$ with $\left|z'_j\right| < \rho \quad (j = 1, \ldots, n-1)$, there must be at least one point q in the set*

$$Z_{z'} := \left\{z : z_j = z'_j \quad (j = 1, \ldots, n-1), \quad 0 < |z_n| < r'\right\},$$

such that $\mathcal{F}$ is not normal at q.

Proof. Since $\{0\}\times\{|z_n| = r'\} \subset Z_0$, it follows from our assumptions that there exist a ρ sufficiently small and $0 < \varepsilon < r'$ such that, setting $S_2 := \overline{P}(0', \rho)\times\Gamma^*$, where

$$\overline{P}(0', \rho) = \left\{z \in \mathbb{C}^{n-1} : |z_j| \le \rho, j = 1, \ldots, n-1;\right\}$$

and

$$\Gamma^* = \left\{z_n \in \mathbb{C} : r' - \varepsilon \le |z_n| \le r' + \varepsilon\right\}.$$

we have $S_2 \subset \Omega$ and $\mathcal{F}$ is normal on S_2. We prove that this $\rho > 0$ yields the conclusion of the theorem. For suppose not. Then there exists some $z_0' = (z_{01}, \ldots, z_{0(n-1)})$ with $\left|z_{0j}'\right| < \rho$ $(j = 1, \ldots, n-1)$ such that $\mathcal{F}$ is normal on $L_{z_0'} = \{z_0'\}\times\{|z_n| \le r'\}$. We can thus find a neighborhood S_1 of $L_{z_0'}$ in Ω such that $\mathcal{F}$ is normal on S_1 indeed, we can take S_1 of the form $S_1 := \overline{P}'(z_0', \rho_0')\times\Gamma$, where

$$\overline{P}'(z_{0j}', \rho_0') = \left\{z \in \mathbb{C}^{n-1} : \left|z_j - z_{0j}'\right| \le \rho_0', j = 1, \ldots, n-1;\right\},$$

and

$$\Gamma := \left\{z_n \in \mathbb{C} : |z_n| \le r' + \varepsilon\right\}.$$

Here $\overline{P}'(z_{0j}', \rho_0') \subset \overline{P}(0', \rho)$. Now let $\{f_j\}$ be any sequence in $\mathcal{F}$. We can find a subsequence $\{f_{j_k}\}$ of $\{f_j\}$ such that f_{j_k} converges uniformly on $S_1 \cup S_2$ to a function g holomorphic on $S_1 \cup S_2$ or to $g \equiv \infty$.

If g is holomorphic, then $|g(z)| \le M < \infty$ on the skeleton Ξ of the polydisk $P(0, \rho_0') \times \{z_n \in \mathbb{C} : |z_n| \le r'\}$ and hence $|f_j| \le M + 1$ on Ξ. But then, by the maximum principle, $|f_j| \le M + 1$ on $P(0, \rho_0') \times \{z_n \in \mathbb{C} : |z_n| \le r'\}$ for $j \ge j_0$. By Montel's theorem, $\{f_j(z)\}$ converges uniformly on $P(0, \rho_0') \times \{z_n \in \mathbb{C} : |z_n| \le r'\}$. This contradicts our assumption that $\mathcal{F}$ is not normal at the origin 0.

Thus we may assume that the limiting function $g \equiv \infty$ on Ξ. If $g \equiv \infty$, then, relabeling a subsequence, we may suppose that $|f_j| \ge j$ on $G := P(z_{0j}', \rho_0') \times \{z_n \in \mathbb{C} : r' - \varepsilon < |z_n| \le r'\}$.

If infinitely many of the $f_j(z)$ are nonvanishing on

$$P(0, \rho_0') \times \{z_n \in \mathbb{C} : |z_n| \le r'\}$$

it follows that $1/f_j$ is holomorphic and $1/|f_j| < 1/j$ on

$$P(0, \rho_0') \times \{z_n \in \mathbb{C} : |z_n| \le r'\}.$$

Therefore the sequence $\{1/f_j(z)\}$ converges uniformly on

$$P(0, \rho_0') \times \{z_n \in \mathbb{C} : |z_n| \le r'\} \text{ to } 0.$$

Hence $\{f_j\}$ converges uniformly on $P(0, \rho_0') \times \{z_n \in \mathbb{C} : |z_n| \le r'\}$ to ∞. This also contradicts our assumption that $\mathcal{F}$ is not normal at the origin 0.

Thus there exist a subsequence of function $\{f_{j_k}(z)\}$ of $\{f_j(z)\}$ and a sequence of points $\{(a_k, b_k)\}$ with $a_k \in P(0', \rho)\backslash P'(z_{0j}', \rho_0')$ and $|b_k| < r' - \varepsilon$ such that $f_{j_k}(a_k, b_k) = 0$. On the other hand, note that $f_{j_k}(z) \to \infty$ as $k \to \infty$ on $P'(z'_{0j}, \rho'_0)$ say $|f_{j_k}(z)| > 1$ on $P'(z'_{0j}, \rho'_0)$ for all $k > k_0$. It follows that for any $z' \in P(0', \rho)$ each $f_{j_k}(z)$ for $k > k_0$ vanishes at some point $(z', z_n(z')) \in \mathbb{C}^n$ with $|z_n(z')| < r' - \varepsilon$ by the Weierstrass preparation theorem. In particular, if we fix $z' = z'_0$ in $\overline{P}'(z'_{0j}, \rho'_0)$ and consider a limit point w^* of $\{w_k(z'_0)\}_{k \geq k_0}$ on $L_{z'_0}$, then $|w^*| < r' - \varepsilon$. We conclude that $\{f_{j_k}(z)\}$ cannot converge uniformly to ∞ on any neighborhood of (z'_0, w^*). This contradicts our assumption that $f_{j_k}(z) \to \infty$ as $k \to \infty$ uniformly on $\overline{P}'(z'_{0j}, \rho'_0)$, and proves the theorem. □

Remark 3.8.2. In general topology, a subset of a topological space is perfect if it is closed and has no isolated points. Denote by $(\mathcal{F})$ a subset of Ω where $\mathcal{F}$ is not locally normal. It follows at once from the definition of the locally normal family that $(\mathcal{F})$ is closed in Ω. From the above theorem we see that $(\mathcal{F})$ cannot have isolated points, hence $(\mathcal{F})$ is perfect.

The points of $(\mathcal{F})$ must go beyond the boundary of the domain Ω. The proof goes by contradiction. Suppose that there are two points $Q \in (\mathcal{F})$ and $Z \in \partial\Omega$ such that $\text{dist}(V, Z) \leq \text{dist}(Q, Z)$ for all $V \in (\mathcal{F})$. We may choose coordinates $z_1, \ldots, z_n$ in $\mathbb{C}^n$ such that $Q = (0)$ and $Z = (\delta,' 0)$. Clearly, $\text{dist}(Q, Z) = \delta$. By Theorem 3.8.1 for $z' = (-\rho/2,' 0) \in \mathbb{C}^n$, there must be at least one point q on the set

$$Z_{z'} : z_1 = -\rho/2, \quad z_j = 0 \quad (j = 2, \ldots, n-1), \quad 0 < |z_n| < r',$$

such that $\mathcal{F}$ is not normal at q. It is plain by inspection that

$$\text{dist}(q, Z) > \delta + \rho/2 > \delta = \text{dist}(Q, Z).$$

This is a contradiction, which proves our assertion.

Remark 3.8.3. There are three definitions of pseudoconvex domains in $\mathbb{C}^n$, $n \geq 2$, which are equivalent (see [220, Chapter 4, p. 112-115]. One of them is the following:

Definition 3.8.4. Let Ω be a domain in $\mathbb{C}^n (n \geq 2)$ and let $P = (a_1, \ldots, a_n) \in \partial\Omega$. We say that Ω is pseudoconvex at the boundary point P if the following holds: under the assumption that there exists $r > 0$ such that the punctured disk

$$Z_a : z_j = a_j \quad (j = 1 \ldots, n-1), \quad 0 < |z_n - a_n| < r$$

is contained in Ω, we have, for any r' satisfying $0 < r' < r$, that there exists $\rho > 0$ such that for each $(z'_1 \ldots, z'_{n-1}) \in \mathbb{C}^{n-1}$ with $\left|'_j - a_j\right| < \rho$, $(j = 1, \ldots, n-1)$. the disk

$$Z_{z'} : z_j = z'_j \quad (j = 1, \ldots, n-1), \quad |z_n - a_n| < r'$$

intersects $\partial\Omega$.

Let $\mathcal{F}$ be a family of holomorphic functions in $\Omega \subset \mathbb{C}^n$. The set Ω' consisting of all points z in Ω such that $\mathcal{F}$ is normal in a neighborhood of z is called the domain of normality of $\mathcal{F}$. Julia's theorem states that if Ω is a pseudoconvex domain, so is Ω'.

In Oka [229] the definition of a normal family of analytic hypersurfaces in a domain in $\mathbb{C}^n$ was given, and it was proved that the domain of normality of such a family in a pseudoconvex domain is also a pseudoconvex domain. This study was developed in his last paper [228].

3.9 Schwick's Normality Criterion

As Royden [252] pointed out, although Marty's result is necessary and sufficient for the relative compactness of a family of holomorphic or meromorphic functions, it may not be easy to apply in certain instances. For example, it is not obvious how Theorem 3.2.3 can be applied to establish normality of the family

$$\mathcal{F} = \left\{ f \in \mathcal{O}(\Omega) : (1 + |f(z)|^2) f^\sharp(z) \le e^{|f(z)|} \right\}.$$

Zalcman's rescalling lemma (Theorem 3.3.1) can be used to prove the following strengthening of the sufficiency part of Marty's normality criterion (Theorem 3.2.3):

Theorem 3.9.1 (see [64, Theorem 3]). *Let $\mathcal{F}$ be a family of holomorphic functions on a domain $\Omega \subset \mathbb{C}^n$ with the property that, for each compact set $K \subset \Omega$, there is a function $h_K : [0, \infty] \to [0, \infty]$, which is bounded in some neighborhood of each $z_0 \in (0, \infty)$, such that*

$$(1 + |f(z)|^2) f^\sharp(z) \le h_K(|f(z)|)$$

for all $f \in \mathcal{F}$ and $z \in K$. Then $\mathcal{F}$ is normal in Ω.

Proof. Since all features of the theorem are local, and translation and scale invariant, it suffices to consider the case when $z_0 = 0$ and $\Omega = B(0, 1)$. If $\mathcal{F}$ is not normal at 0, it follows from Zalcman's rescalling lemma (Theorem 3.3.1) that there exist $f_j \in \mathcal{F}$, $z_j \to 0$, $\rho_j \to 0$, such that the sequence

$$g_j(z) = f_j(z_j + \rho_j z)$$

converges locally uniformly in $\mathbb{C}^n$ to a nonconstant entire function g satisfying $g^\sharp(z) \le g^\sharp(0) = 1$.

It is clear from (3.2) that $g^\sharp$ has the expression

$$g^\sharp(z) = \frac{|Dg(z)|}{1 + |g(z)|^2}$$

which in particular shows that $g^\sharp$ is a continuous function.

Since $g^\sharp(z)$ is continuous on $\mathbb{C}^n$, there exists some δ, $0 < \delta < 1$, such that

$$|g^\sharp(z) - g^\sharp(0)| < \frac{g^\sharp(0)}{2}$$

whenever $z \in B(0, \delta)$. For all such z we have, using the reverse triangle inequality,

$$|g^\sharp(z)| \geq |g^\sharp(0)| - |g^\sharp(z) - g^\sharp(0)| > \frac{g^\sharp(0)}{2} = \frac{1}{2}$$

for all $z \in B(0, \delta)$. Since g is a nonconstant entire function, it follows that there exists $a \in B(0, \delta)$ such that $|g(a)| \neq 0$ and $g^\sharp(a) > 1/2$.

For fixed r, $0 < r < 1$, the hypothesis says that there exists a function $h_r : [0, \infty] \to [0, \infty]$ such that

$$(1 + |f(z)|^2) f^\sharp(z) \leq h_r(|f(z)|)$$

for all $f \in \mathcal{F}$, $z \in \overline{B(0, r)}$. Also there exists some constants $M > 0$ and $\sigma > 0$, say, such that $h_r(x) < M$ for $|x - |g(a)|| < \sigma$. For all j sufficiently large,

$$|z_j + \rho_j a| \leq |z_j| + \rho_j |a| < r;$$

this yields $z_j + \rho_j a \in B(0, r)$. Next we have

$$g_j(a) = f_j(z_j + \rho_j a) \to g(a),$$

which yields $||g_j(a)| - |g(a)|| < \sigma$ for j sufficiently large. So, if j is sufficiently large,

$$g_j^\sharp(a) = f_j^\sharp(z_j + \rho_j a)\rho_j \leq \frac{h_r(|f_j(z_j + \rho_j a)|)}{1 + |f_j(z_j + \rho_j a)|^2}\rho_j =$$

$$\frac{h_r(|g_j(a)|)}{1 + |g_j(a)|^2}\rho_j \leq h_r(|g_j(a)|)\rho_j \leq M\rho_j \to 0.$$

This implies that $g^\sharp(a) = 0$, which is a contradiction to $g^\sharp(a) > 1/2$. This contradiction proves Theorem 3.9.1. □

Remark 3.9.2. Obviously the result has its origin in the sufficiency of Marty's normality criterion (Theorem 3.2.3).

Remark 3.9.3. Set

$$f'(z) = \Big(\frac{\partial f}{\partial z_1}(z), \ldots, \frac{\partial f}{\partial z_n}(z)\Big).$$

It is elementary to see that

$$L_z(\log(1 + |f|^2), v) = \frac{|(f'(z), v)|^2}{(1 + |f(z)|^2)^2}.$$

Appealing to the Cauchy-Schwarz inequality, it is easy to see that

$$|(f'(z), v)|^2 \leq |f'(z)|^2 \, |v|^2$$

with equality if and only if v and $f'(z)$ are linearly dependent.

Then

$$\sup_{|v|=1}(1+|f(z)|^2)^2 L_z(\log(1+|f|^2),v) \le |f'(z)|^2 .$$

If $f'(z) \neq 0$, then the left-hand side is equal to

$$\Big[(1+|f(z)|^2)f^\sharp(z)\Big]^2 = |f'(z)|^2 .$$

If $f'(z) = 0$, then the above equality is trivially true because both sides of it are 0.

So, if $n = 1$, Theorem 3.9.1 coincides with Schwick's theorem [265, Theorem, p. 91], which is the generalization of a normality criterion of Royden [252].

Clearly a result like Theorem 3.9.1 can be useful.

3.10 Grahl and Nevo's Normality Criterion

By Marty's normality criterion (Theorem 3.2.3), the normality of a family $\mathcal{F}$ of holomorphic functions on a domain $\Omega \subset \mathbb{C}^n$ is equivalent to the local boundedness of the corresponding family $\mathcal{F}^\sharp := \big\{f^\sharp(z) : f \in \mathcal{F}, z \in \Omega\big\}$. It is rather surprising that there is also a normality criterion based on a lower bound of the family $\mathcal{F}^\sharp$.

The result which follows is the extension of Grahl and Nevo's normality criterion [86] to a family of holomorphic functions of several complex variables.

Theorem 3.10.1. *Let $\mathcal{F}$ be a family of holomorphic functions on $\Omega \subset \mathbb{C}^n$ and assume that for each compact subset $K \subset \Omega$, there exists a constant $M(K) > 0$ such that*

$$f^\sharp(z) \ge M(K), \quad z \in K,$$

holds for every $f \in \mathcal{F}$. Then $\mathcal{F}$ is normal in Ω.

Proof. The proof will be based again on Zalcman's rescalling lemma (Theorem 3.3.1).

Since normality is a local property, we can restrict all our considerations concerning normal families to the unit ball. So there is no harm in assuming that

$$\mathcal{F} = \big\{f \text{ holomorphic in } \Omega : f^\sharp(z) > \varepsilon \text{ for all } z \in B(0,1)\big\}$$

for some $\varepsilon > 0$.

To obtain a contradiction suppose that $\mathcal{F}$ is not normal at the point 0. If $\mathcal{F}$ is not normal at 0, it follows from Zalcman's rescalling lemma (Theorem 3.3.1) that there exist $f_j \in \mathcal{F}$, $z_j \to 0$, $\rho_j \to 0$, such that the sequence

$$g_j(z) = \rho_j^2 \cdot f_j(z_j + \rho_j z)$$

converges locally uniformly in $\mathbb{C}^n$ to a nonconstant entire function g satisfying $g^\sharp(z) \leq g^\sharp(0) = 1$. Since g is a nonconstant entire function it follows that there exists $a \in \Omega$ such that $|g(a)| > 0$. Hence $|g_j(a)| \neq 0$ for all j sufficiency large. It is easy to see that

$$1 \geq g_j^\sharp(a) = \frac{\max_{|v|=1} |Df_j(z_j + \rho_j a)v| \cdot |g_j(a)|^2}{\rho_j \cdot |f_j(z_j + \rho_j a)|^2 \cdot (1 + |g_j(a)|^2)} \geq$$

$$\frac{f_j^\sharp(z_j + \rho_j a)}{\rho_j} \cdot \frac{|g_j(a)|^2}{1 + |g_j(a)|^2} \geq \frac{\varepsilon}{\rho_j} \cdot \frac{|g_j(a)|^2}{1 + |g_j(a)|^2}.$$

The right-hand side of this inequality tends to infinity as $j \to \infty$, a contradiction. This contradiction shows that $\mathcal{F}$ is normal in Ω. □

Corollary 3.10.2. *For each domain $\Omega \subset \mathbb{C}^n$, there exists a constant $C_\Omega < \infty$ (depending only on Ω) such that*

$$\inf_{z \in \Omega} f^\sharp(z) \leq C_\Omega$$

for each function f holomorphic in Ω.

Proof. Otherwise we could find a sequence $\{f_j\}_{j=1}^\infty$ such that

$$f_j^\sharp(z) > j, z \in \Omega,$$

for each j. By Theorem 3.10.1, $\{f_j\}_{j=1}^\infty$ is normal in Ω. Fix a point $z_0 \in \Omega$. By Marty's criterion (Theorem 3.2.3), there exists a constant $M < \infty$ such that

$$f_j^\sharp(z_0) < M,$$

for all j. But then of course

$$j < f_j^\sharp(z_0) < M$$

for all j. For $j > M$ this gives a contradiction. □

In [8], Alexander has made a study of the volume images of varieties. One of the principal tools in his study is the generalized Crofton's formula. As an application of his results he obtains the following theorem:

Theorem 3.10.3. *Let $\mathcal{F}$ be a family of holomorphic functions on the ball $B \subseteq \mathbb{C}^n$.*

(a) *If the restriction of $\mathcal{F}$ to each complex line through the origin is a normal family, then $\mathcal{F}$ is a normal family.*

(b) *If the restriction of $\mathcal{F}$ to each complex line through the origin is normal at the origin (i.e. in a neighborhood of the origin), then $\mathcal{F}$ is normal on B at the origin.*

It is true that Theorem 3.10.1 can be proved by the method of restriction to a complex lines passing through 0, and this involves calling in the Alexander theorem, a proof of which is not easy. We shall not prove the Alexander theorem here. Instead we refer the interested reader to his [8, Theorem 6.2, p. 248].

Nishino [221] proved that a family $\mathcal{F}$ of analytic functions on a domain Ω in $\mathbb{C}^2$ is normal if the restriction to each coordinate line (of the form $z_1 = \lambda$ or $z_2 = \lambda$) is normal.

It is of great interest to find simple proofs of these theorems.

3.11 Lappan's Normality Criterion

The Zalcman normality criterion has proved to be a useful tool in the study of normal families of holomorphic functions. Zalcman's normality criterion and the one-dimensional Nevanlinna theorem can be used to prove the following theorem, which generalizes those obtained by Lappan [161].

Theorem 3.11.1. *Let $\mathcal{F}$ be a family of holomorphic functions on a domain $\Omega \subset \mathbb{C}^n$ with the property that for each compact set $K \subset \Omega$, there is a function $h'_K(|\cdot|) : \mathbb{C} \to [0, \infty]$, which is finite for at least three point on $\mathbb{C}$, such that*

$$(1 + |f(z)|^2) f^\sharp(z) \leq h'_K(|f(z)|) \tag{3.1}$$

for all $f \in \mathcal{F}$ and $z \in K$. Then $\mathcal{F}$ is normal in Ω.

Let us begin with a definition.

Definition 3.11.2. Let $g(\lambda)$ be a holomorphic function in $\mathbb{C}$. If the equation $g(\lambda) = a$, $a \in \mathbb{C}$ has no simple roots then a called a *totally ramified value.*

Note that an omitted value trivially satisfies this definition, but that it will be useful to distinguish between omitted values and nonomitted totally ramified values.

Theorem 3.11.3 (R. Nevanlinna [269, Theorem 17.3.10., p. 274]). *Let g be an entire (holomorphic in $\mathbb{C}$) function. Then g has at most two totally ramified (finite) values.*

Proof of Theorem 3.11.1. We shall show that the assumption that $\mathcal{F}$ is not normal at point $z_0 \in \Omega$ leads to a contradiction. Since all features of the theorem are local, and translation and scale-change invariant, it suffices to consider the case when $z_0 = 0$ and $\Omega = B(0, 1)$. If $\mathcal{F}$ is not normal at 0, it follows from Zalcman's Lemma 3.3.1 that there exists $f_j \in \mathcal{F}$, $z_j \to 0$, $\rho_j \to 0$, such that the sequence

$$g_j(z) := f_j(z_j + \rho_j z)$$

converges uniformly on compact subsets of $\mathbb{C}^n$ to a nonconstant entire function g satisfying $g^\sharp(z) \le g^\sharp(0) = 1$.

Let $K \subset B(0,1)$ be a closed ball in $\mathbb{C}^n$ about 0. Let $E = E(K) = \{a_1, a_2, a_3\}$ be the set of three distinct values in $\mathbb{C}$ such that the function h'_K is finite on E and let $M = M(K)$ be the maximal value of h'_K on E.

Define the value set by

$$A_{g_j}(a_l) := \{\xi \in \mathbb{C}^n : g_j(\xi) = a_l\} = g_j^{-1}[\{a_l\}].$$

Suppose that $\zeta^l \in A_g(a_l)$. Choose $R > 0$ such that $\zeta^l \in S_R = \{\xi \in \mathbb{C}^n : |\xi| < R\}$. Since normality is a local property, the restriction of a family $\{g_j - a_l\}$ to any open ball $s_n(\zeta^l) := \{\xi \in \mathbb{C}^n : |\xi - \zeta^l| < 1/n\} \subset S_R$ is a normal family. By Hurwitz's theorem (Theorem 2.0.14) $A_{g_j}(a_l) \cap s_n(\zeta^l) \neq \emptyset$ for j sufficiently large since g is not a constant function. It is routine to show that there exists a sequence $\{p_j^l\} \subset S_K$, such that $g_j(p_j^l) = a_l$.

Since $\rho_j \to 0$, we see that $z_j + \rho_j p_j^l \in K$ for j sufficiently large. Now by (3.1), $(1 + |f_j(z_j + \rho_j p_j^l)|^2) f_j^\sharp(z_j + \rho_j p_j^l) \le h'_K(|f_j(z_j + \rho_j p_j^l)|) \le M$ for j sufficiently large, so that

$$g^\sharp(\zeta^l) = \lim_{k\to\infty} g_j^\sharp(p_j^l) = \lim_{k\to\infty} \rho_j f_j^\sharp(z_j + \rho_j p_j^l) \le \lim_{k\to\infty} \rho_j M = 0\,.$$

Thus $g^\sharp(\zeta^l) = 0$.

Since

$$g^\sharp(\zeta^l) = \max_{\{v \in \mathbb{C}^n : |v|=1\}} \frac{|\frac{d}{d\lambda} g(\zeta^l + \lambda \cdot v)|_{\lambda=0}/d\lambda|}{1 + |g(\zeta^l)|^2} = 0$$

it follows that a_l is a finite totally ramifies value for $g(\zeta^l + \lambda \cdot v)$ ($v \in \mathbb{C}^n$ arbitrary (but fixed) and $|v| = 1$).

If $\{\xi \in \mathbb{C}^n : \xi = \zeta^l + \lambda \cdot v\} \cap A_g(a^k) \ni \zeta^k$, then $\zeta^k = \zeta^l + \lambda_k \cdot v$ for some $\lambda_k \in C$. Arguing as above we have

$$g'(\zeta^l + \lambda_k \cdot v) = 0.$$

Hence a_k is a totally ramified value for $g(\zeta^l + \lambda \cdot v)$.

If $\{\xi \in \mathbb{C}^n : \xi = \zeta^l + \lambda \cdot v\} \cap A_g(a^k) = \emptyset$, then a^k is an omitted value for $g(\zeta^l + \lambda \cdot v)$ and hence a totally ramified (finite) value of the function $g(\zeta^l + \lambda \cdot v)$.

Thus a_1, a_2, a_3 are three totally ramified (finite) values for the entire function $g\Big(\zeta^l + \lambda \cdot v\Big)$. By Nevanlinna's theorem, $g\Big(\zeta^l + \lambda \cdot v\Big)$ is constant. Since v was arbitrary the function g is constant, a contradiction. Thereby, the theorem is proved. □

One almost immediate consequence of the (proof of the) above Theorem is the following extension and sharpening of Schwick's extension [265] of a theorem of Royden [252] (see also [64]).

Theorem 3.11.4. *Let $\mathcal{F}$ be a family of holomorphic functions on a domain $\Omega \subset \mathbb{C}^n$ with the property that for each compact set $K \subset \Omega$ there is a function $h''_K(|\cdot|) : \mathbb{C} \to [0, \infty]$, which is finite somewhere on $\mathbb{C}$, such that*

$$(1 + |f(z)|^2) f^\sharp(z) \leq h''_K(|f(z)|)$$

for all $f \in \mathcal{F}$ and $z \in K$. Then $\mathcal{F}$ is normal in Ω.

3.12 Mandelbrojt's Normality Criterion

Mandelbrojt [184] provided a boundedness criterion for a family $\mathcal{F}$ of zero-free holomorphic functions on a region $\Omega \subset \mathbb{C}$ to be normal: For each compact $K \subset \Omega$, $f \in \mathcal{F}$ define

$$M(f, K) := \max\{|f(z)|/|f(w)| : z, w \in K\}$$

and

$$L(f, K) := \max\{\log|f(z)|/\log|f(w)| : z, w \in K\}$$

if $f(K) \cap T = \emptyset$; otherwise $L(f, K) := +\infty$. Then $\mathcal{F}$ is normal if and only if for each K the set $\{\min\{M(f, K), L(f, K)\} : f \in \mathcal{F}\}$ is bounded. That criterion can be considered as a variation of Montel's theorem.

A similar result holds in several variables:

Theorem 3.12.1. *[[63, Theorem 1.1]] Let $\mathcal{F}$ be a family of zero-free holomorphic functions in a domain $\Omega \subset \mathbb{C}^n$. Then $\mathcal{F}$ is normal in Ω if and only if for each point $z_0 \in \Omega$ there exists a ball $\overline{B(z_0, r_0)} \subset \Omega$ such that the set $\{\min\{M(f, \overline{B(z_0, r_0)}), L(f, \overline{B(z_0, r_0)})\} : f \in \mathcal{F}\}$ is bounded.*

We shall need the following preliminary result:

Theorem 3.12.2. *If a family of holomorphic functions $\mathcal{F}$ is normal in a domain $\Omega \subset \mathbb{C}^n$, then about each point $z^0 \in \Omega$ there is a ball $B(z^0, r) \subseteq \Omega$ such that either*

$$|f(z)| < 2 \quad or \quad \frac{1}{|f(z)|} < 2$$

holds for all $z \in B(z^0, r)$, $f \in \mathcal{F}$, and conversely.

The partitioning of the family $\mathcal{F}$ into two normal families works equally well if the constant 2 in the above is replace by an arbitrary positive constant.

Proof of Theorem 3.12.2. If w' and w'' are in the finite plane, then

$$\chi(w', w'') = \frac{|w' - w''|}{\sqrt{1 + |w'|^2}\sqrt{1 + |w''|^2}}.$$

Observe that the *chordal distance* between the points w' and w'' such that $|w'| = 1$ and $|w''| = 2$ is $\chi(w', w'') \geq 1/\sqrt{10}$. Then, for any two points w', w'' that satisfy either the two relations $|w'| \leq 1$ and $|w''| \geq 2$, or the two relations $|w'| \geq 1$ and $|w''| \leq 1/2$, it must be true that $\chi(w', w'') \geq 1/\sqrt{10}$.

Indeed, the function

$$g(t) = \frac{1-t}{\sqrt{2+2t^2}}, 0 < t < 1,$$

visibly satisfies

$$g \text{ is decreasing on the interval } [0, 1] \tag{3.1}$$

as it is the product of the two positive decreasing functions $1 - t$ and $1/\sqrt{2+2t^2}$. Now for complex numbers w' and w'', with $|w'| < 1 < 2 < |w''|$,

$$\chi(w', w'') \geq \frac{|w'| - |w''|}{\sqrt{(1+|w'|^2)(1+|w''|^2)}} >$$

$$\frac{|w''| - 1}{\sqrt{2(1+|w''|^2)}} = g\left(\frac{1}{|w''|}\right) > g(1/2) \text{ [due to (3.1)] } = \frac{1}{\sqrt{10}}.$$

From the symmetry of the chordal metric in its two variables, it is easy to see that

$$\chi(w', w'') > 1/\sqrt{10} \tag{3.2}$$

holds if w' belongs to the complements of the closed unit disk and w'' belongs to the open disk of radius $1/2$.

Let z_0 be arbitrary but fixed point in Ω. If, say, $|f(z_0)| \leq 1$ then $|f(z)| < 2$ for $z \in B(z_0, r)$, $f \in \mathcal{F}$, (otherwise, $\chi(f(z), f(z_0)) > \chi(1, 2)$), a contradiction with the inequality (3.2). However if $|f(z_0)| > 1$, then $|f(z)| > 1/2$ (otherwise, $\chi(f(z), f(z_0)) > \chi(1, 1/2) = \chi(1, 2)$), a contradiction with the inequality (3.2). Therefore, if $z \in B(z_0, r)$, either

$$|f(z)| < 2 \quad \text{or} \quad \frac{1}{|f(z)|} < 2, \quad f \in \mathcal{F}.$$

It follows that $\mathcal{F}$ can be expressed as the union of two families:

$$\mathcal{J} = \{f \in \mathcal{F}, |f(z_0)| \leq 1, |f(z)| < 2, z \in B(z_0, r)\},$$

$$\mathcal{H} = \{f \in \mathcal{F}, \frac{1}{|f(z_0)|} < 1, \frac{1}{|f(z)|} < 2, z \in B(z_0, r)\}.$$

□

Proof of Theorem 3.12.1. ⇒ Fix a point z_0 in Ω and consider a ball

$$B(z_0, r) \subset \Omega.$$

Suppose that $\mathcal{F}$ is normal in Ω but the set

$$\{\min\{M(f,\overline{B(z_0,r_0)}),L(f,\overline{B(z_0,r_0)})\}:f\in\mathcal{F}\},$$

for some $r_0<r$, is unbounded. Then there exists a sequence $\{f_j\}\subseteq\mathcal{F}$ such that

$$\min\{M(f_j,\overline{B(z_0,r_0)}),L(f_j,\overline{B(z_0,r_0)})\}>j \text{ for all } j\in\mathbb{N}. \tag{3.3}$$

By hypothesis $\mathcal{F}$ is normal, and therefore the following two cases exhaust all the possibilities for sequence $\{f_j\}$:

(1) the sequence $\{f_j\}$ has a subsequence $\{f_{j_k}\}$ which converges uniformly on $\overline{B(z_0,r_0)}$ to a holomorphic function f;

(2) the sequence $\{f_j\}$ has a subsequence $\{f_{j_k}\}$ which converges uniformly on $\overline{B(z_0,r_0)}$ to ∞.

Since $\mathcal{F}$ is a family of zero-free holomorphic functions in a domain Ω, by Hurwitz's theorem (Theorem 2.0.14) f is either nowhere zero or identically equal to zero.

Therefore, the following three cases exhaust all the possibilities for sequence $\{f_j\}$:

(a) the sequence $\{f_j\}$ has a subsequence $\{f_{j_k}\}$ which converges uniformly on $\overline{B(z_0,r_0)}$ to the holomorphic function $f\equiv 0$;

(b) the sequence $\{f_j\}$ has a subsequence $\{f_{j_k}\}$ which converges uniformly on $\overline{B(z_0,r_0)}$ to a holomorphic function f which is zero-free on $\overline{B(z_0,r_0)}$;

(c) the sequence $\{f_j\}$ has a subsequence $\{f_{j_k}\}$ which converges uniformly on $\overline{B(z_0,r_0)}$ to ∞.

Since $j_k\geq k$, it follows readily from (3.3) that

$$\min\{M(f_{j_k},\overline{B(z_0,r_0)}),L(f_{j_k},\overline{B(z_0,r_0)})\}>k \text{ for all } k\in\mathbb{N}. \tag{3.4}$$

In case (a) (respectively in case (c)) we have $|f_{j_k}(z)|<1/2$ (respectively $|f_{j_k}(z)|>2$) for all $z\in\overline{B(z_0,r)}$ and all $k\in\mathbb{N}$ sufficiently large. Hence $\log|f_{j_k}(z)|$ is a negative (respectively positive) pluriharmonic function in $B(z_0,r)$. Pluriharmonic functions form a subclass of the class of harmonic functions in $B(z_0,r)$ (obviously proper for $n>1$). So by Harnack's inequality there exists some constant $C=C(\overline{B(z_0,r_0)},B(z_0,r))$, $C\in(1,\infty)$, such that

$$\frac{\log|f_{j_k}(z)|}{\log|f_{j_k}(w)|}\leq C \text{ for all } z \text{ and } w \text{ in } \overline{B(z_0,r_0)},$$

and hence $L(f_{j_k},\overline{B(z_0,r_0)})\leq C$ for all $k\in\mathbb{N}$ sufficiently large.

In case (b) we have $\lim_{k\to\infty} |f_{j_k}(z)| = f(z)$ for all $z \in \overline{B(z_0, r)}$. It follows that

$$\lim_{k\to\infty} \frac{|f_{j_k}(z)|}{|f_{j_k}(w)|} = \left|\frac{f(z)}{f(w)}\right|$$

uniformly for $z, w \in \overline{B(z_0, r_0)}$. The function $f(z)/f(w)$ is holomorphic on $\overline{B(z_0, r_0)} \times \overline{B(z_0, r_0)}$, so it follows that $M(f, \overline{B(z_0, r_0)})$ is bounded. Since $M(f_{j_k}, \overline{B(z_0, r_0)}) \to M(f, \overline{B(z_0, r_0)})$ as $k \to \infty$, we conclude that

$$M(f_{j_k}, \overline{B(z_0, r_0)})$$

is also bounded for all $k \in \mathbb{N}$ sufficiently large.

Hence the set of quantities $\min\{M(f_{j_k}, \overline{B(z_0, r_0)}), L(f_{j_k}, \overline{B(z_0, r_0)}), \quad k \in \mathbb{N}, \}$ is bounded, which is a contradiction to (3.4).

$\Leftarrow$ Fix a point z_0 in Ω, consider a ball $B(z_0, r_0) \subset \Omega$, and define the families $\mathcal{J}$ and $\mathcal{H}$ by

$$\mathcal{J} = \{f \in \mathcal{F}, |f(z_0)| \leq 1\},$$
$$\mathcal{H} = \{f \in \mathcal{F}, |f(z_0)| > 1\}.$$

It will be shown that $\mathcal{J}$ is normal in $\mathcal{O}(\Omega)$ and that $\mathcal{H}$ is normal in $C(\Omega, \mathbb{C}^\infty)$.

To prove that the family $\mathcal{J} := \{f \in \mathcal{F} : |f(z_0)| \leq 1\}$ is normal, it is sufficient to show that each sequence $\{f_j\} \subset \mathcal{J}$ contains a subsequence converging locally uniformly in $B(z_0, r_0)$ to a holomorphic function or to ∞. The following two cases exhaust all the possibilities:

(a) there exists a subsequence $\{f_{j_k}\}$ such that for any $k \in \mathbb{N}$ the function $\log |f_{j_k}|$ does not vanish in $\overline{B(z_0, r_0)}$;

(b) for each $j \in \mathbb{N}$ there exists $z_j \in \overline{B(z_0, r_0)}$ such that $\log |f_j(z_j)| = 0$.

In case (a) we have that $|f_{j_k}| < 1$ in $\overline{B(z_0, r_0)}$ for all elements of the sequence. Such a subsequence is normal in $B(z_0, r_0)$ by Montel's theorem, and hence we are done in case (a).

In case (b) we have $L(f_j, B(z_0, r_0)) = +\infty$ for all $j \in \mathbb{N}$. Therefore, according to the hypothesis, $M(f_j, \overline{B(z_0, r_0)}) < C$ for all $j \in \mathbb{N}$ and some constant $C > 0$. It follows that $|f_j| < C$ in $B(z_0, r_0)$ for all $j \in \mathbb{N}$, which means that $\{f_j\}$ is a normal family in $B(z_0, r_0)$ and hence finishes the proof in case (b).

If $f \in \mathcal{H}$, then $1/f$ is holomorphic on Ω because f never vanishes. Also, $1/f$ never vanishes and $1/|f(z_0)| < 1$. Hence reasoning similar to that in the above proof shows that $\widetilde{\mathcal{H}} = \{1/f : f \in \mathcal{H}\}$ is also normal in $\mathcal{O}(B(z_0, r_0))$. So if $\{f_j\}$ is a sequence in $\mathcal{H}$ there is a subsequence $\{f_{j_k}\}$ and an analytic function h on $B(z_0, r_0)$ such that $\{1/f_{j_k}\}$ converges in $\mathcal{O}(B(z_0, r_0))$ to h. By the generalized Hurwitz theorem (Theorem 2.0.14), either $h \equiv 0$ or h never vanishes. If $h \equiv 0$ it is easy to see that $f_{j_k}(z) \to \infty$ uniformly on compact subsets of $B(z_0, r_0)$. If h never vanishes, then $1/h$ is analytic and it follows that $f_{j_k}(z) \to 1/h(z)$ uniformly on compact subsets of $B(z_0, r_0)$.

It follows that $\mathcal{J}$ and $\mathcal{H}$ are normal at z_0 so that the union $\mathcal{F}$ is normal in $B(z_0, r_0)$. Since normality is a local property, $\mathcal{F}$ is a normal family in Ω. This completes the proof of the theorem. □

Remark 3.12.3. It should be pointed out that the above theorem is not true if the condition "for each point $z_0 \in \Omega$ there exists a ball $B(z_0, r_0) \subset \Omega$ such that the the set of quantities $L(f, B(z_0, r_0))$, $f \in \mathcal{F}$, is bounded" is replaced by the condition "the corresponding family of functions given by $|g(z)|/|g(w)|$ is locally bounded on $\Omega \times \Omega$" (cf. [260, Theorem 2.2.8]).

To see this, consider the family $\mathcal{F} = \{z^j\}_{j=1}^{\infty}$ of holomorphic functions. If we take $\mathcal{A} = \{z \in \mathbb{C} : 1/2 < |z| < 1\}$, then $\mathcal{F}|_{\mathcal{A}}$ is a set of bounded (by 1) zero-free holomorphic functions in $\mathcal{A}$ so Montel's theorem guarantees that $\mathcal{F}$ is normal. It is plain by inspection that the family $\left\{\frac{|z|^j}{|w|^j}\right\}_{j=1}^{\infty}$ is not locally bounded on $\mathcal{A} \times \mathcal{A}$, while $\left\{\frac{\log|z|^j}{\log|w|^j}\right\}_{j=1}^{\infty}$ is a locally bounded family on $\mathcal{A} \times \mathcal{A}$. Hence Theorem 2.2.8 in [260] is not true.

Remark 3.12.4. As in the proof of Marty's normality criterion (Theorem 3.2.3), the necessary part of Theorem 3.12.1 also can be proved as follows. Suppose that $\mathcal{F}$ is normal in Ω, but $L(f_j, B(z_0, r_0))$ is unbounded. Hence $B(z_0, r_0)$ is compactly included in Ω, the map $\mathcal{F} \ni f \mapsto M(f, B(z_0, r_0)) \in \mathbb{R}$ is continuous and explodes only in a neighborhood of $\{f \equiv 0\}$ and $\{f \equiv \infty\}$. So it is sufficient to consider a sequence $\{f_j\}$ converging locally uniformly in Ω (and hence uniformly in $B(z_0, r_0)$) either to 0 or to ∞. For these cases one uses argument based on the Harnak inequality in order to obtain a contradiction.

3.13 Zalcman-Pang's Lemma

Marty's criterion (Theorem 3.2.3) in conjunctions with the ideas of Lawrence Zalcman and Xue-cheng Pang yields the following theorem.

Theorem 3.13.1. *A family $\mathcal{F}$ of functions holomorphic on $\Omega \subset \mathbb{C}^n$ is not normal at some point $z_0 \in \Omega$ if and only if there exist sequences $f_j \in \mathcal{F}$, $z_j \to z_0$, $r_j \to 0$, such that the sequence*

$$g_j(z) := \frac{f_j(z_j + r_j z)}{r_j^{\alpha}} \quad (0 \leq \alpha < 1 \quad \textit{arbitrary})$$

converges locally uniformly in $\mathbb{C}^n$ to a nonconstant entire function g satisfying $g^{\sharp}(z) \leq g^{\sharp}(0) = 1$.

Remark 3.13.2. Zalcman's lemma (Theorem 3.3.1) is the case $\alpha = 0$ of the result given above.

Of central importance in the proof of Theorem 3.13.1 is the following lemma.

Lemma 3.13.3. *Let f be a holomorphic function on the unit ball $B = \{z \in \mathbb{C}^n : |z| < 1\}$, and α be a real number with $0 \leq \alpha < 1$. Suppose that $j \geq 3$ and*

$$\max_{|z|\leq 1/j} \frac{(1-j|z|)^{1+\alpha}(1+|f(z)|^2)f^\sharp(z)}{(1-j|z|)^{2\alpha}+|f(z)|^2} > 1.$$

Then there exists a point ξ^*, $|\xi^*| < 1/j$, and a real number ρ, $0 < \rho < 1$, such that

$$\max_{|z|\leq 1/j} \frac{(1-j|z|)^{1+\alpha}\rho^{1+\alpha}(1+|f(z)|^2)f^\sharp(z)}{(1-j|z|)^{2\alpha}\rho^{2\alpha}+|f(z)|^2} =$$

$$\frac{(1-j|\xi^*|)^{1+\alpha}\rho^{1+\alpha}(1+|f(\xi^*)|^2)f^\sharp(\xi^*)}{(1-j|\xi^*|)^{2\alpha}\rho^{2\alpha}+|f(\xi^*)|^2} = 1.$$

Proof. To prove this lemma, we consider the function

$$\varphi(t,z) := \frac{(1-j|z|)^{1+\alpha}t^{1+\alpha}(1+|f(z)|^2)f^\sharp(z)}{(1-j|z|)^{2\alpha}t^{2\alpha}+|f(z)|^2}.$$

It is easy to see that $\varphi(t,z)$ is defined and continuous on $[0,1) \times \{z \in \mathbb{C}^n : |z| \leq 1/j\}$. The continuous function $(1+|f(z)|^2)f^\sharp(z)$ attains a maximum as z varies over the compact ball $\{z \in \mathbb{C}^n : |z| \leq 2/3\}$. Then there exists a constant $M > 0$ such that

$$\varphi(t,z) \leq (1-j|z|)^{1-\alpha}t^{1-\alpha}M \qquad (z \in \overline{B(0,2/3)}). \tag{3.1}$$

It follows that $\varphi(t,z)$ is continuous on $[0,1] \times \{z \in \mathbb{C}^n : |z| \leq 1/j\}$ and $\varphi(0,z) = 0$ on $\{z \in \mathbb{C}^n : |z| \leq 1/j\}$, $j > 3$.

By Weierstrass's theorem there exists z_1^*, $|z_1^*| < 1/j$ such that

$$\varphi(1,z_1^*) = \max_{|z|\leq 1/j} \varphi(1,z) > 1.$$

Since $\varphi(0,z_1^*) = 0$ and $\varphi(1,z_1^*) > 1$, by the intermediate value theorem for continuous functions on $[0,1]$, there exists ρ_1, $0 < \rho_1 < 1$, such that $\varphi(\rho_1,z_1^*) = 1$. We are done if $\max_{|z|\leq 1/j}\varphi(\rho_1,z) = \varphi(\rho_1,z_1^*) = 1$; if not, then $\max_{|z|\leq 1/j}\varphi(\rho_1,z) > 1$ and we repeats our discourse for $\varphi(\rho_1,z)$.

Again, by Weierstrass's theorem there exists z_2^*, $|z_2^*| < 1/j$ such that

$$\varphi(\rho_1,z_2^*) = \max_{|z|\leq 1/j} \varphi(\rho_1,z) > 1.$$

Since $\varphi(0,z_2^*) = 0$ and $\varphi(\rho_1,z_2^*) > 1$, by the intermediate value theorem for continuous functions on $[0,1]$ there exists ρ_2, $0 < \rho_2 < 1$, such that $\varphi(\rho_1\rho_2,z_2^*) = 1$. We are done if $\max_{|z|\leq 1/j}\varphi(\rho_1\rho_2,z) = \varphi(\rho_1\rho_2,z_2^*) = 1$; if not, then we repeats our discourse for $\varphi(\rho_1\rho_2,z)$ again, and so on.

If we stop after a finite number of steps, we are done; if not, we repeat this process, ad infinitum, to obtain points ρ_m, $0 < \rho_m < 1$, and z_m^*, $|z_m^*| < 1/j$, such that

$$\max_{|z|\leq 1/j} \varphi(\rho_1 \dots \rho_m, z) = \varphi(\rho_1 \dots \rho_m, z_m^*) > 1\,. \tag{3.2}$$

$$\varphi(\rho_1 \dots \rho_m \rho_{m+1}, z_m^*) = 1\,. \tag{3.3}$$

The sequence $\{x_m := \rho_1 \dots \rho_m\}$ is a bounded and decreasing. Then the greatest lower bound of the set $\{x_m : m \in N\}$, say ρ, is the limit of $\{x_m\}$. The sequence $\{z_m^*\}$ contains a subsequence, again denoted by $\{z_m^*\}$, such that $\lim_{m\to\infty} z_m^* = \xi^*$. From (3.2) follows that $0 < \rho < 1$ and $|\xi^*| < 1/j$. (This follows at once from Weierstrass's theorem that every bounded sequence has a convergent subsequence.)

From (3.2) and (3.3), we have

$$\begin{aligned} &\lim_{m\to\infty} \max_{|z|\leq 1/j} \varphi(\rho_1 \dots \rho_m, z) = \lim_{m\to\infty} \varphi(\rho_1 \dots \rho_m, z_m^*) = \\ &\lim_{m\to\infty} \varphi(\rho_1 \dots \rho_m \rho_{m+1}, z_m^*) = \varphi(\rho, \xi^*) = 1\,. \end{aligned} \tag{3.4}$$

We claim that

$$\max_{|z|\leq 1/j} \lim_{m\to\infty} \varphi(\rho_1 \dots \rho_m, z) = \lim_{m\to\infty} \max_{|z|\leq 1/j} \varphi(\rho_1 \dots \rho_m, z). \tag{3.5}$$

By Weierstrass's theorem

$$\max_{|z|\leq 1/j} \lim_{m\to\infty} \varphi(\rho_1 \dots \rho_m, z) = \max_{|z|\leq 1/j} \varphi(\rho, z) = \varphi(\rho, \eta),$$

where $|\eta| \leq 1/j$.

There is a sequence $\{w_m\}$ contained in $\{|z| \leq 1/j\}$ such that

$$\varphi(\rho_1 \dots \rho_m, \eta) \leq \max_{|z|\leq 1/j} \varphi(\rho_1 \dots \rho_m, z) = \varphi(\rho_1 \dots \rho_m, w_m).$$

Since $\{|z| \leq 1/j\}$ is compact, there is an infinite subsequence of $\{w_m\}$, again denoted by $\{w_m\}$, and ς, $|\varsigma| \leq 1/j$ such that $w_m \to \varsigma$ as $m \to \infty$. Therefore

$$\lim_{m\to\infty} \varphi(\rho_1 \dots \rho_m, \eta) \leq \lim_{m\to\infty} \max_{|z|\leq 1/j} \varphi(\rho_1 \dots \rho_m, z) = \lim_{m\to\infty} \varphi(\rho_1 \dots \rho_m, w_m)$$

and

$$\varphi(\rho, \eta) \leq \lim_{m\to\infty} \max_{|z|\leq 1/j} \varphi(\rho_1 \dots \rho_m, z) = \varphi(\rho, \varsigma) \leq \varphi(\rho, \eta).$$

We may conclude the desired claim (3.4). Combining (3.4) and (3.5), we have

$$\max_{|z|\leq 1/j} \varphi(\rho, z) = \varphi(\rho, \xi^*) = 1 \quad (|\xi^*| < 1/j).$$

This is the desired equality. The proof of the lemma is complete. □

Proof of Theorem 3.13.1. Assume $\mathcal{F}$ is not normal in point $z_0 \in \Omega$. To simplify matters, we assume that $z_0 = 0$ and all functions under consideration are holomorphic on the unit ball $\{z \in \mathbb{C}^n : |z| < 1\}$. By Marty's criterion (Theorem 3.2.3), $\mathcal{F}$ contains functions f_j, $j \in \mathbb{N}$, satisfying $\max_{|z|<1/(2j)} f_j^\sharp(z) > 2^{1+\alpha} j^{3(1+\alpha)}$. Hence there exists a ξ_j with $|\xi_j| < 1/j$ such that

$$\max_{|z|\le 1/j} (1 - j|z|)^{1+\alpha} f_j^\sharp(z) = (1 - j|\xi_j|)^{1+\alpha} f_j^\sharp(\xi_j) \ge j^{3(1+\alpha)}.$$

Since $1 + |f_j(\xi_j)|^2 > (1 - j|\xi_j|)^{2\alpha} + |f_j(\xi_j)|^2$, we have

$$\frac{(1 - j|\xi_j|)^{1+\alpha}(1 + |f_j(\xi_j)|^2) f_j^\sharp(\xi_j)}{(1 - j|\xi_j|)^{2\alpha} + |f_j(\xi_j)|^2} > (1 - j|\xi_j|)^{1+\alpha} f_j^\sharp(\xi_j) > j^{3(1+\alpha)}. \quad (3.6)$$

Hence

$$\max_{|z|\le 1/j} \frac{(1 - j|z|)^{1+\alpha}(1 + |f_j(z)|^2) f_j^\sharp(z)}{(1 - j|z|)^{2\alpha} + |f_j(z)|^2} > 1.$$

According to Lemma 3.13.3, there exists ξ_j^*, $|\xi_j^*| < 1/j$, and ρ_j, $0 < \rho_j < 1$, such that

$$\max_{|z|\le 1/j} \frac{(1 - j|z|)^{1+\alpha} \rho_j^{1+\alpha} (1 + |f_j(z)|^2) f_j^\sharp(z)}{(1 - j|z|)^{2\alpha} \rho_j^{2\alpha} + |f_j(z)|^2} =$$

$$\frac{(1 - j|\xi_j^*|)^{1+\alpha} \rho_j^{1+\alpha} (1 + |f_j(\xi_j^*)|^2) f_j^\sharp(\xi_j^*)}{(1 - j|\xi_j^*|)^{2\alpha} \rho_j^{2\alpha} + |f_j(\xi_j^*)|^2} = 1.$$

Using this equation and (3.6), we obtain

$$1 = \frac{(1 - j|\xi_j^*|)^{1+\alpha} \rho_j^{1+\alpha} (1 + |f_j(\xi_j^*)|^2) f_j^\sharp(\xi_j^*)}{(1 - j|\xi_j^*|)^{2\alpha} \rho_j^{2\alpha} + |f_j(\xi_j^*)|^2} \ge$$

$$\frac{(1 - j|\xi_j|)^{1+\alpha} \rho_j^{1+\alpha} (1 + |f_j(\xi_j)|^2) f_j^\sharp(\xi_j)}{(1 - j|\xi_j|)^{2\alpha} + |f_j(\xi_j)|^2} \ge$$

$$(1 - j|\xi_j|)^{1+\alpha} \rho_j^{1+\alpha} f_j^\sharp(\xi_j) \ge \rho_j^{1+\alpha} j^{3(1+\alpha)} \quad (|\xi_j| < 1/j).$$

It follows

$$\left(\frac{1}{j}\right)^3 \ge \rho_j \to 0. \quad (3.7)$$

Put

$$r_j = (1 - j|\xi_j^*|)\rho_j \to 0.$$

Set

$$h_j(z) = \frac{f_j(\xi_j^* + r_j z)}{r_j^\alpha}.$$

We claim that appropriately chosen subsequences $z_k = \xi_{j_k}$, $\rho_k = r_{j_k}$, and $g_k = h_{j_k}$ will do. First of all, $h_j(z)$ is defined on $|z| < \frac{1}{j\rho_j}$, hence on $|z| < j$, since

$$|\xi_j^* + r_j z| \leq |\xi_j^*| + r_j|z| \leq |\xi_j^*| + r_j \frac{1 - j|\xi_j^*|}{jr_j} = \frac{1}{j}.$$

By the invariance of the Levi form under biholomorphic mappings, we have

$$L_z(\log(1 + |h_j|^2), v) = L_{\xi_j^* + r_j z}(\log(1 + |h_j|^2), r_j v)$$

and hence

$$h_j^\sharp(z) = r_j h_j^\sharp(\xi_j^* + r_j z).$$

Since $r_j = (1 - j|\xi_j^*|)\rho_j$ a simple computations shows that

$$h_j^\sharp(z) = \frac{r_j r_j^{-\alpha}(1 + |f_j(\xi_j^* + r_j z)|^2) f_j^\sharp(\xi_j^* + r_j z)}{1 + r_j^{-2\alpha}|f_j(\xi_j^* + r_j z)|^2} =$$

$$\frac{(1 - j|\xi_j^*|)^{1+\alpha}\rho_j^{1+\alpha}(1 + |f_j(\xi_j^* + r_j z)|^2) f_j^\sharp(\xi_j^* + r_j z)}{[(1 - j|\xi_j^*|)/(1 - j|\xi_j^* + r_j z|)]^{2\alpha}(1 - j|\xi_j^* + r_j z|)^{2\alpha}\rho_j^{2\alpha} + |f_j(\xi_j^* + r_j z)|^2} =$$

$$\frac{(1 - j|\xi_j^*|)^{1+\alpha}(1 - j|\xi_j^* + r_j z|)^{1+\alpha}\rho_j^{1+\alpha}(1 + |f_j(\xi_j^* + r_j z)|^2) f_j^\sharp(\xi_j^* + r_j z)}{(1 - j|\xi_j^* + r_j z|)^{1+\alpha}\left[\left(\frac{1 - j|\xi_j^*|}{1 - j|\xi_j^* + r_j z|}\right)^{2\alpha}(1 - j|\xi_j^* + r_j z|)^{2\alpha}\rho_j^{2\alpha} + |f_j(\xi_j^* + r_j z)|^2\right]}$$

Bearing in mind Lemma 3.13.3, it is easy to see that $h_j^\sharp(0) = 1$. Since[1]

$$\frac{1}{1 + 1/j} \leq \frac{1 - j|\xi_j^*|}{1 - j|\xi_j^* + r_j z|} \leq \frac{1}{1 - 1/j}$$

we have

$$\left(\frac{1}{1 + 1/j}\right)^{2\alpha}\left[(1 - j|\xi_j^* + r_j z|)^{2\alpha}\rho_j^{2\alpha} + |f_j(\xi_j^* + r_j z)|^2\right] \leq$$

$$\left(\frac{1 - j|\xi_j^*|}{1 - j|\xi_j^* + r_j z|}\right)^{2\alpha}(1 - j|\xi_j^* + r_j z|)^{2\alpha}\rho_j^{2\alpha} + |f_j(\xi_j^* + r_j z)|^2 \leq$$

[1] In fact, for $|z| < j$, $\rho_j < 1/j^3$, and $r_j = (1 - j|\xi_j^*|)\rho_j$ the triangle inequality for complex numbers and their absolute values $|a| - |b| \leq |a + b| \leq |a| + |b|$ $(a, b \in C)$ implies

$$\frac{1}{1 + 1/j} \leq \frac{1}{1 + j^2\rho_j} \leq \frac{1}{1 + j\rho_j|z|} =$$

$$\frac{1}{1 + j\frac{r_j}{1 - j|\xi_j^*|}|z|} \leq \frac{1 - j|\xi_j^*|}{1 - j|\xi_j^* + r_j z|} = \frac{1}{\frac{1}{1 - j|\xi_j^*|} - j\left|\frac{\xi_j^*}{1 - j|\xi_j^*|} + \frac{r_j}{1 - j|\xi_j^*|}z\right|} \leq$$

$$\frac{1}{1 - j\frac{r_j}{1 - j|\xi_j^*|}|z|} = \frac{1}{1 - j\rho_j|z|} \leq \frac{1}{1 - j^2\rho_j} \leq \frac{1}{1 - 1/j}.$$

$$\left(\frac{1}{1-1/j}\right)^{2\alpha}\left[(1-j|\xi_j^*+r_jz|)^{2\alpha}\rho_j^{2\alpha}+|f_j(\xi_j^*+r_jz)|^2\right].$$

From the above inequalities and Lemma 3.13.3, we infer that

$$h_j^\sharp(z)\le$$

$$(1+1/j)^{2\alpha}\cdot\left(\frac{1-|\xi_j^*|}{1-j|\xi_j^*+r_jz|}\right)^{1-\alpha}\cdot\frac{(1-j|\xi_j^*|)^{1+\alpha}\rho_j^{1+\alpha}(1+|f_j(\xi_j^*)|^2)f_j^\sharp(\xi_j^*)}{(1-j|\xi_j^*|)^{2\alpha}\rho_j^{2\alpha}+|f_j(\xi_j^*)|^2}=$$

$$(1+1/j)^{2\alpha}\cdot\left(\frac{1-j|\xi_j^*|}{1-j|\xi_j^*+r_jz|}\right)^{1-\alpha}\cdot 1\le$$

$$(1+1/j)^{2\alpha}\cdot\left(\frac{1}{1-1/j}\right)^{1-\alpha}\quad(|z|<j).$$

For every $m\in N$ the sequence $\{h_j\}_{j>m}$ is normal in $|z|<m$ by Marty's criterion (Theorem 3.2.3). The well-known Cantor diagonal process yields a subsequence $\{g_k=h_{j_k}\}$ which converges uniformly on every ball $|z|<R$. The limit function g satisfies $g^\sharp(z)\le\limsup_{j\to\infty}h_j^\sharp(z)\le 1=g^\sharp(0)$. Clearly g is nonconstant because $g^\sharp(0)\ne 0$. Conversely, suppose that there exist sequences $f_j\in\mathcal{F}$, $z_j\to z_0$, $r_j\to 0$ such that the sequence

$$g_j(z)=\frac{f_j(z_j+r_jz)}{r_j^\alpha}$$

converges locally uniformly in $\mathbb{C}^n$ to a nonconstant entire function g satisfying $g^\sharp(z)\le g^\sharp(0)=1$, but $\mathcal{F}$ is normal. Without restriction, we may suppose that $z_0=0$. By Marty's criterion (Theorem 3.2.3), there exists a constant $M>0$ such that

$$\max_{|z|<1/j_0}\frac{f_j^\sharp(z)}{r_j^\alpha}<M.$$

Since $z_j\to 0$, $r_j\to 0$, for $|z|<1/j_0$, there exists j_1 such that for all $j>j_1$

$$|z_j+\rho_jz|\le|z_j|+\rho_j|z|\le|z_j|+\rho_j/j_0<1/j_0.$$

Therefore

$$g_j^\sharp(z)=\frac{r_jf_j^\sharp(z_j+r_jz)}{r_j^\alpha}=\frac{f_j^\sharp(z_j+r_jz)}{r_j^\alpha}r_j\le Mr_j\to 0.$$

But then $g^\sharp(0)=0$, a contradiction, since $g^\sharp(0)=1$. This contradiction concludes the proof of the theorem. □

By similar arguments as those used to prove Theorem 3.13.1 we can obtain the following:

Theorem 3.13.4. *Let $\mathcal{F}$ be a family of zero-free holomorphic functions in a domain $\Omega\subset\mathbb{C}^n$. The statement of Theorem 3.13.1 remains valid if $0\le\alpha<1$ is replaced with $-\infty<\alpha<\infty$.*

Proof of Theorem 3.13.4. Assume $\mathcal{F}$ is not normal at the point $z_0 \in \Omega$. To simplify matters we assume that $z_0 = 0$ and all functions under consideration are holomorphic on the unit ball $\{z \in \mathbb{C}^n : |z| < 1\}$. The notations used below are those in the proof of Lemma 3.13.3 and Theorem 3.13.1.

First we restrict our attention to the case where $0 \leq \alpha < \infty$. The case $0 \leq \alpha < 1$ is proved in Theorem 3.13.1. The proof of the case where $1 \leq \alpha < \infty$ is, mutatis mutandis, the same as that of Theorem 3.13.1.

In fact, since f is zero-free we have

$$\varphi(\rho, z) \leq \frac{(1 - j|z|)^{1+\alpha}\rho^{1+\alpha}(1 + |f(z)|^2)f^{\sharp}(z)}{|f(z)|^2}.$$

The continuous function $(1 + |f(z)|^2)f^{\sharp}(z)/|f(z)|^2$ attains a maximum as z varies over the compact ball $\overline{B(0, 2/3)}$. Then there exists a constant $M > 0$ such that

$$\varphi(\rho, z) \leq (1 - j|z|)^{1+\alpha}\rho^{1+\alpha}M \quad (z \in \overline{B(0, 2/3)}).$$

It follows that $\varphi(\rho, z)$ is continuous on $[0, 1] \times \{z \in \mathbb{C}^n : |z| \leq 1/j\}$ and $\varphi(0, z) = 0$ on $\{z \in \mathbb{C}^n : |z| \leq 1/j\}$, $j > 3$.

The same kind of argument as in the proof of Lemma 3.13.3 yields that the statement of Lemma 3.13.3 remains valid if f is zero-free and $0 \leq \alpha < 1$ is replaced with $1 \leq \alpha < \infty$. The rest of the proof of the theorem is the same as in Theorem 3.13.1. The case $0 \leq \alpha < \infty$ is proved. Thus we have proved the first part of the theorem.

It remains to consider the case $-\infty < \alpha < 0$. Since a family $\{1/f, f \in \mathcal{F}\}$ conforms to the hypotheses of Theorem 3.13.4, the earlier argument shows that there exist sequences $1/f_j$, $z_j \to z_0$, $r_j \to 0$, such that the sequence

$$g_j(z) := \frac{1}{r_j^{\alpha} f_j(z_j + r_j z)} \quad (0 \leq \alpha < \infty \text{ arbitrary})$$

converges locally uniformly in $\mathbb{C}^n$ to a nonconstant entire function g satisfying $g^{\sharp}(z) \leq g^{\sharp}(0) = 1$. By Hurwitz's theorem either $g \equiv 0$ or g never vanishes. Since $g^{\sharp}(0) = 1$, it is easy to see that if g never vanishes, then $1/g$ is an entire function in $\mathbb{C}^n$. It follows $r^{\alpha} f_j \to 1/g$ uniformly in $\mathbb{C}^n$.

Since Levi form vanishes for any pluriharmonic function,

$$\begin{aligned} L_z(\log(1 + |1/g|^2), v) &= \\ &L_z(\log(1 + |g|^2), v) - 2L_z(\log|g|, v) = \\ &\qquad L_z(\log(1 + |g|^2), v). \end{aligned}$$

Therefore

$$g^{\sharp}(z) = (1/g)^{\sharp}(z).$$

For every $z \in \mathbb{C}^n$ we have $g^{\sharp}(z) \leq g^{\sharp}(0) = 1$, hence

$$(1/g)^{\sharp}(z) \leq (1/g)^{\sharp}(0) = 1.$$

The theorem now follows from the observation that the above means that for all $-\infty < \alpha < 0$, the sequence $f_j(z_j + r_j z)/r_j^\alpha$ converges locally uniformly in $\mathbb{C}^n$ to a nonconstant entire function $1/g$ satisfying $(1/g)^\sharp(z) \leq (1/g)^\sharp(0) = 1$.

The proof of the other implication is similar to that of Theorem 3.13.1, so we omit the details. □

4

Normal Functions in $\mathbb{C}^n$

Prologue: In the papers [288] and [226], Yosida and Noshiro introduce the notion of normal function (although they do not use that terminology). This name was given to these functions much later in another pioneering paper in this vein [164]. In that article Lehto and Virtanen show that the notion of a normal meromorphic function is closely related to some of the most important problems of the boundary behavior of meromorphic functions.

After the appearance of the paper [164], one of the main directions of research was simply to test various properties of bounded holomorphic functions in order to establish the same properties for normal functions, or to construct examples of normal functions that do not have this property.

The most important work related to normal functions was published by MacLane [181], who considered the general question of the asymptotic values of holomorphic functions. It should be noted here that there is a natural division in the study of normal functions into normal meromorphic functions (which may not have asymptotic values) and normal holomorphic functions; the study of the latter is more fruitful, since a holomorphic function always has at least one asymptotic value.

Normal functions of several complex variables and their boundary behavior began to be studied in the papers [61], [67], and [58]. Dovbush defined normal functions in these articles in terms of the Bergman or Carathéodory metrics. The Kobayashi metric is more convenient and the right definition of a normal function in several variables. That these ideas can be generalized to complex spaces was developed in another pioneering paper in this vein [47]. However, for the case of strictly pseudoconvex domains, which Dovbush was interested in, the Bergman, Carathéodory and Kobayashi metrics are "equivalent" in the sense that the space of normal functions defined in terms of any of these three metrics is the same.

After the appearance of the paper of Cima and Krantz [47], there were a large number of works on this subject and a new rubric in the Mathematics Subject Classification System -MSC2010 appeared: "32A19—Bloch functions, normal functions of several complex variables".

DOI: 10.1201/9781032669861-4

4.1 Definitions and Preliminaries

In Chapter 3, all the results were proved in standard n-dimensional complex Euclidean space $\mathbb{C}^n$, but in essence its complex structure was not used in the proofs. Nevertheless, for the study of holomorphic functions, we need those objects which are intrinsically related to the domain in question and which reflect more intimately the complex structure of $\mathbb{C}^n$.

We begin with the intrinsic metrics on domains in $\mathbb{C}^n$. Here we present the basic definitions and some well-known facts. It is intended that this material be used mainly for reference.

By a *domain* we mean a connected open subset Ω of $\mathbb{C}^n$. We denote by $D(\Omega)$ (resp. $B_n(\Omega)$) the family of holomorphic maps from Ω into the unit disk D (resp. the unit ball B_n in $\mathbb{C}^n$) and by $\Omega(D)$ (resp. $\Omega(B_n)$) the family of holomorphic maps from D (resp. B_n) into Ω. Furthermore, $T(\Omega) = \{(z, v) \in \Omega \times \mathbb{C}^n\}$ means the holomorphic tangent bundle on Ω.

The infinitesimal metric

$$ds_D(z, v) := \frac{|v|}{1 - |z|^2}$$

is called the *Poincaré metric* on D.

The Poincaré metric on a disk in the complex plane has the important property that holomorphic maps from the disk to itself are distance decreasing in this metric by the Schwarz-Pick Lemma.

Definition 4.1.1. The infinitesimal Carathéodory metric on Ω is the function $C_\Omega : T(\Omega) \to \mathbb{R}^+ \cup \{0\}$, defined by [244]

$$F^C_\Omega(z, v) := \sup \{|f_*(v)| : f \in \mathcal{O}(\Omega, D), f(z) = 0\}$$

where $f_* = df$ is the differential of f at z.

(The condition $f(z) = 0$ is superfluous [244].) The Carathéodory distance is defined by [40]

$$C_\Omega(z, w) := \sup_{f \in D(\Omega)} \rho(f(z), f(w)), \quad z, w \in \Omega,$$

where ρ is the Poincaré distance in the unit disk; for points $z, w \in D$ it is equal to

$$\rho(z, w) := \ln \frac{|1 - \overline{w}z| + |z - w|}{|1 - \overline{w}z| - |z - w|},$$

It was shown by Reiffen [244] that if $\gamma : [0, 1] \to \Omega$ is a C^1 curve in Ω with Carathéodory length defined by

$$L_C(\gamma) := \sup_{0=t_0<t_1<\cdots<t_k=1} \sum_{i=1}^{k} C_\Omega(\gamma(t_{i-1}, \gamma(t_i))),$$

then

$$L_C(\gamma) = \int_0^1 F_\Omega^C(\gamma(t), \gamma'(t))dt.$$

This infinitesimal metric was first investigated in depth by Reiffen [244], who showed that it gives the same length for arcs as the Carathéodory metric, and is thus the inner metric corresponding to the Carathéodory metric. Set

$$C_\Omega^i(z, w) = \inf \int_0^1 F_\Omega^C(\gamma(t), \gamma'(t))dt,$$

where the infimum is taken over all smooth paths $\gamma : [0,1] \to \Omega$ connecting z to w. This metric, which is properly called the *Carathéodory-Reiffen metric*, is, in its integrated form, equivalent to the Carathéodory metric. Since it is the inner metric derived from the Carathéodory metric, it is always at least as large as the Carathéodory metric, but it is actually larger in many cases. The Carathéodory-Reiffen metric is also non-increasing under holomorphic maps, both in its integrated and in its differential form.

Although Carathéodory states in one of his papers that he is not sure how useful his metric will be in the future study of functions of several complex variables, it has found numerous applications both by him and by Cartan.

Kobayashi [132, 136], however, introduced a new metric for complex manifolds which has proved even more useful. In 1971 Royden [254] gave the following infinitesimal definition of the Kobayashi metric.

Definition 4.1.2. The infinitesimal Kobayashi metric on Ω is the function $F_\Omega^K : T(\Omega) \to \mathbb{R}^+ \cup \{0\}$, defined by

$$F_\Omega^K(z, v) :=$$

$$\inf \{\alpha : \alpha > 0 \text{ and } \exists f \in \mathcal{O}(D, \Omega) \text{ with } f(0) = z \text{ and } f'(0) = v/\alpha\}.$$

Obviously $F_\Omega^K(z, v) \geq 0$ and $F_\Omega^K(z, cv) = |c| F_\Omega^K(z, v)$ for all complex numbers c. The following Schwarz-Pick property is also an immediate consequence of the definition (see [253] or [224, Theorem 1.2.6, p. 6]).

Proposition 4.1.3. *Let $\Omega_j, j = 1, 2$, be bounded domains in $\mathbb{C}^n$. If $f \in \mathcal{O}(\Omega_1, \Omega_2)$, then*

$$F_{\Omega_2}^K(f(z), f_*(x)v) \leq F_{\Omega_1}^K(z, v) \quad \textit{for all } (z, v) \in \Omega_1 \times \mathbb{C}^n.$$

In particular, if f is biholomorphic, then $F_{\Omega_2}^K(f(x), f_*(z)v) = F_{\Omega_1}^K(z, v)$.

Corollary 4.1.4. *We have that $k_D(w, z) = \frac{|z|}{1-|w|^2}$ for all (w, z) in $D \times \mathbb{C}$.*

Proof. Now we prove the proposition. Take any $v \in \mathbb{C}^n$. Let $h : D(r) \to \Omega_1$ be a holomorphic mapping such that $h(0) = z$ and $h'(0) = v$. Then $f \circ h : D(r) \to \Omega_2$ is a holomorphic mapping such that $(f \circ h)'(0) = f_*(v)$. Hence,

$F^K_{\Omega_2}(f(z), f_*(v)) \le 1/\alpha$. Since h is arbitrary, we have $F_{\Omega_2}(f(z), f_*(v)) \le F^K_{\Omega_1}(z, v)$.

Moreover, if f is biholomorphic, we see that

$$F^K_{\Omega_1}(z, v) = F_{\Omega_1}\left(f^{-1} \circ f(z), f_*^{-1}(f_*(v))\right) \le F^K_{\Omega_2}(f(z), f_*(v)).$$

Hence we obtain $F^K_{\Omega_1}(z, v) = F^K_{\Omega_2}(f(z), f_*(v))$. □

Proposition 4.1.3 and Corollary 4.1.4 give

$$|f_*(v)| \le \frac{|f_*(v)|}{1 - |f(z)|^2} \le F^K_\Omega(z, v) \quad \text{for all } f \text{ in } \mathcal{O}(\Omega, D). \tag{4.1}$$

Therefore the number $F^C_K(z, v)$ is finite and bounded by $F^K_\Omega(z, v)$.

Given any w in D, we use the map $\zeta \mapsto (\zeta - w)/(1 - \zeta\bar{w})$ in $\mathcal{O}(D, D)$ to obtain

$$\frac{|z|}{1 - |w|^2} \le F^C_D(w, z) \le F^K_D(w, z) = \frac{|z|}{1 - |w|^2} \quad \text{for all } z \text{ in } \mathbb{C} \tag{4.2}$$

using (4.1), and Corollary 4.1.4. These observations verify the well known fact that the functions F^C_D and F^K_D are respectively the infinitesimal Poincarè metric on D.

Remark 4.1.5. In general, given $z \in \Omega$ and $v \in \mathbb{C}^n$, there is not a holomorphic map $\varphi : D \to \Omega$ such that $F_K(z, v) = |v| / |\varphi'(0)|$ (see [143]). When there exists one, it may not be unique (see [143]). Lempert has shown that when Ω is strongly convex (all principal curvatures of the boundary positive) then the extremal function will exist and be unique (see [170]).

Royden's definition has its classical source in the Schottky-Landau theorem, according to which for any holomorphic map f of the disk $D_r = \{|z| < r\}$ into to the Riemann sphere $\widehat{\mathbb{C}}$ with three points a, b, c removed, with $f'(0) \neq 0$, the radius r is bounded, from below by a constant depending only on a, b, c and the values of $f(0)$, $f'(0)$.

In [254] Royden proved that F^K_Ω has the contraction property under holomorphic maps: if $\varphi : \Omega \to \Omega'$ is a holomorphic map then for each $v \in \mathbb{C}^n$

$$F^K_{\Omega'}(\varphi(z), \varphi_* v) \le F^K_\Omega(z, v).$$

The Kobayashi distance is defined as follows [134, Chapter 3, p. 50]: Let $z, w \in \Omega$. Consider all finite sequences of points $z = p_0, p_1, \ldots, p_{k-1}, p_k = w$ of Ω such that there exist points $z_1, \ldots, z_k$, $w_1, \ldots, w_k$ of D and mappings $f_1, \ldots, f_k \in \Omega(D)$ satisfying $f(z_i) = p_{i-1}$ and $f(w_i) = p_i$, $i = 1, \ldots, k$. Set

$$K_\Omega(z, w) := \inf \sum_{i=1}^{k} \rho(z_i, w_i)$$

where the infimum is taken over all choices of points and mappings.

Royden [254] showed that K_Ω is actually the integrated form of F_K. That is, given $z, w \in \Omega$,

$$K_\Omega(z,w) := \inf_\gamma \int_0^1 F_\Omega^K(\gamma(t), \gamma'(t))dt,$$

where the infimum is taken over all piecewise differentiable curves $\gamma : [0,1] \to \Omega$ joining z to w.

Remark 4.1.6. The Carathéodory distance and metric do not satisfy such a relationship.

Definition 4.1.7. If the Kobayashi metric K_Ω is a distance and if the topology defined by K_Ω is equivalent to the relative topology of Ω in $\mathbb{C}^n$, then the domain Ω is said to be *hyperbolic*.

By Theorem IV.2.2 [76], any bounded domain in $\mathbb{C}^n$ is hyperbolic. In $\mathbb{C}^n$, the fact that K_Ω is a distance suffices to ensure that Ω is hyperbolic [76, Proposition IV.2.3].

A variant of the definition of hyperbolicity is given in Kobayashi's book [134, Chapter 3, Section 2, p. 60].

It is known that the Carathéodory metric is the smallest of the class of metrics which generalize the Schwarz lemma and that the new metric of Kobayashi is the largest. These metrics are particularly useful in the hyperbolic case, that is, when they are actually metrics rather than pseudometrics. The advantage of the Kobayashi metric lies in that, being larger, it will be an actual metric for some manifolds on which the Carathéodory metric is only a pseudometric.

The classical infinitesimal metric of Carathéodory and infinitesimal metric of Kobayashi share two basic properties:

(1) they agree with the Poincare metric on the unit disk;

(2) they are contractible with respect to holomorphic mappings.

Among the metrics with these two properties, the infinitesimal Caratheodory metric is the smallest and the infinitesimal Kobayashi metric is the largest.

Let us also remark that F_Ω^C is continuous on the tangent bundle $T(\Omega) = \mathbb{C}^n$ and F_Ω^K is upper semicontinuous on $T(\Omega)$ see [254].

The notation F_Ω without a superscript C or K refers to either metric unless specified otherwise; also we shall usually drop the notational dependence on Ω unless there are different domains involved.

The metric F is biholomorphically invariant which follows immediately from the following theorem.

Theorem 4.1.8 (Schwarz lemma). *Let Ω_1, Ω_2 be bounded domains in $\mathbb{C}^n$ and let F_1, F_2 both be one of the above metrics on Ω_1, Ω_2 respectively—as defined above. Suppose that $f : \Omega_1 \to \Omega_2$ is a holomorphic map. Then one has $f^*(F_2) \leq F_1$*

For a proof, the reader is referred to Kobayashi [134].

Remark 4.1.9. If Ω is bounded, then there exists a positive constant R such that $\Omega \subset B_R = \{z \in \mathbb{C}^n, |z| < R\}$. By the Schwarz lemma (Theorem 4.1.8),

$$F_\Omega(z,v) \geq F_{B_R}(z,v) \quad \text{for all } (z,v) \in \Omega \times \mathbb{C}^n.$$

Since

$$F_{B_R}(z,v) = \frac{[(R^2 - |z|^2)|v|^2 + |(z,v)|^2]^{1/2}}{R^2 - |z|^2} \geq |v|/R,$$

we have

$$F_\Omega(z,v) \geq |v|/R \quad \text{for all } (z,v) \in \Omega \times \mathbb{C}^n. \tag{4.3}$$

In function theory of several variables, the Bergman metric has played an important role as a canonically defined invariant metric. It is also a Kähler metric. Let Ω be a bounded domain in $\mathbb{C}^n$ with coordinate system $z_1, \ldots, z_n$. The set $H(\Omega)$ ol all holomorphic functions on Ω which are square integrable forms a separable complex Hilbert space. Let $h_0, h_1, h_2, \ldots$ be a complete orthonormal basis for $H(\Omega)$. Then

$$B(z,w) := \sum_{j=0}^{\infty} h_j(z)\overline{h}_j(\overline{w})$$

is a holomorphic function in z and $\overline{w}$, or a holomorphic function on $\Omega \times \overline{\Omega}$, where $\overline{\Omega}$ is the domain conjugate to Ω, and is called the kernel function of Bergman. Define

$$B_\Omega(z,v) := \sum_{k,l=1}^{n} \frac{\partial^2 \log B(z,z)}{\partial z_k \partial \overline{z}_l} v_k \overline{v}_l.$$

This form does not depend on the choice of the coordinates in $\mathbb{C}^n$. It defines an Hermitean and, even more, a Kählerian metric on Ω—called the Bergman metric. Let us also remark that B_Ω is continuous on the tangent bundle $T(\Omega)$. For more details, see [133].

We take

$$F_\Omega^B(z,v) = (B_\Omega(z,v))^{1/2}$$

in the notation of [52].

Remark 4.1.10. It is easy to see that the Bergman metric is invariant under biholomorphic maps of Ω, but it does not enjoy the contraction property for holomorphic maps. Moreover, it turns out that the Bergman pseudometric is not monotone with respect to inclusions (cf. [25]).

Remark 4.1.11. Let Ω be a bounded domain in $\mathbb{C}^n$. Then the matrix

$$T(z,\overline{z}) := \left(\frac{\partial^2 \log B(z,z)}{\partial z_k \partial \overline{z}_l} \right)_{k,l=1,\ldots,n}$$

determines an invariant Hermitian metric on Ω. If $f = (f_1, \ldots, f_n) : \Omega \to \mathbb{C}^n$ is holomorphic and bounded by a constant M (i.e., $f\overline{f} \le M^2$), then C. H. Look [177, 178] proved in 1957 that

$$\frac{\partial f}{\partial z}\frac{\partial \overline{f}}{\partial \overline{z}} \le M^2 T(z, \overline{z}). \tag{4.4}$$

When $f = (f_1, 0, \ldots, 0)$ and $M = 1$, the above inequality obviously implies

$$\sup_{f_1 \in H^\infty(\Omega)} |df_1| \le F_\Omega^B, \tag{4.5}$$

where the lefthand side is exactly the infinitesimal Carathéodory metric defined eight years later by Reiffen [244]. Twenty years later Hahn [106] reproduced the proof of (4.4) and noted that the infinitesimal Carathéodory metric is not greater than the Bergman metric.

In the class of all pseudodifferential metrics on complex manifolds satisfying the Schwarz Lemma with respect to holomorphic mappings and agreeing with the Poincaré metric on the unit disk, the Caratheodory metric is known to be the smallest and the Kobayashi metric is the largest. The Bergman metric does not belong to this class, since it does not satisfy the required Schwarz Lemma. Nevertheless, as shown in [104, 177, 178], on bounded domains in $\mathbb{C}^n$ (and even on more general complex manifolds) the Bergman metric is always larger than the Caratheodory metric.

It is known that the boundary behavior of the Bergrnan metric F_B and the Kobayashi metric F_K on a strictly pseudo-convex domains $\Omega \subset \mathbb{C}^n$ is the same [52, 85]. As a consequence there exists a constant $c_\Omega > 0$ which might depend on Ω, such that

$$F_B(z, v) < c_\Omega F^K(z, v) \tag{4.6}$$

for all $z \in \Omega$, $v \in \mathbb{C}^n$.

Nevertheless, one has in contrast to this situation the following

Theorem 4.1.12 (Diederich and Fornaess, see [53, Theorem p. 257]). *There exists a pseudoconvex domain $\Omega \subset \mathbb{C}^3$ with smooth C^∞ boundary such that the quotient*

$$\Phi(z, v) := F_B(z, v)/F^K(z, v)$$

is unbounded from above on $\Omega \times \mathbb{C}^3$.

For the needed details on the Bergman metric we refer the reader to Bergman [24] or [145]. All relevant properties of the Kobayashi metric are contained in Kobayashi [134].

4.1.1 Homogeneous domains.

Let Ω be a general domain in $\mathbb{C}^n$. If Ω' is another domain and $g \colon \Omega \to \Omega'$ is a holomorphic mapping with holomorphic inverse, then we say that g is an *isomorphism*. If $\Omega = \Omega'$, g is called an *automorphism* of Ω. The set of all automorphisms of Ω is a group under composition of mappings and we denote it by $\mathrm{Aut}(\Omega)$ (which sometimes, indeed generically, consists only of the identity element).

The domain Ω is said to be *homogeneous* if $\mathrm{Aut}(\Omega)$ acts transitively on Ω; i.e., there exists for any pair of points $z_1, z_2 \in \Omega$ an automorphism $g \in \mathrm{Aut}(\Omega)$ such that $g(z_l) = z_2$.

It follows that the unit disk $D \subset \mathbb{C}$ is a homogeneous domain. By Riemann's famous theorem, every proper simply connected domain Ω in $\mathbb{C}$ (i.e. Ω is a holomorphically simply connected open set in $\mathbb{C}$ that is not equal to all of $\mathbb{C}$) is biholomorphically equivalent to D; i.e., there are $h \in \mathrm{Hol}(D, \Omega)$ and $g \in \mathrm{Hol}(\Omega, D)$ such that

$$h(D) = \Omega \quad , \quad g(\Omega) = D \quad , \quad h \circ g = id_\Omega \quad , \quad g \circ h = id_D \,.$$

Hence for every proper, simply connected domain $\Omega \subset \mathbb{C}$, we can describe explicitly the qroup $\mathrm{Aut}(\Omega)$ by the formula

$$\mathrm{Aut}(\Omega) = \left\{ h \circ m \circ h^{-1} : m \in \mathrm{Aut}(D) \right\}.$$

Let us recall that the mapping h can be expressed in terms of the Green's function of the domain Ω. In particular, each proper simply connected domain of $\mathbb{C}$ is homogeneous.

If Ω is bounded, or isomorphic with a bounded domain, then Ω has an invariant Hermitian metric, the Bergman metric.

In tracing the history of the classification theory of homogeneous bounded domains in $\mathbb{C}^n$, one must mention the pioneering work by H. Poincaré [237]. In 1907, he proved that the ball $\{|z_1|^2 + |z_2|^2 < 1\}$ and the polidisk $\{|z_1|^2 < 1, |z_2|^2 < 1\}$ in $\mathbb{C}^2$ are not holomorphically equivalent to each other, but they are both connected and simply connected. One sometimes adds the misleading comment that there is no Riemann mapping theorem in several variables. We next describe two positive mapping theorems. It is not hard to see that the only plane domains $\Omega \neq \mathbb{C}$ for which the automorphism group is transitive are those biholomorphic to the disk. Bun Wong [286] showed that the corresponding statement is true for strictly pseudoconvex domains in $\mathbb{C}^n$, $n > 1$. Later, Rosay [250] generalized it to the following statement.

Theorem 4.1.13. *Let $\Omega \subset \mathbb{C}^n$ be an open set with C^2 boundary. If Ω has a transitive automorphism group, then Ω is biholomorphic to the ball.*

It should be remarked that bounded homogeneous domains are always (without any assumptions on the boundary) domains of holomorphy, i.e. pseudoconvex (see, [133, Theorem 6.3, p. 280]). There are, of course, many such domains without smooth boundary.

H. Cartan gave a complete classification of homogeneous bounded domains in $\mathbb{C}^2$ and $\mathbb{C}^3$, which was published in É. Cartan's paper [41]. Cartan found all the bounded homogeneous domains in the spaces $\mathbb{C}^2$ and $\mathbb{C}^3$. It was shown that in the space $\mathbb{C}^2$ any bounded homogeneous domain can be biholomorphically mapped either onto the ball $\{|z_1|^2 + |z_2|^2 < 1\}$ or onto the polydisk $\{|z_1|^2 < 1, |z_2|^2 < 1\}$.

In the space $\mathbb{C}^3$ any bounded homogeneous domain can be biholomorphically mapped onto one of the following domains:

(1) the ball $\{z \in \mathbb{C}^3 : |z_1|^2 + |z_2|^2 + |z_3|^2 < 1\}$;

(2) the domain $\{z \in \mathbb{C}^3 : |z_1|^2 + |z_2|^2 < 1, |z_3|^2 < 1\}$;

(3) the polydisk $\{z \in \mathbb{C}^3 : |z_1|^2 < 1, |z_2|^2 < 1, |z_3|^2 < 1\}$;

(4) a bounded domain which is obtained by a biholomorphic mapping from the domain $\left\{y_3 > \sqrt{|y_1|^2 + |y_2|^2}\right\}$. (Here, as usual, $z_k = x_k + \imath y_k$, $k = 1, 2, 3$.)

One can show that all of the above-mentioned domains are not only homogeneous, but also symmetric. (A domain is called symmetric if the group $\mathrm{Aut}(\Omega)$ acts transitively on Ω and if there exists an automorphism with a single fixed point in Ω whose square is the identity map).

In this connection E. Cartan [41] raised the problem: does there exist in the space $\mathbb{C}^n$ for $n \geq 4$ a bounded homogeneous nonsymmetric domain? An answer to this problem was given by I. I. Pjateckii-Shapiro, who, in his paper [236] published in 1959, constructed the first example of a bounded homogeneous nonsymmetric domain. It was shown that similar domains exist in the space $\mathbb{C}^n$ of all dimensions, beginning with the fourth. Moreover, in every one of the spaces $\mathbb{C}^4$, $\mathbb{C}^5$, $\mathbb{C}^6$ there appears a finite set of biholomorphically inequivalent homogeneous bounded nonsymmetric domains, but already in the space $\mathbb{C}^7$, the set of such domains attains a continuous cardinality. On the other hand, the set of bounded symmetric domains is finite for any space.

Bounded complex homogeneous domains are of great interest from many points of view. This arises from the fact that they form a comparatively broad class of domains in $\mathbb{C}^n$, for which it has been possible to obtain a whole series of interesting essentially-many-dimensional results.

Having got a good many of the basic definitions under control, we are ready to examine the concept of normal function in some detail. In the next section we shall study holomorphic normal functions in several complex variables.

In several complex variables there is not a single standard region like D in which to consider biholomorphic mappings. Cartan also pointed out that some classical results of univalent function theory do not have analogs in several complex variables.

4.2 Normal Functions in $\mathbb{C}^n$

Much of the one dimensional theory of normal meromorphic functions is inspired by the classical theory of conformal mappings, a very special case.

The concept of normal functions was introduced by Noshiro [226] following from earlier related work of Yosida [288]. His definition reads as follows: a meromorphic function $f : D \to \overline{\mathbb{C}}$ is said to be normal if $\{f \circ g : g \in \mathrm{Aut}(D)\}$ is a normal family. Noshiro's definition was restricted to the unit disk and was later extended by Lehto and Virtanen in their seminal paper of 1957 [164].

The Riemann mapping theorem asserts that every simply connected domain $\Omega \subset \mathbb{C}$ with $\Omega \neq \mathbb{C}$ admits a unique conformal mapping f onto the unit disk D with the properties $f(\zeta) = 0$ and $f'(\zeta) > 0$ for an arbitrarily prescribed point $\zeta \in \Omega$. Because the inverse function is necessarily analytic, it is equivalent to say that D can be mapped conformally onto Ω. The Carathéodory extension theorem says (in a special case) that each conformal mapping of a Jordan domain Ω onto the Jordan domain D can be extended to a homeomorphism of $\overline{\Omega}$ onto $\overline{D}$. The modern proof of Riemann's theorem is based on the theory of normal families. A full discussion and proof of Riemann's theorem was given in Section 3.7.

Lehto and Virtanen have introduced the following definition: A meromorphic function $f(z)$ is called *normal* in a simply connected domain $\mathcal{D}$, if the family $\{f(S(z))\}$ is normal, where $z' = S(z)$ denotes an arbitrary conformal mapping of $\mathcal{D}$ onto itself[1].

Normal functions have been studied by many authors; there is now a large volume of literature on the subject (see [33] for a survey of twenty-two characterization of normal functions). The survey article by A. J. Lohwater's [180] may be consulted for further summary accounts of the subject up to 1973.

One can naturally introduce the definition of normal holomorphic functions in bounded homogeneous domains in $\mathbb{C}^n$ which is analogous to the definition of normal function given by Noshiro [226] for the unit disk.

Definition 4.2.1. Let $\mathcal{D}$ be a homogeneous bounded domain in $\mathbb{C}^n$. We say that a holomorphic function $f \colon \mathcal{D} \to \mathbb{C}$ is normal if the family

$$\mathcal{F} = \{f \circ g : g \in \mathrm{Aut}(\mathcal{D})\}$$

is normal, where $\mathrm{Aut}(\mathcal{D})$ denotes the holomorphic automorphism group of $\mathcal{D}$.

Remark 4.2.2. If $f(z) \in \mathcal{O}(\mathcal{D})$ omits two values, then the family $\mathcal{F}$ also omits these values. By Montel's normality criterion (Theorem 3.5.2) $\mathcal{F}$ is normal. Now, from Definition 4.2.1 we see that $f(z)$ is normal. Therefore bounded holomorphic functions are normal.

The hypotheses imposed on $f \circ g$ for all $g \in \mathrm{Aut}(\mathcal{D})$ will lead to notable conclusions about f.

An important result of Noshiro [226, Theorem 1, p. 150] characterizes normal functions in terms of the spherical derivative (see also Lehto-Virtanen [164]).

---

[1]In the case of the unit disk, Noshiro [226] said such a function belongs to the class (A) (instead of normal function) and obtained some interesting results.

Theorem 4.2.3. *A meromorphic function $f(z)$ in the unit disk $D \subset \mathbb{C}$ is normal if and only if*

$$\sup_{z \in D} (1 - |z|^2) f^\sharp(z) < \infty, \tag{4.1}$$

where $f^\sharp(z)$ is the spherical derivative.

That is, f is normal if there exists a constant $C > 0$ such that

$$\frac{|f'(z)|}{1 + |f(z)|^2} < C \frac{1}{1 - |z|^2} \tag{4.2}$$

for all $z \in D$.

Theorem 4.2.3 is a central result in the geometric theory of normal meromorphic functions, with important applications.

Remark 4.2.4. From the proof (see [225, Theorem 7, p. 87]), it is easily seen that a nonconstant function, meromorphic in $|z| < 1$, is normal if and only if $\{f(S(0))\}$ is bounded for all S.

Normal meromorphic functions admit the following characterization in terms of the spherical derivative: A nonconstant meromorphic $f(z)$ is normal in a domain $\Omega \subset \mathbb{C}$, which then necessarily is of hyperbolic type, if and only if there exists a finite constant C so that

$$\frac{|f'(z)|\,|dz|}{1 + |f(z)|^2} < C d\sigma(z)$$

where $d\sigma(z)$ denotes the element of length in the hyperbolic metric of Ω (see [164, p. 47]).

As a direct application of the Marty normality criterion (Theorem 3.4), we obtain that Theorem 4.2.3 can be extended to bounded homogeneous domains in $\mathbb{C}^n$:

Theorem 4.2.5. *Let $\mathcal{D}$ be a bounded homogeneous domain in $\mathbb{C}^n$. Let F be either the infinitesimal Carathéodory, Kobayashi, or Bergman metrics on $\mathcal{D}$. A holomorphic function $f \colon \mathcal{D} \to \mathbb{C}$ is normal if and only if there exists a constant C such that*

$$L_z(\log(1 + |f|^2), v) \leq C(F(z, v))^2 \tag{4.3}$$

for all $z \in \mathcal{D}$ and all $v \in \mathbb{C}^n$.

Proof. Suppose that f is normal in $\mathcal{D}$ (i.e. the family

$$\mathcal{F} = \{f \circ g : g \in \mathrm{Aut}(\mathcal{D})\}$$

is normal). Let $z_0 \in \mathcal{D}$ be a fixed point. Of course the singleton $\{z_0\}$ is compact. By Marty's criterion (Theorem 3.2.5) for the compact subset $\{z_0\} \subset \mathcal{D}$ there exists a constant $C(z_0)$ such that

$$L_{z_0}(\log(1 + |f \circ \varphi|^2), v) \leq C(z_0)|v|^2 \tag{4.4}$$

for all $\varphi \in \text{Aut}(\mathcal{D})$ and all $v \in \mathbb{C}^n$. From the above inequality, taking into account (4.3) and (4.5), we obtain

$$L_{z_0}(\log(1 + |f \circ \varphi|^2), v) \leq C(z_0)R^2(F(z_0, v))^2 \tag{4.5}$$

for all $\varphi \in \text{Aut}(\mathcal{D})$ and all $v \in \mathbb{C}^n$.

Since the Levi form is invariant relative to biholomorphic mappings, we have

$$L_{z_0}(\log(1 + |f \circ \varphi|^2), v) = L_{\varphi(z_0)}(\log(1 + |f|^2), \varphi_* v), \tag{4.6}$$

where $\varphi_* = d\varphi$ is the differential of the mapping φ at z_0. Now (4.5) and (4.6) together give

$$L_{\varphi(z_0)}(\log(1 + |f|^2), \varphi_* v) \leq C(z_0)R^2(F(z_0, v))^2. \tag{4.7}$$

Since $Aut(\mathcal{D})$ is transitive on $\mathcal{D}$ and

$$F(z_0, v) = F(\varphi(z_0), \varphi_* v),$$

when this is substituted in (4.7) we obtain (4.3) with $C = C(z_0)R^2$.

Conversely, suppose that (4.3) holds. Then, since the Levi form is invariant relative to biholomorphic mappings, we have

$$L_z(\log(1 + |f \circ \varphi|^2), v) \leq C(F(\varphi(z), \varphi_* v))^2 \tag{4.8}$$

for all $\varphi \in Aut(\mathcal{D})$ and all $(z, v) \in \Omega \times \mathbb{C}^n$. The function $F(z, v)/|v|$ is at least upper semicontinuous on $\Omega \times \mathbb{C}^n \setminus \{0\}$ and therefore assumes its maximum, say $M(K)$, on any compact subset $K \times \{|v| = 1 : v \in \mathbb{C}^n\} \subset \Omega \times \{\mathbb{C}^n \setminus \{0\}\}$. Hence one obtains from (4.8)

$$L_z(\log(1 + |f \circ \varphi|^2), v) \leq C \cdot M(K)|v|^2 \tag{4.9}$$

for all $\varphi \in Aut(\mathcal{D})$ and all $(z, v) \in K \times \mathbb{C}^n$. It follows from Marty's normality criterion (Theorem 3.2.5) that $\mathcal{F} = \{f \circ g : g \in \text{Aut}(\mathcal{D})\}$ is normal, and so, by Definition 4.2.1, f is normal. The proof is complete. □

Remark 4.2.6. Later, in a refined form, this theorem appeared in Funahashi [81, Theorem 1, p. 80], Hahn [99, 2.7 Lemma, p. 721], [101, Theorem 1, p. 113].

As function theory on two biholomorphic domains is the same, one of the main questions in complex analysis is to classify domains up to biholomorphic equivalence. In one variable, there is the rather striking theorem due to Riemann: If $U \subset \mathbb{C}$ is a nonempty simply connected domain such that $U \neq \mathbb{C}$, then U is biholomorphic to D.

In one variable, a topological property on U is enough to classify a whole class of domains. It is one of the reasons why studying the disk is so important in one variable and why many theorems are stated for the disk only.

There is simply no such theorem in several complex variables. The unit ball and the polydisk are not biholomorphically equivalent
(see H. Poincaré [237]). Both are simply connected (have no holes), and they are the two most obvious generalizations of the disk to several variables. They are homeomorphic, that is to say, topology does not see any difference between them.

This motivates the following definition. We want a definition that includes the classical definition of normal function and also applies to arbitrary bounded domain in $\mathbb{C}^n$.

In view of Theorem 4.2.5 the definition of normal function is extended in the obvious way to arbitrary domain in $\mathbb{C}^n$.

Definition 4.2.7. Let Ω be a bounded domain in $\mathbb{C}^n$ and let F be either the infinitesimal Carathéodory, Kobayashi, or Bergman metrics on Ω. A holomorphic function $f\colon \Omega \to \mathbb{C}$ [one writes $f \in \mathcal{O}(\Omega)$] is called *normal* if there exists a positive constant C such that

$$L_z(\log(1 + |f|^2), v) \leq C \cdot (F(z, v))^2 \tag{4.10}$$

for all $(z, v) \in \Omega \times \mathbb{C}^n$.

Definition 4.2.8. We call $f \in \mathcal{O}(\Omega)$ a C-normal (respectively K-normal and B-normal) function on $\Omega \subset \mathbb{C}^n$ if in Definition 4.2.7 $F = F_\Omega^C$ (respectively $F = F_\Omega^K$ and $F = F_\Omega^B$).

However, for the case of strongly pseudoconvex domains the Bergman, Kobayashi, and Caratheodory metrics are "equivalent" in the sense that normal functions defined in terms of either of the three metrics are the same. This assertion follows from well-known estimates on the asymptotic behavior of these metrics (see [74], [84]).

In the case of strongly pseudoconvex domains $\Omega \subset \mathbb{C}^n$, we will shorten the above terminology and say simply that f is in class $\mathcal{N}$, or equivalently f is a normal function in Ω.

Denote by $N^C(\Omega)$ (respectively $N^K(\Omega)$ and $N^B(\Omega)$) a class of all C-normal (respectively K-normal and B-normal) functions in a domain $\Omega \subset \mathbb{C}^n$.

Since $F_\Omega^C \leq F_\Omega^K$ and $F_\Omega^C \leq F_\Omega^B$ from this we obtain the inclusions

$$N^C(\Omega) \subseteq N^K(\Omega) \quad \text{and} \quad N^C(\Omega) \subseteq N^B(\Omega).$$

Remark 4.2.9. Suppose that f is a holomorphic function from Ω to $\mathbb{C}$ such that $\mathbb{C} \setminus f(\Omega)$ contains two points a_1 and a_2. Then f is K-normal. First, the distance decreasing property of the Kobayashi metric yields

$$F_{\mathbb{C}\setminus\{a_1,a_2\}}^K(f(z), f'(z) \cdot \xi) \leq F_\Omega^K(z, \xi) \quad \text{for all } z \in \Omega \text{ and all } \xi \in \mathbb{C}^n.$$

Further, making use of the homogeneity and the continuity of $F_{\mathbb{C}\setminus\{a_1,a_2\}}^K$ and the spherical metric as well as the fact that

$$|1|_{sph,z}/F_{\mathbb{C}\setminus\{a_1,a_2\}}^K(z, 1) \to 0 \quad \text{as } z \to a_j, j = 1, 2.$$

one readily finds a constant C such that

$$|\eta|_{sph,z} \leq C \cdot F^K_{\mathbb{C}\setminus\{a_1,a_2\}}(z,\eta) \quad \text{for all } z \in \mathbb{C} \setminus \{a_1, a_2\} \text{ and all } \eta \in \mathbb{C}. \quad (4.11)$$

Combining (4.10) and (4.11) gives the assertion. The above proof is taken from [119].

Another deduction of this result is given in [47, p. 308].

Let $\mathbf{G}$ denote the family of all functions holomorphic on D which omit the values 0 and 1. According to Montel's theorem [292, p. 218], the family $\mathbf{G}$ is normal. Marty's theorem [187] yields a non-negative constant L such that, for any $g \in \mathbf{G}$,

$$L_0(\log(1 + |g|^2), 1) \leq L\,.$$

The constant L in the inequality above does not depend on $g \in \mathbf{G}$. The exact value L is not known.

Hence for $\varphi : D \to \Omega$ such that $\varphi(0) = z$ and $d\varphi(0) = \alpha v$, $\alpha > 0$, we have

$$L_0(\log(1+|f\circ\varphi|^2), 1) = L_z(\log(1+|f|^2), d\varphi(0)\cdot 1) = L_z(\log(1+|f|^2), \alpha v) \leq L,$$

It follows that

$$L_z(\log(1 + |f|^2), v) \leq L\alpha^{-2}.$$

Since the left-hand side of the above inequality does not depend on α and $F^K_\Omega(z, v) = \inf_\alpha \{\alpha^{-1}\}$ over all maps with the normalization $\varphi(0) = z$ and $d\varphi(0) = \alpha v$, with $\alpha > 0$, we conclude that

$$L_z(\log(1 + |f|^2), v) \leq L\left(F^K_\Omega(z, v)\right)^2 .$$

By Definition 4.2.7, f is a K-normal function in Ω. To prove this fact, we have only to use the idea in the proof of Proposition 1.4 in [47].

Remark 4.2.10. Bloch functions are precisely the holomorphic functions on D that are globally $d\sigma$-to-euclid Lipschitz. Similarly, a meromorphic function on D is a normal function f that are globally ρ-to-s Lipschitz. Finally, a meromorphic function on $\mathbb{C}$ is a Yosida function if $f^\sharp(z)$ is uniformly bounded on $\mathbb{C}$, so Yosida functions are euclid-to-s Lipschitz functions [20, Example 6.1, p. 345].

Notice that if $f \in \mathcal{F}$, $\mathcal{F}$ is a normal family in Ω, and $z, w \in \Omega$ then

$$s(f(z), f(w)) \leq C \cdot K_\Omega(z, w)$$

as follows from the definition of the integrated distance. It follows that the family $\mathcal{F}$ is uniformly Lipschitz on compact subsets of Ω with respect to the Euclidean distance and the spherical distance.

Remark 4.2.11. From Definition 4.2.7 and the definitions of the Levi form and the infinitesimal Carathéodory metric, it is easy to see that a natural generalization of the class of bounded holomorphic functions is the class of normal holomorphic functions.

Remark 4.2.12. Geometrically, Definition 4.2.7 means that the spherical derivatives of $f(z)$ in various directions can be estimated in terms of holomorphic disks in Ω passing through the point z in these directions. In particular, in strictly pseudoconvex domains this means that the spherical derivatives along complex normals grow like $O(1/\delta(z))$ and along complex tangential directions, like $O(1/\sqrt{\delta(z)})$. Here and later $\delta(z)$ is the Euclidean distance from the point z to $\partial\Omega$.

Next we prove the equivalence of Definition 4.2.7 with some others.

Remark 4.2.13. In case Ω is a homogeneous domain, Theorem 4.2.5 shows that Definition 4.2.7 coincides with Definition 4.2.1.

Remark 4.2.14. For the unit disk Theorem 4.2.3 coincides with Noshiro's classical theorem. Indeed, the metric F_D coincides with the Poincaré metric on D. A straightforward calculation shows that inequality (4.3) reduces to

$$(1-|z|^2)f^\sharp(z) \leq C < \infty$$

for all $z \in D$. Taking sup on both sides over $z \in D$, we obtain the desired inequality (4.1).

The following simple result will prove useful in later sections.

Proposition 4.2.15. *Let* $\Omega_1 \subseteq \mathbb{C}^{n_1}$*;* $\Omega_2 \subseteq \mathbb{C}^{n_2}$ *be domains. Let* $\varphi : \Omega_1 \to \Omega_2$ *be holomorphic. Let* $f : \Omega_2 \to \mathbb{C}$ *be* K*-normal. Then* $f \circ \varphi$ *is* K*-normal.*

Proof. The map $\varphi : \Omega_1 \to \Omega_2$ is distance decreasing in the Kobayashi metric. □

One obvious question is whether the normality of function remains a hereditary property under projections. The answer is yes.

Corollary 4.2.16. *Let* $\Omega \subset \mathbb{C}^n$ *be a domain and let* $f : \Omega \to \mathbb{C}$ *be* K*-normal. Let* $p \subseteq \mathbb{C}^n$ *be a complex linear subspace. Assume that* $p \cap \Omega \equiv \widetilde{\Omega}$ *is linearly biholomorphic to a domain in some* $\mathbb{C}^k$, $k \leq n$*. Then* $f|_{\widetilde{\Omega}}$ *is* K*-normal.*

Definition 4.2.7 has an obvious analogue for a complex manifold. The same applies to the infinitesimal Carathéodory and Kobayashi metrics. The Carathéodory metric is a continuous metric defined on every complex manifold (sometimes trivial). The distance-decreasing property is the most prominent of all, in that it produces useful applications. On the other hand, since the definition of this metric depends on the bounded holomorphic functions, it is identically zero on any compact complex manifold, or compact complex manifolds with subsets taken away, when the set (that was removed) happens to be necessarily a removable singularity set for bounded holomorphic functions. The advantage of the infinitesimal Kobayashi pseudometric lies in that, being larger, it will be an actual metric for some manifolds on which the infinitesimal Carathéodory pseudometric is only a pseudometric. Therefore, in the definition of normal functions on manifolds, the Kobayashi pseudometric is used.

We first formulate a sufficient condition for the non-normality of a holomorphic function.

Lemma 4.2.17. *Let* $f(z)$ *be a holomorphic function in* $\Omega \subset \mathbb{C}^n$, *and let* $\{z_j\}$ *and* $\{w_j\}$ *be two sequences of points in* Ω *such that* $\lim_{j\to\infty} K_\Omega(z_j, w_j) = 0$, *if*

$$\lim_{j\to\infty} f(z_j) = a \text{ and } \lim_{j\to\infty} f(w_j) = b \quad (a \neq b),$$

then $f(z)$ *is not a* K*-normal function.*

Proof. Suppose, reasoning by contradiction, that f is K-normal. By the very definition of K-normal function, there exist a positive constant C such that

$$L_z(\log(1+|f|^2), v) \leq C(F_\Omega^K(z,v))^2 \text{ for all } (z,v) \in \Omega \times \mathbb{C}^n.$$

Let $\gamma : [0,1] \to \Omega$ be any differentiable curve from z_j to w_j. Then $f \circ \gamma : [0,1] \to \mathbb{C}$ is a curve from $f(z_j)$ to $f(w_j)$. It is elementary to see that

$$L_z(\log(1+|f|^2), v) = \frac{|Df(z)v|^2}{(1+|f(z)|^2)^2} \geq 0,$$

where

$$Df(z)v := \nabla f \cdot v .$$

Hence

$$\frac{|df(\gamma(t))/dt|^2}{(1+|f(\gamma(t))|^2)^2} = L_{\gamma(t)}(\log(1+|f|^2), \gamma'(t)).$$

Therefore

$$\int_0^1 \frac{|f'(\gamma(t))|}{1+|f(\gamma(t))|^2} dt \leq \sqrt{C} \int_0^1 F_\Omega(\gamma(t), \gamma'(t)) dt.$$

Taking the infimum of the above inequality over all differentiable curves from z_j to w_j, and applying Theorem 1 from [254] one has

$$s(a,b) = \lim_{j\to\infty} s(f(z_j), f(w_j)) \leq \sqrt{C} \lim_{j\to\infty} K_\Omega(z_j, w_j) = 0.$$

Clearly, when $j \to \infty$ the right-hand side of the above inequality tends to zero, while the term on the left-hand side of the inequality tends to $s(a,b)$. This is a contradiction, since $a \neq b$. This contradiction completes the proof. □

Next let us introduce a little notation. If $l \subseteq \mathbb{C}^n$ is a complex line, $l = \{a\zeta + b : \zeta \in \mathbb{C}\}$, then $B \cap l$ is a complex disk of Euclidean radius $r = r_l$, $0 \leq r < 1$. The Kobayashi metric for the standard disk $D(0,r) \subseteq \mathbb{C}$ is given by

$$F_{D(0,r)}^K(z,\xi) = F_D^K(z/r, \xi/r)$$

Let $\psi_l : D(0,r) \to B \cap l$ denote the standard rigid identification of $D(0,r)$ with $B \cap l$. Notice that if $\varphi \in \mathrm{Aut}(B)$, then $\varphi(B \cap l) = B \cap l'$ for some other complex line l' see [257].

Proposition 4.2.18. *Let $f : B \to \mathbb{C}$ be holomorphic. Then the following are equivalent:*

(1) *f is normal;*

(2) *For any family $\{\Phi_\alpha\}_{\alpha \in A}$ of automorphisms of B it holds that*

$$\{f \circ \Phi_\alpha\}_{\alpha \in A}$$

is a normal family;

(3) *There is a constant $C = C(f)$ such that for any complex line q and $f_q = f|_{B \cap q} \circ \psi_q$ it holds that*

$$\frac{|f_q'(\zeta)|}{1 + |f_q(\zeta)|^2} \leq C \cdot F_K^{D(0,r)}(\zeta, 1) = C \cdot \frac{1}{r^2 - |\zeta|^2} \quad \textit{for all} \quad \zeta \in D(0,r).$$

Proof. **(2)** $\Rightarrow$ **(3)**: Pick a complex line q. Without loss of generality we may assume that $q = \{\beta \cdot 1 : \beta \in \mathbb{C}\}$ where $1 = (1, 0, \ldots, 0)$.

Let $\{\varphi_\alpha\}_{\alpha \in A}$ be automorphisms of D. Then each φ_α can be identified in a natural way with an automorphism $\widetilde{\varphi}_\alpha$ of $q \cap B$. And each $\widetilde{\varphi}_\alpha$ is the restriction to $q \cap B$ of an element $\Phi_\alpha \in \text{Aut } B$. By **(2)**, $\{f \circ \Phi_\alpha\}$ is a normal family. Therefore $\{f \circ \widetilde{\varphi}_\alpha\}$ is a normal family. Hence $\{f_q \circ \varphi_\alpha\}$ is a normal family. By Remark 4.2.14, we may conclude that **(3)** holds for each complex line q. That it holds uniformly over such lines follows from an easy argument by contradiction.

(3) $\Rightarrow$ **(1)**: The inequality in **(3)** is invariant under automorphisms of B and so is the inequality (1.1) which defines normality. So we need only check normality at $z = 0$ with $\xi = 1$. But then the two inequalities are identical.

(1) $\Rightarrow$ **(2)**: The inequality (4.10) is clearly invariant under automorphisms of B. Therefore it simply amounts to an equicontinuity condition for the family $\{f \circ \Phi_\alpha\}$ as maps from B equipped with the Kobayashi metric to $\widehat{\mathbb{C}}$ equipped with the spherical metric. By a variant of the Ascoli-Arzela theorem (see [287]), $\{f \circ \Phi_\alpha\}$ is a normal family. □

The following theorem is well known.

Theorem 4.2.19 (Hartogs extension theorem). *If the function f is holomorphic everywhere in the domain $D \subset \mathbb{C}^n$ $(n > 1)$ except perhaps for a set $K \subset\subset D$ such that K does not separate the domain (i.e., such that $D \backslash K$ is connected), then f extends holomorphically to the whole domain D.*

The condition that K does not separate the domain in the above theorem is essential: Let $K = \left\{|z| = \frac{1}{2}\right\}$ be a sphere and D is the unit ball B in $\mathbb{C}^n$, $n > 1$; then the function f that is equal to 0 in $\left\{|z| < \frac{1}{2}\right\}$ and to 1 in $\left\{\frac{1}{2} < |z| < 1\right\}$ is holomorphic in $B \backslash K$ but does not extend holomorphically to B.

The question then naturally arises: Can normal functions on $D \backslash K$ be extended to normal functions on D?

Let r be a positive number and $B_r(a)$ the open ball of radius r with center $a \in \mathbb{C}^n$, i.e.,

$$B_r(a) = \{z \in \mathbb{C}^n : |z - a| < r\}.$$

As an abbreviation we denote by $'z = (z_1, \dots, z_{n-1})$ the projection of $z \in \mathbb{C}^n$ into $\mathbb{C}^{n-1}$ and by $'D$ the projection of D into $\mathbb{C}^{n-1}$ (i.e., the set of all $'z$ for $z \in D$).

Theorem 4.2.20. *If the function f is C-normal on $B \setminus \overline{B}_{1/2}('0, 1/2) \subset \mathbb{C}^n$ $(n > 1)$, then f can be extended to a function that is normal on B.*

Proof. Set $\Omega = B \setminus \overline{B}_{1/2}('0, 1/2)$. Since f is a C-normal function on Ω by hypothesis, from the definition of C-normal function, there exists a positive constant C_0 such that

$$L_z(\log(1 + |f(z)|^2), v) \leq C_0(F_\Omega^C(z, v))^2 \tag{4.12}$$

for all $z \in \Omega$ and all $v \in \mathbb{C}^n$.

By the Hartogs extension theorem any g holomorphic in Ω such that $|g| < 1$ extends holomorphically to the whole ball B and this extension $\widetilde{g}$ can take only those values at points of $\overline{B}_{1/2}('0, 1/2)$ that g takes in Ω. Hence $|\widetilde{g}| < 1$ in B. It follows that

$$F_\Omega^C(z, v) \equiv F_B^C(z, v)$$

for all $z \in \Omega$ and all $v \in \mathbb{C}^n$.

Therefore from (4.12) we obtain

$$L_z(\log(1 + |f(z)|^2), v) \leq C_0(F_B^C(z, v))^2 \tag{4.13}$$

for all $z \in \Omega$ and all $v \in \mathbb{C}^n$.

By the Hartogs extension theorem f extends holomorphically to the whole ball B. Denote this extension also by f. Since f is holomorphic in B we have $|f| < C_1$ on $B_{3/4}(0)$. It follows at once that

$$L_w(\log(1 + |f(w)|^2), v) \leq C_1^2(F_{B_{3/4}(0)}^C(w, v))^2 \tag{4.14}$$

for all $w \in B_{3/4}(0)$ and all $v \in \mathbb{C}^n$.

Under the map

$$\varphi : z \longrightarrow \frac{4}{3}z, \quad \varphi : \mathbb{C}^n \to \mathbb{C}_w^n,$$

the ball $B_{3/4}(0)$ goes to the unit ball B. It is easy to see that $\varphi_*(z) \equiv 4/3 \cdot I$ for all $z \in B_{3/4}(0)$. Here I is the identity matrix, i.e., the square matrix of size $n \times n$ with 1s on the main diagonal and zeros elsewhere.

Since the Carathéodory infinitesimal metric is invariant with respect to biholomorphic transformations, we conclude that

$$F^C_{B_{3/4}(0)}(w, v) = F^C_B(\varphi^{-1}(w), \varphi_*^{-1}(w)v)$$

for all $w \in B_{3/4}(0)$ and all $v \in \mathbb{C}^n$.

From (4.14) we obtain

$$L_{\varphi^{-1}(w)}(\log(1 + \left|f(\varphi^{-1}(w))\right|^2), v) \leq C_1^2(F^C_B(\varphi^{-1}(w), \varphi_*^{-1}(w)v))^2$$

for all $w \in B_{3/4}$ and all $v \in \mathbb{C}^n$. It follows that

$$\frac{1}{|\det \varphi(w)_*|^2} L_{\varphi^{-1}(z)}(\log(1 + \left|f(\varphi^{-1}(w))\right|^2), \varphi_*^{-1}(w)v) \leq$$

$$C_1^2(F^C_B(\varphi^{-1}(w), \varphi_*^{-1}(w)v))^2$$

for all $w \in \overline{B}_{2/3}(0)$ and all $v \in \mathbb{C}^n$. If we set $z = \varphi^{-1}(w)$ and $v = \varphi_*^{-1}(w)v$, then from the inequality above we have

$$L_z(\log(1 + |f(z)|^2), v) \leq \frac{16}{9} C_1^2(F^C_B(z, v))^2 \tag{4.15}$$

for all $z \in \overline{B}_{1/2}(0) \equiv \varphi^{-1}(\overline{B}_{2/3}(0))$ and all $v \in \mathbb{C}^n$.

From (4.13) and (4.15), we obtain

$$L_z(\log(1 + |f(z)|^2), v) \leq C(F^C_B(z, v))^2$$

for all $z \in B$ and all $v \in \mathbb{C}^n$. Here $C = \max\left\{C_0, 16C_1^2/9\right\}$.

Since, for the domain the unit ball, the Bergman, Kobayashi, and Carathéodory metrics are "equal" up to a constant factor we can say simply that f is in the class $\mathcal{N}$, or equivalently f is a normal function in all of B. □

Remark 4.2.21. We now look at the Behnke-Sommer's theorem. To state it, we define an m-dimensional holomorphic surface S in $\mathbb{C}^n$ to be the image of some domain $G \subset \mathbb{C}^m (m < n)$ under a nondegenerate holomorphic mapping

$$\varphi : G \to \mathbb{C}^n. \tag{4.16}$$

In particular, for $m = 1$, S is a holomorphic curve, and if $G \subset \mathbb{C}$ is a disk and φ is continuous in $\overline{G}$, then $S = \varphi(\overline{G})$ is called a *holomorphic disk*. Recall that a mapping (4.16) is said to be *nondegenerate* if the rank of the Jacobian matrix $\left(\frac{\partial \varphi_\mu}{\partial z_j}\right)$ is equal to m at all points of G. The surface (4.16) is said to be *bounded* if the set $S = \varphi(G)$ is bounded in $\mathbb{C}^n$.

Next we shall say that a sequence of sets M_j converges to a set M (notation: $M_j \to M$) if, for any $\varepsilon > 0$, there is a j_0 such that for all $j \geq j_0$ we have

$$M_j \subset M^{(\varepsilon)} \text{ and } \quad M \subset M_j^{(\varepsilon)}$$

where $M^{(\varepsilon)}$ and $M_j^{(\varepsilon)}$ denote the ε-dilations of the sets M and M_j, respectively (i.e., the union of all polydisks $P(z, \varepsilon)$ with centers at points of the sets).

Theorem 4.2.22 (Behnke-Sommer). *Let* S_j *be a sequence of bounded holomorphic surfaces which, together with the boundaries* ∂S_j, *are contained in a domain* $D \subset \mathbb{C}^n$. *If* S_j *converges to some set* S, ∂S_j *converges to a set* Γ, *and* $\Gamma \subset\subset D$, *then any function* $f \in \mathcal{O}(D)$ *extends holomorphically to some neighborhood of the set* S.

The Behnke-Sommer theorem is sometimes called the continuity principle. Relatively speaking, it states that the property of a function to be holomorphic in a neighborhood of the holomorphic surfaces S_j is also preserved for the limit set of these surfaces.

The special case of the Behnke-Sommer theorem where the S_j are closed subsets of the complex lines $'z = 'a_j$ and S is a closed subset of the line $'z = 'a = \lim_{j\to\infty} 'a_j$ was already proved by Hartogs, and is known as Hartogs's continuity theorem.

Suppose that a domain $D \subset \mathbb{C}^n$ in a neighborhood U of a boundary point a is given by the condition

$$D \cap U = \{z \in U : \varphi(z) < 0\}$$

where $\varphi \in C^2(U)$ and $\nabla_z\varphi = \left(\frac{\partial\varphi}{\partial z_1}, \ldots, \frac{\partial\varphi}{\partial z_n}\right)_z \neq 0$ for all $z \in U$; a function φ with these properties is said to *locally define* the domain D in the neighborhood U.

Definition 4.2.23. A domain D with a C^2 boundary in a neighborhood of a point a is said to be *pseudoconvex* at a if, in a neighborhood of a, there exists a local defining function φ of the domain D such that

$$L_a(\varphi, \omega) \geq 0 \quad \text{for all } \omega \in T_a^c(\partial D),$$

and *strictly pseudoconvex* if

$$L_a(\varphi, \omega) > 0 \quad \text{for all } \omega \in T_a^c(\partial D), \omega \neq 0$$

Definition 4.2.24. A domain $D \subset \mathbb{C}^n$ is *not holomorphically extendable* at a point $a \in \partial D$, if there exist a neighborhood $U = U(a) \subset \mathbb{C}^n$ and a holomorphic function $f : U \cap D \to \mathbb{C}$ that does not extend holomorphically to a neighborhood of a. We say that D is *holomorphically non-extendable* if it is not holomorphically extendable at each point of its boundary.

Theorem 4.2.25 (Levi-Krzoska). *If a domain* Ω *with a* C^2 *boundary in a neighborhood of a boundary point* a *is strictly pseudoconvex at this point, then* Ω *is not holomorphically extendable at* a. *Conversely, if* Ω *is not holomorphically extendable at a point* a, *then it is pseudoconvex at* a.

Corollary 4.2.26. *Let* $\varphi \in C^2$ *be a real function in a neighborhood of a point* $a \in \mathbb{C}^n, \varphi(a) = 0$, *and suppose that the Levi form* $L_a(\varphi, \omega)$ *on the complex tangent plane* $T_a^c(S)$, *where* $S = \{\varphi(z) = 0\}$, *has at least one negative* *eigenvalue (i.e., there exists a vector* $\omega \in T_a^c(S)$ *such that* $H_a(\varphi, \omega) < 0$). *Then any function* f *that is holomorphic in the part of a neighborhood of* S *where* $\varphi < 0$ *has a holomorphic extension to a neighborhood of the point* a.

Suppose that the domain Ω at some boundary point a can be tangent from the exterior to a complex hypersurface $A = \{f(z) = 0\}$; this means that $f(a) = 0$, but in some neighborhood U_a the surface A lies outside of Ω. Then Ω is not holomorphically extendable at the point a (the function $1/f$ cannot be extended), and hence, by the previous theorem, is pseudoconvex at this point. This sufficient condition for psuedoconvexity is similar to the classical geometric criterion for geometric convexity (instead of A we must take a real hyperplane).

Until now we have considered domains with boundaries of class C^2; however the concept of local pseudoconvexity can also be formulated in the general case. For this let us agree to say that a domain Ω at a boundary point a can be tangent from the interior by a family of holomorphic disks if there is a family of holomorphic disks $S_t \subset\subset D, 0 < t \leq t_0$, converging as $t \to 0$ to a disk S such that $S_t \to S, \partial S_t \to \partial S$, where $\partial S \subset\subset \Omega$, and S contains the point a. We take the absence of this property as the definition of local pseudoconvexity in the case of domains with arbitrary boundaries.

Definition 4.2.27. A domain Ω is said to be *pseudoconvex* at a boundary point a, if at a it cannot be tangent from the interior by a family of holomorphic disks.

In particular, if $\partial\Omega$ is C^2 in a neighborhood of the point a and is pseudoconvex at this point in the sense of Definition 4.2.23, then it is pseudoconvex in the sense of Definition 4.2.27.

In fact, in the proof of the Levi-Krzoska theorem it is proved that if $\partial\Omega$ is not pseudoconvex at a in the sense of Definition 4.2.23, then at a it can be tangent from the interior by a family of holomorphic disks. We stress that the notion of strict pseudoconvexity implies that the boundary is C^2-smooth and does not extend to arbitrary domains.

Theorem 4.2.28. *If a domain* Ω *in* $\mathbb{C}^n$ *is not pseudoconvex at a point* $a \in \partial\Omega$, *then any function* f *that is* C*-normal in* Ω *has a* C*-normal extension to some larger domain* Ω' *which contains* a.

Proof. Since Ω is not pseudoconvex at a point a in the boundary then there exists a ball $B_r(a)$ such that any function holomorphic in Ω extends holomorphically to $B_r(a)$. Any $g \in \mathcal{O}(\Omega)$ extends holomorphically to a domain $\Omega_1 = \Omega \cup B_r(a)$ and this extension $\widetilde{g}$ can take only those values at points of $\Omega \cup B_r(a)$ that g takes in Ω. It follows that

$$F_{\Omega_1}^C(z, v) = F_\Omega^C(z, v)$$

for all $z \in \Omega$ and all $v \in \mathbb{C}^n$.

The function f extends holomorphically to a domain $\Omega_1 = \Omega \cup B_r(a)$ and this extension $\widetilde{f}$ defines a function $L_z(\log(1+|\widetilde{f}|^2), v)$ which is continuous on $\Omega_1 \times \mathbb{C}^n$. Since the function $F^C_{\Omega_1}(z, v) > 0$ and is continuous on $\Omega_1 \times \{\mathbb{C}^n \setminus \{0\}\}$ it follows that the function

$$L_z(\log(1+|\widetilde{f}|^2), v)/F^C_{\Omega_1}(z, v)$$

is continuous on $\Omega_1 \times \{\mathbb{C}^n \setminus \{0\}\}$. Since any continuous function is bounded on any compact set which belongs to the domain of definition it follows that there exists some positive constant C_1 such that

$$L_z(\log(1+|\widetilde{f}|^2), v)/F^C_{\Omega_1}(z, v) < C_1$$

for all $z \in \overline{B_{r/2}(a)} \setminus \Omega$ and all v such that $|v| = 1$.

By hypothesis,

$$L_z(\log(1+|\widetilde{f}|^2), v) < C \cdot F^C_{\Omega_1}(z, v) \text{ for all } z \in \Omega \text{ and all } v \in \mathbb{C}^n.$$

Set $\Omega' = B_{r/2}(a) \cup \Omega$. It easy to see that $F^C_{\Omega_1}(z, v) = F^C_{\Omega'}(z, v)$. From the above inequalities we obtain

$$L_z(\log(1+|\widetilde{f}|)^2, v) < K \cdot F^C_{\Omega'}(z, v) \text{ for all } z \in \Omega' \text{ and all } v \in \mathbb{C}^n$$

where $K = \max\{C, C_1\}$. □

Remark 4.2.29. A direct consequence of the definitions is that, if f is normal in B, then there is a positive constant C such that a function

$$V(z) = \log\left((1+|f(z)|^2)(1-|z|^2)^{C(n+1)}\right)$$

is plurisubharmonic in B. For an arbitrary holomorphic function f this may not be the case.

If $V(z_1, \ldots, z_p)$ is plurisubharmonic in $\mathbb{C}^p \subset \mathbb{C}^n$, $n > p$, then V is also plurisubharmonic in $\mathbb{C}^n$.

4.3 Algebraic Operation in Class of Normal Function

Let Ω be a bounded domain in $\mathbb{C}^n$ and let F_Ω be either the infinitesimal Carathéodory, Kobayashi, or Bergman metrics on Ω.

Let $f \colon \Omega \to \mathbb{C}$ be a holomorphic function in Ω. Suppose that there exists a positive constant C such that the Levi form of $\log(1+|f|^2)$ at $z \in \Omega$ satisfies

$$L_z(\log(1+|f|^2), v) \leq C \cdot (F_\Omega(z, v))^2 \tag{4.1}$$

for all $(z, v) \in \Omega \times \mathbb{C}^n$.

Recall that

$$L_z(\log(1+|f|^2), v) = \frac{|Df(z)v|^2}{(1+|f(z)|^2)^2},$$

where

$$Df(z)v := \sum_{k=1}^{n} \frac{\partial f}{\partial z_k}(z)v_k.$$

Lemma 4.3.1. *Let Ω be a bounded domain in $\mathbb{C}^n$ and let $m \geq 2$ be an integer. If $f(z) \in \mathcal{O}(\Omega)$ satisfy the inequality (4.1), then $h(z) = (f(z))^m$ also satisfies the inequality (4.1).*

Proof. The proof consists of direct verification that the function $h(z)$ satisfies the definition of a normal function. It is elementary to see that

$$L_z(\log(1+|h|^2), v) = \frac{|f|^{2m-2}\,|D(f(z))v|^2}{(1+|f(z)|^{2m})^2} =$$

$$\frac{|Df(z)v|^2}{(1+|f|^2)^2}\frac{|f|^{2m-2}\,(1+|f|^2)^2}{(1+|f(z)|^{2m})^2} = \frac{|Df(z)v|^2}{(1+|f|^2)^2}\frac{(1/\,|f|+|f|)^2}{(1/|f(z)|^m+|f(z)|^m)^2} \leq$$

$$\frac{|Df(z)v|^2}{(1+|f|^2)^2};$$

this can be seen from the following elementary inequality[2]:

$$\frac{1}{|f|} + |f| \leq \frac{1}{|f|^m} + |f|^m.$$

By hypothesis f is normal, i.e.,

$$L_z(\log(1+|f|^2), v) \leq C(F_\Omega(z,v))^2$$

for some positive constant C. From the inequalities above we obtain

$$L_z(\log(1+|h|^2), v) \leq C(F_\Omega(z,v))^2.$$

Thus $h = f^m$ is normal. The proof of the lemma is complete. □

In their paper concerning normal meromorphic functions, Lehto and Virtanen [5, p. 53] remark that the sum of a "normal" function and a bounded function (which is necessarily normal) is a normal function.

[2]The inequality to be proved may be written in the form

$$1 + |f|^{2m} - |f|^{m-1} - |f|^{m+1} \geq 0.$$

It is clear that

$$1 + |f|^{2m} - |f|^{m-1} - |f|^{m+1} = (|f|^{m-1} - 1)(|f|^{m+1} - 1) \geq 0.$$

Lemma 4.3.2. *Let Ω be a bounded domain in $\mathbb{C}^n$ and let $f(z) \in \mathcal{O}(\Omega)$ satisfy the inequality (4.1). Suppose that $g(z) \in \mathcal{O}(\Omega)$ is such that*

$$|g(z)| \leq m, z \in \Omega,$$

where m is finite constant.

Then the function $h(z) = f(z) - g(z)$ also satisfies the inequality (4.1) with some constant $C' = C'(C, m)$.

Proof. The proof consists of direct verification that the function $h(z)$ satisfies the inequality (4.1). It is elementary to see that

$$L_z(\log(1 + |h|^2), v) = \frac{|D(f(z) - g(z))v|^2}{(1 + |f(z) - g(z)|^2)^2} \leq$$

$$\frac{|Df(z)v|^2}{(1 + |f|^2)^2} \frac{(1 + |f|^2)^2}{(1 + |f(z) - g(z)|^2)^2} + \frac{|Dg(z)v|^2}{(1 + |f(z) - g(z)|^2)^2} \leq$$

$$\left[4\frac{|Df(z)v|^2}{(1 + |f|^2)^2} + \frac{|Dg(z)v|^2}{1 + |g(z)|^2}\right](1 + |g(z)|^2)^2;$$

this can be seen from the following elementary inequalities[3]:

$$\frac{1}{2(1 + |g|^2)} < \frac{1 + |f - g|^2}{1 + |f|^2} < 2(1 + |g|^2).$$

Since $|g| \leq m$, it follows from the definition of the infinitesimal Carathéodory metric that

$$L_z(\log(1 + |g|^2), v) \leq m^2 (F_\Omega^C(z, v))^2 .$$

Since the infinitesimal Carathéodory metric is not greater than the Bergman metric or the infinitesimal Kobayashy metric, we have

$$L_z(\log(1 + |g|^2), v) \leq m^2 (F_\Omega(z, v))^2 .$$

By hypothesis f is normal, i.e.,

$$L_z(\log(1 + |f|^2), v) \leq C(F_\Omega(z, v))^2$$

for some positive constant C. From the inequalities above, we obtain

$$L_z(\log(1 + |h|^2), v) \leq \left[4C + m^2\right](1 + m^2)^2 (F_\Omega(z, v))^2.$$

It follows that the function $h(z) = f(z) - g(z)$ satisfies the inequality (4.1) with constant

$$C' = \left[4C + m^2\right](1 + m^2)^2$$

The proof of the lemma is complete. □

[3]The right inequality follows from the fact that

$$1 + |f - g|^2 < 1 + |f|^2 + 2|fg| + |g|^2 + (|f| - |g|)^2 + 2|fg|^2 < 2(1 + |g|^2)(1 + |f|^2),$$

while the left inequality is obtained from the right one by replacing f by $f + g$.

Lemma 4.3.3. *Let Ω be a bounded domain in $\mathbb{C}^n$ and let $f(z) \in \mathcal{O}(\Omega)$ satisfy the inequality (4.1), Suppose that $g(z) \in \mathcal{O}(\Omega)$ is such that*

$$\frac{1}{m} < |g(z)| < m,$$

where $m > 0$ is a finite constant. Then the function $h(z) = f(z) \cdot g(z)$ also satisfies the inequality (4.1) with some constant $C' = C'(C, m)$.

Proof. The proof consists of direct verification that the function $h(z)$ satisfies the inequality (4.1). It is easy to see that

$$\frac{1}{m^2} \leq \frac{1 + |f(z)|^2}{1 + |g(z)f(z)|^2} \leq m^2.$$

It is elementary to see that

$$L_z(\log(1 + |h|^2), v) = \frac{|D(f(z)g(z))v|^2}{(1 + |f(z)g(z)|^2)^2} \leq$$

$$\frac{|Df(z)v|^2}{(1 + |f(z)|^2)^2} \frac{|g(z)|^2 (1 + |f(z)|^2)^2}{(1 + |f(z)g(z)|^2)^2} + \frac{|f(z)g(z)|^2 |Dg(z)v|^2}{|g(z)|^2 (1 + |f(z)g(z)|^2)^2} \leq$$

$$\left[m^2 |g(z)|^2 \frac{|Df(z)v|^2}{(1 + |f(z)|^2)^2} + \frac{|Dg(z)v|^2}{2 |g(z)|^2 (1 + |g(z)|^2)}\right] (1 + |g(z)|^2)^2.$$

Arguing as in Lemma 4.3.2, we can show that

$$L_z(\log(1 + |h|^2), v) \leq \left[m^4 C + \frac{m^4}{2}\right] (1 + m^2)^2 (F_\Omega(z, v))^2.$$

If follows that the function $h(z) = f(z) \cdot g(z)$ satisfies the inequality (4.1) with constant

$$C' = \left[m^4 C + \frac{m^4}{2}\right] (1 + m^2)^2.$$

The details are omitted here. The proof of the lemma is complete. □

Remark 4.3.4. Let $f, g \in \mathcal{O}(\Omega)$ such that $|g| < |f|$. Then g/f satisfies the inequality (4.1) with constant 1 (as a bounded holomorphic function in Ω) and therefore

$$\frac{|f(z)Dg(z)v - gDf(z)v|^2}{|f|^4 (1 + |g/f|^2)^2} < (F_\Omega(z, v))^2.$$

Theorem 4.3.5. *Let Ω be a bounded domain in $\mathbb{C}^n$ and $m \in \mathbb{N}$. If a family $\mathcal{F} \subset \mathcal{O}(\Omega)$ is normal, $q, g \in \mathcal{O}^\infty(\Omega)$, $|g| >$ const, then the family*

$$G = \{gf^m + q, f \in \mathcal{F}\}$$

is also normal.

Proof. Since normality is a local property, it sufficient to prove that G is a normal family in the unit ball. Fix a point z_0 in Ω and consider a ball $\overline{B(z_0, r)} \subset \Omega$. Marty's characterization of normal families yields a non-negative constant M such that, for any $f \in \mathcal{F}$,

$$L_z(\log(1 + |f|^2), v) < M\,|v|^2 \text{ for all } z \in \overline{B(z_0, r)}, v \in \mathbb{C}^n.$$

Arguing as in the lemmas, we conclude that there is a positive constant C such that for any $f \in \mathcal{F}$,

$$L_z(\log(1 + |gf^m + q|^2), v) < C\,|v|^2 \text{ for all } z \in \overline{B(z_0, r)}, v \in \mathbb{C}^n\,.$$

From Marty's criterion it now follows that $G = \{gf^m + q, f \in \mathcal{F}\}$ is a normal family. □

4.4 An Extension Theorem for Bloch and Normal Functions

As an extension of the big Picard theorem, Lehto and Virtanen showed [164, Theorem 9] that isolated singularities are removable for normal meromorphic functions. Järvi [119] has given a generalization of this result for functions defined in subdomains of $\mathbb{C}^n$.

Set $D^* = D \setminus \{0\}$, and let $f : D^* \to \widehat{\mathbb{C}}$ be normal. As noted above, f extends to a function f^* meromorphic in D [164, p. 62]. Since $F_D(z, \eta)$ and $F_{D^*}(z, \eta)$ are comparable near ∂D, f^* is normal in D. Moreover, the order of normality of f^* does not deviate too much from that of f. More precisely, we have

Lemma 4.4.1. *Given a positive number K, there is a positive number K' such that any function f normal in D^* with $C_f \leq K$ extends to a function f^* normal in D with $C_{f^*} \leq K'$.*

Proof. We begin with a quick proof of the Lehto-Virtanen extension theorem. Let f be normal in D^*. Recall that

$$F_{D^*}(z, \xi) = \text{ hyperbolic metric of } D^* = \frac{|\xi|}{|z| \log(1/|z|)}.$$

Hence the hyperbolic area of $D^*(r) = \{z \in \mathbb{C} : 0 < |z| < r\}$ is finite for every $r < 1$. By (4.1), we have

$$\iint_{D^*(r)} \frac{|f'(z)|^2}{(1 + |f(z)|^2)^2} dxdy < \infty$$

By the big Picard theorem, the singularity at 0 is inessential.

Fix $K > 0$, and let $f : D^* \to \widehat{\mathbb{C}}$ be normal with $C_f \leq K$. Again, let f^* stand for the extended function. Since the spherical metric is invariant under rotations of the sphere, we may assume that $f^*(0) = 0$. Clearly it is sufficient to exhibit an $R, 0 < R < 1$, depending only on K, such that $|f^*(z)| < 1$ for $z \in D(R) = \{z \in \mathbb{C} : |z| < R\}$. By assumption,

$$\frac{|f'(z)|}{1 + |f(z)|^2} \leq K \cdot \frac{1}{|z| \log(1/|z|)} \quad \text{for all } z \in D^* . \tag{4.1}$$

Set $\gamma_r = \{z \in \mathbb{C} : |z| = r\}$, $0 < r < 1$, and let $s\left(f^*\left(\gamma_r\right)\right)$ denote the spherical length of $f^*\left(\gamma_r\right)$. We claim that $R = e^{-8K}$ does the job. Suppose, on the contrary, that $|f^*(z)| \geq 1$ for some $z \in D(R)$. Pick out $z_0 \in D(R)$ such that $|f^*\left(z_0\right)| = 1$ and $|f^*(z)| < 1$ for $|z| < |z_0|$. A simple estimate based on (4.1) gives $s\left(f^*\left(\gamma_{|z_0|}\right)\right) < \pi/4$. Since the spherical distance of 0 and $f\left(z_0\right)$ is $\pi/2$, $f^*\left(\gamma_{|z_0|}\right)$ lies in the half plane $\operatorname{Re} \overline{f\left(z_0\right)} z > 0$. Therefore, the winding number of $f^*\left(\gamma_{|z_0|}\right)$ with respect to 0 (in $\mathbb{C}$) is 0. This contradiction with $f^*(0) = 0$ completes the proof. □

The next lemma is readily deduced by elementary considerations of the Kobayashi metric. We omit the proof.

Lemma 4.4.2. *Let* $f : \left(D^*\right)^n \to \widehat{\mathbb{C}}$ *be normal.*

(1) *For every* $k \in \{1, \ldots, n\}$ *and every* $\left(a_1, \ldots, a_{n-1}\right) \in \left(D^*\right)^{n-1}$, *the map*

$$z \to f\left(a_1, \ldots, a_{k-1}, z, a_k, \ldots, a_{n-1}\right), \quad D^* \to \widehat{\mathbb{C}},$$

is a normal function.

(2) *For every* $k \in \{1, \ldots, n\}$ *and every* $a \in D^*$, *the map*

$$\left(z_1, \ldots, z_{n-1}\right) \to f\left(z_1, \ldots, z_{k-1}, a, z_k, \ldots, z_{n-1}\right), \quad \left(D^*\right)^{n-1} \to \widehat{\mathbb{C}},$$

is a normal function.

(3) *For every* $a = \left(a_1, \ldots, a_n\right) \in \partial D^n$ *with* $a_i \neq 0, i = 1, \ldots, n$, *the map*

$$z \to f\left(a_1 z, \ldots, a_n z\right), \quad D^* \to \widehat{\mathbb{C}},$$

is a normal function.

Further, the orders of normality of all these functions are bounded above by that of f.

Let $\Omega \subset \mathbb{C}^n$ be a domain and let $V \subset \Omega$ be an analytic subvariety of codimension one. The singularities of V are said to be *normal crossings* provided $\Omega \setminus V$ is locally biholomorphic to $\left(D^*\right)^k \times D^{n-k}$ for some $k \in \{0, \ldots, n\}$. Our main theorem reads as follows.

Theorem 4.4.3. *Let $\Omega \subset \mathbb{C}^n$ be a domain and let $V \subset \Omega$ be an analytic subvariety of codimension one, whose singularities are normal crossings. Suppose $f : \Omega \setminus V \to \widehat{\mathbb{C}}$ is normal. Then f extends to a holomorphic mapping $f^* : \Omega \to \widehat{\mathbb{C}}$.*

Proof. Since the problem is of a local nature and the inclusion mapping is distance-decreasing (in the Kobayashi metrics), we may assume that $\Omega = D^n$ and $D^n \setminus V = (D^*)^n$.

The proof will be by induction on n. The case $n = 1$ is part of Lemma 4.4.1. So let $n \geq 2$ and assume that the extension is possible for $1, \ldots, n-1$. Let $a = (a_1, \ldots, a_n) \in V \setminus \{0\}$, and choose $k \in \{1, \ldots, n\}$ such that $a_k \neq 0$.

Consider the function

$$f_{a_k} : (z_1, \ldots, z_{n-1}) \to f(z_1, \ldots, z_{k-1}, a_k, z_k, \ldots, z_{n-1}),$$

mapping $(D^*)^{n-1} \to \widehat{\mathbb{C}}$. It follows from Lemma 4.4.2 that f_{a_k} is normal. Hence, by the induction hypothesis, f_{a_k} admits a holomorphic extension $f^*_{a_k} : D^{n-1} \to \widehat{\mathbb{C}}$. We set $f^*(a) = f^*_{a_k}(a_1, \ldots, a_{k-1}, a_{k+1}, \ldots, a_n)$. We will show that the extended mapping is holomorphic on $D^n \setminus \{0\}$. By the Riemann extension theorem, it suffices to prove that f^* is continuous, i.e., $\mathrm{Cl}(f; a)$, the cluster set of f at any $a \in V \setminus \{0\}$, reduces to a singleton (of course this also shows that the extension does not depend on the choice of k).

So let a, k and $f^*(a)$ be as above. Pick $\varepsilon > 0$. Set

$$\chi(z, w) = \frac{|z - w|}{\left(1 + |z|^2\right)^{1/2} (1 + |w|^2)^{1/2}} \qquad \text{for } z, w \in \widehat{\mathbb{C}}$$

and $B(a, \delta) = \{z \in \mathbb{C}^n : |z - a| < \delta\}$. By Lemma 4.4.2, the family

$$\left\{z \to f(b_1, \ldots, b_{k-1}, z, b_k, \ldots, b_{n-1}) : (b_1, \ldots, b_{n-1}) \in (D^*)^{n-1}\right\}$$

is equicontinuous at a_k. Hence there exists a positive δ such that

$$\chi\left(f(z), f(z_1, \ldots, z_{k-1}, a_k, z_{k+1}, \ldots, z_n)\right) < \varepsilon/2$$

and

$$\chi\left(f(z_1, \ldots, z_{k-1}, a_k, z_{k+1}, \ldots, z_n), f^*(a)\right) < \varepsilon/2$$

for $z = (z_1, \ldots, z_n) \in B(a, \delta) \setminus V$. Thus $\chi(f(z), f^*(a)) < \varepsilon$ for $z \in B(a, \delta) \setminus V$. It follows that $\mathrm{Cl}(f; a) = f^*(a)$.

It remains to extend f to 0. First, we infer from Lemmas 4.4.2 and 4.4.1 that $f(z, \ldots, z)$ tends to a limit, say w_0, as $z \to 0$. It suffices to show that $\mathrm{Cl}(f; 0) = w_0$. Let $\varepsilon > 0$. Consider the mappings described in Lemma 4.4.2 (**3**), or rather their counterparts for the restrictions of f to the hyperplanes of the form $\{z_n = a\}, a \in D^*$. By Lemma 4.4.1, all of them extend holomorphically to 0. Moreover, it follows from Lemmas 4.4.2 and 4.4.1 that the extensions constitute an equicontinuous family at 0 ($\in \mathbb{C}$). Therefore we find a positive δ such that $\chi(f(z), f^*(0, \ldots, 0, z_n)) < \varepsilon/3$, $\chi(f^*(0, \ldots, 0, z_n), f(z_n, \ldots, z_n)) < \varepsilon/3$ and $\chi(f(z_n, \ldots, z_n), w_0) < \varepsilon/3$ for $z = (z_1, \ldots, z_n) \in B(0, \delta) \setminus V$. Therefore $\chi(f(z), w_0) < \varepsilon$ for $z \in B(0, \delta) \setminus V$. Hence $\mathrm{Cl}(f; 0) = w_0$. This completes the proof. □

Remark 4.4.4. Set $V = \{(z_1, z_2) \in \mathbb{C}^2 \mid z_1 z_2 (z_1 - z_2) = 0\}$ and consider the mapping $f : D^2 \setminus V \to \widehat{\mathbb{C}}, (z_1, z_2) \to z_1/z_2$. Since f omits the values 0,1 and ∞ in $D^2 \setminus V$, it is normal by Remark 4.2.9. Yet f does not extend to a holomorphic map $D^2 \to \widehat{\mathbb{C}}$. Accordingly, we cannot dispense with some restrictions on the singularities of V in Theorem 4.4.3.

Remark 4.4.5. One may ask whether the extended function is normal in Ω (provided Ω is hyperbolic). However, it seems that this need not be the case even in dimension one.

Remark 4.4.6. Results related to Theorem 1 can be found in [128], [135] and [152]. Cf. in particular [128, Theorem 2].

In the counterexample discussed above the function involved is the restriction to $D^2 \setminus V$ of a meromorphic function, i.e., a function with "indeterminacies". This is always the case as shown by

Theorem 4.4.7. *Let $\Omega \subset \mathbb{C}^n$ be a domain and let V be a subvariety of Ω. Suppose $f : \Omega \setminus V \to \widehat{\mathbb{C}}$ is normal. Then f extends to a meromorphic function in Ω.*

Proof. Denote by $S(V)$ the set of singular points of V. By Theorem 4.4.3, f extends to a holomorphic mapping of $\Omega \setminus S(V)$ into $\widehat{\mathbb{C}}$. Thus f can be regarded as a meromorphic function in $\Omega \setminus S(V)$. Since $\dim S(V) \leq n-2$, [251, p. 144], f extends to a function meromorphic in Ω, [251, p. 149]. □

Classically, a holomorphic function on the disk $D \subseteq \mathbb{C}$ is called a *Bloch function* if

$$\sup_{z \in D} \left(1 - |z|^2\right) |f'(z)| < \infty$$

(see [10]).

In several complex variables Bloch functions were studied on the ball and on bounded homogeneous domains in [102, 107, 276, 277]. Generic smoothly bounded domains in $\mathbb{C}^n$, $n > 1$, are not homogeneous (for example, strictly pseudo convex domains are generically rigid, see [143]).

Definition 4.4.8. ([148, p. 146]) A holomorphic function defined on a bounded domain $\Omega \subset \mathbb{C}^n$, $n \geq 1$, is said to be a *Bloch function* $f \in \mathcal{B}(\Omega)$, if

$$|f_*(z) \cdot v| \leq C \cdot F_\Omega^K(z, v), \qquad \text{for all } z \in \Omega, v \in T_z(\Omega),$$

where $f_*(z)$ is the mapping from $T_z(\Omega)$ to $T_{f(P)}(\mathbb{C})$ induced by f.

Remark 4.4.9. Bloch functions form, in general, a proper subclass of normal functions.

Remark 4.4.10. It is easily checked, using the explicit formula for the Kobayashi metric given in [143], that this definition coincides with the more classical definitions on the disk and ball.

To prove our theorem, we need some elementary results involving Hausdorff measures. A good exposition of the theory of Hausdorff measures can be found in Federer's monograph [72]. We begin by introducing the necessary terminology and notation.

Definition 4.4.11. Let μ be a nonnegative measure on X. For any function $f : X \to [0, +\infty]$, the upper integral of f is defined as

$$\int^{*} f d\mu = \inf \left\{ \int g d\mu : \text{g is integrable and } g \geq f \right\}.$$

By convention (throughout this chapter), $\inf \varnothing = +\infty$.

We state the following elementary properties of the upper integral:

1) If $f : X \to [0, +\infty]$ is such that $\int^{*} f d\mu = 0$, then $f = 0$ almost everywhere.

Let $f_n : X \to [0, +\infty]$ for $n = 1, 2, \cdots$. Then

2) $\int^{*} \liminf f_n d\mu \leq \liminf \int^{*} f_n d\mu$
3) $\int^{*} \sum f_n d\mu \leq \sum \int^{*} f_n d\mu$.

Let E be an arbitrary subset of a metric space. Consider a covering of E by at most countably many balls B_j (it is immaterial whether these are open or closed) of radii r_j, respectively, and compute the sum Σr_j^{α}, where α is a fixed positive integer. The infimum of such sums, $m_{\alpha}(E)$, will to some extent have to reflect the property that E is a set of "dimension α"; however, this quantity is too coarse (for, say, the unit ball in $\mathbf{R}^m$ it equals 1 for all $\alpha \leq m$). Therefore we first fix $\varepsilon > 0$, and define the quantity

$$H_{\varepsilon}^{\alpha}(E) := \inf \left\{ c_{\alpha} \sum_{j} r_j^{\alpha} : E \subset \cup B_j, r_j < \varepsilon \right\}.$$

For integer $\alpha = m$ we take the constant $c_k > 0$ equal to the volume of the unit ball in $\mathbf{R}^m$ (in particular, $c_0 = 1, c_1 = 2, c_2 = \pi$, etc.); for noninteger α we take it to be the corresponding expression with the gamma function (for us the constants do not play a role at all, but this normalization is commonly adopted); the index α will always be nonnegative.

As ε decreases, the quantities $H_{\varepsilon}^{\alpha}(E)$ monotone increase, hence have a (finite or infinite) limit:

$$H^{\alpha}(E) := \lim_{\varepsilon \to 0} H_{\varepsilon}^{\alpha}(E).$$

This number is called the *Hausdorff measure of order* (dimension) α (or the H^α-measure, or simply the α-measure) of E.

If $f : X \to Y$ is a mapping of metric spaces, we write

$$\mathrm{Lip}(f) = \sup_{a \neq b} \frac{d_Y(f(x), f(x'))}{d_X(x, x')} \leq +\infty,$$

and we say that f satisfies a *Lipschitz condition of order 1* (f ϵL_1) if $\mathrm{Lip}(f) < +\infty$.

If $f : X \to Y$ is a continuous map between metric spaces that uniformly satisfies a Lipschitz condition (i.e. $d_Y(f(x), f(x')) \leqslant C d_X(x, x')$ for some constant C and all $x, x' \in X$), then $H_\alpha(f(E)) \leqslant C^\alpha H_\alpha(E)$ for all $E \subset X$. In particular, under projections $\mathbf{R}^N \to \mathbf{R}^m \subset \mathbf{R}^N$, Hausdorff measures do not increase. This property readily follow from the definition.

A subset A of a metric space X is said to be of type $L(p, c)$, for $0 \leq c < +\infty$, if there exists $\delta > 0$ such that $H^p(B) \leq c\delta^p(B)$ for all $B \subset A$ with diameter smaller than δ.

Let $\widetilde{H}^p$ denote the restriction of H^p to the Borel sets, and write

$$\int^* f(x) d^p x = \int^* f(x) d\widetilde{H}^p(x) = \int^* f d\widetilde{H}^p.$$

Lemma 4.4.12. *Let X and Y be metric spaces, and let f map X into Y, with* $\mathrm{Lip}(f) = \lambda < +\infty$. *Let $\alpha \geq 0, \beta > 0$, and suppose that $f(X)$ is of type $L(\beta, \mathrm{c})$ and $H^{\alpha+\beta}(X) < +\infty$. Then*

$$\int^* H^\alpha(f^{-1}(y)) d^\beta y \leq c\lambda^\beta H^{\alpha+\beta}(A).$$

Proof. Federer proved a more general version of Lemma 4.4.12 see [72, p.243]. We give an elementary proof of the version above, formulated as follows: Pick δ as in the definition of $L(\beta, \mathrm{c})$. Let $A \subset X$ with $\delta(A) < \delta/(\lambda+1)$, and let $\mathrm{B} = \mathrm{f}(A)$. Then $\delta(\overline{\mathrm{B}}) \leq \lambda\delta(A) < \delta$, and therefore

$$\int^* \delta^\alpha(A \cap f^{-1}(Y)) \mathrm{d}^\beta Y \leq \delta^\alpha(A) \widetilde{H}^\beta(\overline{\mathrm{B}}) \leq \mathrm{c}\lambda^\beta \delta^{\alpha+\beta}(A).$$

Let $\varepsilon < \delta/(\lambda+1)$, and pick a covering $\{A_n\}$ of X such that

$$\delta(A_n) < \varepsilon \quad \text{and} \quad \sum \delta^{\alpha+\beta}(A_n) \leq H_\varepsilon^{\alpha+\beta}(X) + \varepsilon.$$

Then, by the above inequality and property 3) of the upper integral,

$$\int^* H_\varepsilon^\alpha\left(f^{-1}(y)\right) d^\beta y \leq c\lambda^\beta \Sigma \delta^{\alpha+\beta}(A_n) \leq c\lambda^\beta H_\varepsilon^{\alpha+\beta}(X) + c\lambda^\beta \varepsilon,$$

and therefore

$$\int^* H^\alpha\left(f^{-1}(y)\right) d^\beta y \leq \liminf \int^* H_{1/n}^\alpha\left(f^{-1}(y)\right) d^\beta y \leq c\lambda^\beta H^{\alpha+\beta}(X).$$

□

Corollary 4.4.13. *Let X be a metric space, and let $a \in X$. Then*

$$\int_{[0,+\infty)}^{*} H^{\alpha}\left(S_r(a)\right) d^1 r \leq H^{\alpha+1}(X) \quad \text{for } \alpha \geq 0,$$

where $S_r(a) = \{x \in X : d_X(x,a) = r\}$.

Proof. Consider the function $f : X \to [0,+\infty)$ given by $f(x) = d(x,a)$. □

Corollary 4.4.14. *Let $H^1(X) = 0$, and let $a \in X$. Then $S_r(a)$ is empty for H^1-almost all r.*

Corollary 4.4.15. *If $H^{n+1}(X) = 0$, then the topological dimension of X is at most n.*

Corollary 4.4.16. *Let A be an arbitrary subset of R^n, let $\alpha \geq 0$, and let $\pi_k : \mathrm{R}^n \to \mathrm{R}^k$ denote the projection onto the first k coordinates.*

(i) If $H^{k+\alpha}(A) = 0$, then $H^{\alpha}\left(A \cap \pi_k^{-1}(X)\right) = 0$ for H^k-almost all $X \in \mathrm{R}^k$.

(ii) If $H^{k+\alpha}(A) < +\infty$, then $H^{\alpha}\left(A \cap \pi_k^{-1}(x)\right) < +\infty$ for H^k-almost all $x \in \mathrm{R}^k$.

Lemma 4.4.17. *Let Y be an arbitrary subset of $\mathbb{C}^n$ with $0 \notin Y$, and let $\alpha \geq 0$. If $H^{2k+\alpha}(Y) = 0$, then there exists a complex $(\mathrm{n}-k)$-plane P through 0 such that*

$$H^{\alpha}(Y \cap \mathrm{P}) = 0.$$

Remark 4.4.18. If $\alpha > 0$, the hypothesis $0 \notin Y$ is inessential. If $\alpha = 0$, the conclusion asserts that $Y \cap \mathrm{P} = \varnothing$.

Proof. We use downward induction on k. If $k = n$, the lemma is trivial. Now suppose the lemma is true for $k+1$, and let Y be given as above. Then the Hausdorff $(2k+2+\alpha)$-measure of Y also vanishes; therefore we can pick a complex $(\mathrm{n}-k-1)$-plane Q through 0 such that $H^{\alpha}(Y \cap Q) = 0$. After rotating the coordinate system, we may assume that

$$Q = \{z_1 = \cdots = z_{k+1} = 0\}.$$

Let

$$\pi : \mathbb{C}^n \setminus Q \to P^k = \text{ complex projective } k\text{-space}.$$

be defined by $\pi(z_1, \cdots, z_n) = [z_1, \cdots, z_{k+1}]$. Set

$$Y_m = \left\{z \in Y : \|z_1\|^2 + \cdots + \|z_{k+1}|^2 \geq 1/m\right\} \quad \text{for m} = 1, 2, \cdots.$$

Since $\pi \mid Y_m$ satisfies a Lipschitz condition of order 1 and P^k is a compact $2k$-dimensional Riemannian manifold, we conclude by Lemma 4.4.1 that

$$H^{\alpha}\left(Y_m \cap \pi^{-1}(w)\right) = 0 \quad \text{for } \left(H^{2k}\right)\text{-almost all } w \in P^k.$$

Since $Y = (Y \cap Q) \cup (\bigcup_m Y_m)$, there exists $w_0 \in P^k$ such that

$$H^\alpha \left(Y \cap \pi^{-1}(w_0)\right) = 0,$$

and $\mathrm{P} = \pi^{-1}(w_0)$ is our desired $(n-k)$-plane. □

The following lemma generalizes some elementary results from the theory of functions of one complex variable.

Lemma 4.4.19. *Let U be open in $\mathbb{C}^n$, and let E be a closed subset of U. Let f be an analytic function on $U \setminus E$. Then f can be extended to an analytic function on U if one of the following conditions is satisfied: (i) $H^{2n-2}(E) = 0$; (ii) f is bounded and $H^{2n-1}(E) = 0$; (iii) f can be extended to a continuous function on U, and $H^{2n-1}(E) < +\infty$; (iv) for some α in the range $0 < \alpha \le 1$, $f \in L_\alpha$ and $H^{2n-1+\alpha}(E) = 0$.*

Proof. The proof of Lemma 4.4.19 for $n = 1$ is elementary: case (i) is a tautology; cases (ii) and (iii) are proved in [28]; and case (iv) can easily be established by the methods of [28]. We now prove Lemma 4.4.19 for $n > 1$, utilizing the result for $n = 1$. In all cases, $H^{2n}(E) = 0$; therefore E is nowhere dense. Consider an arbitrary point p of E, and assume without loss of generality that $p = 0^n$. We first consider cases (i) and (ii). By Lemma 4.4.17, we can pick a complex line L through 0^n so that $H^l(E \cap L) = 0$. Rotate coordinates so that $L = \{z_1 = \cdots = z_{n-1} = 0\}$. By Corollary 4.4.14 of Lemma 4.4.12, we can pick an open disk d about 0 in $\mathbb{C}$ such that $0^{n-1} \times \overline{d} \subset U$ and $\left(0^{n-1} \times \partial d\right) \cap E = \varnothing$. Since E is closed in U, there exists an open neighborhood W of 0^{n-1} in $\mathbb{C}^{n-1}$ such that $W \times \overline{d} \subset U$ and $(W \times \partial d) \cap E = \varnothing$. Define

$$g(z) = \frac{1}{2\pi i} \int_{\partial d} \frac{f(z_1, \cdots, z_{n-1}, \zeta)}{\zeta - z_n} d\zeta \quad \text{for } z \in W \times d.$$

Consider the complex lines $L_w = w \times C^L$ for $w \in w$. For case (i), let $F = \{w \in W : E \cap L_w \neq \varnothing\}$; and for case (ii), let $f = \left\{w \in W : H^1(E \cap L_w) > 0\right\}$.

Consider Corollary 4.4.16 of Lemma 4.4.12. Thus the set of points $w \in W$ such that

$$g \not\equiv f \text{ on } W \times d \setminus E$$

is an open subset of F, and hence it is empty. Therefore g extends $f|_{(W \times d \setminus E)}$. Since g is analytic, these cases are proved. Now consider cases (iii) and (iv). We can assume that f is a continuous function on all of U. Using the original coordinates, we define g as above, where W and d are chosen so that $W \times \overline{d} \subset U$. For case (iv), let $F = \left\{w \in W : H^{1+\alpha}(E \cap L_W) > 0\right\}$, and for case (iii), let $F = \left\{w \in W : H^l(E \cap L_w) = +\infty\right\}$. Proceeding as before, we conclude that

$$g = f|_{W \times d}.$$

Hence $f|_{W \times d}$ is analytic with respect to z_n and is similarly analytic with respect to each $z_k (1 \le k \le n)$. Therefore f is analytic in $W \times d$. □

We begin with a short proof of a result announced by Poletskiı and Shabat [238, Theorem 2.3, p. 79]. They do not give any proof in their survey article, but just announce that a proof can be based on a variation of reasoning by Campbell, Howard and Ochiai [34, Theorem 1, p. 106]. We base our proof on the earlier argument of Campbell and Ogawa [35, Proposition 2, p. 40] (on which the argument of Campbell, Howard and Ochiai is based, too).

Lemma 4.4.20. *Let Ω be a domain in $\mathbb{C}^n, n \geq 1$. Let $E \subset \Omega$ be closed in Ω such that $H^{2n-2}(E) = 0$. Then*

$$K_{\Omega \setminus E}(z, z') = K_{\Omega}(z, z')$$

for all $z, z' \in \Omega \setminus E$.

Proof. We use here the standard notation

$$\mathcal{O}(D, \Omega) = \{f : D \to \Omega : f \text{ is holomorphic }\}.$$

First we prove that if $\mathcal{O}(D, \Omega \setminus A)$ is dense in $\mathcal{O}(D, \Omega)$ then the Kobayashi pseudo-distance on Ω restricts to that on $\Omega \setminus A$.

Let z and z' be two points of Ω not in A. Let $r = K_{\Omega}(z, z')$. Choose $\varepsilon > 0$, and let $f_i \in \mathcal{O}(D, \Omega)$, $i = 1, \cdots, m$ be holomorphic maps such that $f_1(0) = z, f_i(a_i) = f_{i+1}(0), f_m(a_m) = z'$ for points $a_1, \cdots, a_m \in D$ satisfying $\sum_{i=1}^{m} d_D(0, a_i) \leq r + \varepsilon$.

Suppose that g_i is a map of D into $\Omega \setminus A, i = 1, \cdots, m$, and we put $\zeta_i = g_i(0), w_i = g_i(a_i), \zeta_0 = z, \zeta_{m+1} = z', \xi_i = f_i(a_i) = f_{i+1}(0), \xi_0 = z$.

Then

$$K_{\Omega \setminus A}(z, z') \leq \sum_{i=0}^{m} K_{\Omega \setminus A}(\zeta_i, \zeta_{i+1}) \leq \sum_{i=0}^{m} \left[K_{\Omega \setminus A}(\zeta_i, w_i) + K_{\Omega \setminus A}(w_i, \zeta_{i+1})\right]$$

$$\leq \sum_{i=0}^{m} d_D(0, a_i) + \sum_{i=0}^{m} K_{\Omega \setminus A}(w_i, \zeta_{i+1}) \leq r + \varepsilon + \sum_{i=0}^{m} K_{\Omega \setminus A}(w_i, \zeta_{i+1}).$$

Now, if the g_i are chosen close to the f_i in $\mathcal{O}(D, \Omega)$, then the points $w_i = g_i(a_i)$ and $\zeta_{i+1} = g_{i+1}(0)$ will both be close to $\xi_i = f_i(a_i) = f_{i+1}(0)$ and hence close to each other. Since $K_{\Omega \setminus A}$ is continuous, $K_{\Omega \setminus A}(w_i, \zeta_{i+1})$ will be small. Choose the g_i so close to the f_i in $\mathcal{O}(D, \Omega)$ that $\sum_{i=0}^{m} K_{\Omega \setminus A}(w_i, \zeta_{i+1}) < \varepsilon$. We obtain $K_{\Omega - A}(z, z') \leq r + 2\varepsilon = d_X(z, z') + 2\varepsilon$. Finally, letting $\varepsilon \to 0$ we obtain $K_{\Omega - A}(z, z') \leq K_{\Omega}(z, z')$. Since the other inequality $K_{\Omega \setminus A}(z, z') \geq K_{\Omega}(z, z')$ is always satisfied, we obtain $K_{\Omega \setminus A}(z, z') = K_{\Omega}(z, z')$.

Recall also that $\mathcal{O}(D, \Omega)$ is equipped with the usual compact-open topology. Write

$$M = \{f \in \mathcal{O}(D, \Omega) : f(D) \subset\subset \Omega\}.$$

It is clear that M is dense in $\mathcal{O}(D, \Omega)$. Take $g \in M$ arbitrarily. Then the mapping $G : D \times \Omega \to D \times \Omega$,

$$G(t, b) = (t, g(t) - b)$$

is clearly a holomorphic injection. It follows, at least from [72, 2.10.45, p. 202], that $H^{2n}(D \times E) = 0$. But then $H^{2n}(G(D \times E)) = 0$. Since the projections do not increase Hausdorff measure, we find a sequence $c_k \to 0, c_k \in \mathbb{C}^n, k = 1, 2, \ldots$, such that $c_k \neq g(t) - b$ for all $t \in D, b \in E$ and $k = 1, 2, \ldots$ Define for each $k = 1, 2, \ldots$ holomorphic mappings $g_k : D \to \mathbb{C}^n$,

$$g_k(t) = g(t) - c_k.$$

Since $g(D) \subset\subset \Omega$, there is a positive integer N such that $g_k(D) \subset\subset \Omega$ for all $k \geq N$. It is clear that $g_k \to g$ in $\mathcal{O}(D, \Omega)$ as $k \to \infty$. Moreover, one easily sees that $g_k(D) \subset \Omega \setminus E$ for all $k = 1, 2, \ldots$. Hence the assertion follows. □

Theorem 4.4.21. *Let Ω be a domain in $\mathbb{C}^n, n \geq 1$, and let $E \subset \Omega$ be closed in Ω such that $H^{2n-2}(E) = 0$. If $f : \Omega \setminus E \to \mathbb{C}$ is a Bloch function (respectively if $f : \Omega \setminus E \to \overline{\mathbb{C}}$ is a normal meromorphic function), then f has a Bloch extension $f^* : \Omega \to \mathbb{C}$ (respectively a normal extension $f^* : \Omega \to \overline{\mathbb{C}}$).*

Proof. By Lemma 4.4.19, f has a holomorphic extension f^* to Ω. Using then Lemma 4.4.20 above, the continuity of the Kobayashi pseudodistance K_Ω and also the continuity of f^*, we see that f^* is Bloch. To prove the normal function case, take $z_0 \in E$ arbitrarily. Choose $R > 0$ such that $B^{2n}(z_0, R) \subset\subset \Omega$ and take $z^* \in B^{2n}\left(z_0, \frac{R}{2}\right) \setminus E$ arbitrarily. Then either $f_{B^{2n}\left(z^*, \frac{R}{2}\right) \setminus E}$ or $\frac{1}{f}|_{B^{2n}\left(z^*, \frac{R}{2}\right) \setminus E}$ is a holomorphic function, and has thus again by Lemma 4.4.19 a holomorphic extension g to $B^{2n}\left(z^*, \frac{R}{2}\right)$. Thus f has a spherically continuous meromorphic extension f^* to Ω. The normality of f^* follows then again by using Lemma 4.4.20, the continuity of the Kobayashi pseudodistance K_Ω and the spherical continuity of f^*. □

Remark 4.4.22. The above result supplements the result of [247, Corollary 3.2, p. 148]. For a partial generalization, where the assumption "$H^{2n-2}(E) = 0$" is replaced by the weaker assumption "$H^{2n-2}(E \cap K) < \infty$ for each $K \subset \Omega$ compact", see Theorem 4.4.26 below.

For convenience of the reader, we recall first three basic results which we need in the proof of Theorem 4.4.26 below.

Lemma 4.4.23 (see [72, 2.10.25, p. 188] or [268, Corollary 4, p. 114]). *Let $A \subset \mathbb{C}^n, n \geq 2$. (a) If $H^{2n-2}(A) < \infty$, then for each $j, 1 \leq j \leq n$, and for H^{2n-2}-almost all $Z_j \in \mathbb{C}^{n-1}$, we have that $H^0(A(Z_j)) < \infty$.*

(b) If $H^{2n-1}(A) = 0$, then for each $j, 1 \leq j \leq n$, and for H^{2n-2}-almost all $Z_j \in \mathbb{C}^{n-1}$, we have that $H^1(A(Z_j)) = 0$.

The proof of the next lemma follows from the properties of the Kobayashi metric, just as in [23, Lemma 2.2, p. 925], see [5, Proposition 1.6 and Corollary 1.7, p. 309] and [12, Lemma 2, p. 1173].

Lemma 4.4.24. *Let Ω be a domain in $\mathbb{C}^n, n \geq 2$, and $1 \leq j \leq n$. Suppose that $f : \Omega \to \mathbb{C}$ is Bloch (respectively $f : \Omega \to \overline{\mathbb{C}}$ is normal). If $Z_j \in \mathbb{C}^{n-1}$ is* *such that* $\Omega(Z_j)$ *is a nonempty domain in* $\mathbb{C}$*, then the holomorphic function* $f_{Z_j} : \Omega(Z_j) \to \mathbb{C}$ *(respectively the meromorphic function* $f_{Z_j} : \Omega(Z_j) \to \overline{\mathbb{C}}$*),*

$$f_{Z_j}(z_j) = f(z_j, Z_j)$$

is Bloch (respectively normal).

Lemma 4.4.25. *([118, Lemma 3.4, p. 299]). Let* Ω *be a domain in* $\mathbb{C}^n, n \geq 2$*. Let* $E \subset \Omega$ *be closed in* Ω *and such that for each* $j, 1 \leq j \leq n$*, and for* H^{2n-2}*-almost all* $Z_j \in \mathbb{C}^{n-1}$ *the section* $E(Z_j)$ *is totally disconnected. Let* $f : \Omega \setminus E \to \mathbb{C}$ *be a holomorphic function. If for each* $j, 1 \leq j \leq n$*, and for* H^{2n-2}*-almost all* $Z_j \in \mathbb{C}^{n-1}$ *the function* $f_{Z_j} : (\Omega \setminus E)(Z_j) \to \mathbb{C}$*,*

$$f_{Z_j}(z_j) = f(z_j, Z_j)$$

has a holomorphic (respectively a meromorphic) extension $f^*_{Z_j}$ *to* $\Omega(Z_j)$*, then* f *has a holomorphic (respectively a meromorphic) extension* f^* *to* Ω*.*

A holomorphic function $f : \Omega \to \mathbb{C}$ is called *quasi-Bloch* if there is a constant $C = C(f)$ such that

$$|f(z) - f(z')| \leq C K_\Omega(z, z')$$

for all $z, z' \in \Omega$.

On the other hand, the assumption "$H^{2n-2}(E) < \infty$" is sufficient for Bloch functions (and for normal functions, too):

Theorem 4.4.26. *Let* Ω *be a domain in* $\mathbb{C}^n, n \geq 1$*. Let* $E \subset \Omega$ *be closed in* Ω *and such that* $H^{2n-2}(E \cap K) < \infty$ *for all* $K \subset \Omega$ *compact. If* f *is Bloch (respectively normal) in* $\Omega \setminus E$*, then* f *has a holomorphic (respectively meromorphic) extension* f^* *to* Ω*.*

Proof. Because of Lemma 4.4.23 (a), Lemma 4.4.24 and Lemma 4.4.25 it is sufficient to give the proof for $n = 1$. But then the exceptional set E is locally finite in Ω. Since Bloch functions are quasi-Bloch, the result follows from [248, Theorem 2.10]. The normal function case is proved similarly, see [246, Theorem 3.5, p. 927]. □

The results of this section are due to Järvi [119] and Riihentaus [248].

4.5 Schottky's Theorem in $\mathbb{C}^n$

The center-piece of this chapter is the extension to $\mathbb{C}^n$ of the following remarkable result due to Friedrich Schottky: if f is holomorphic in D and assumes neither of the values 0 or 1, then f is bounded in $D(r)$ by a bound depending on r and $|f(0)|$ only, for each $r < 1$. The principal application is the almost immediate fact that a family of holomorphic functions on a common domain, none of which has 0 or 1 in its range, if bounded at a point, is uniformly bounded in a neighborhood of that point. The compactness theorems then come into play with astounding consequences. Of course "0" and "1" here are convenient normalization: any two distinct complex numbers would serve as well.

A century ago, Picard [232] showed that, in any neighborhood of an essential singularity, a complex function attains every complex value, with at most one exception, an infinite number of times. Picard's proof was non-elementary, in that it made use of the theory of modular functions. The first elementary proof was given twenty-five years later by Schottky [262], and was based on a theorem which now bears his name. Subsequently, Montel [200] used Schottky's theorem and the notion of a normal family to give what has become the standard elementary proof of Picard's theorem.

Classically, there are at least three approaches to Schottky's theorem. The most sophisticated of these uses the theory of modular functions [78, Chapter 5]. A second approach, via Bloch's theorem, is described in §§1 − 3 of Chapter XII of [49]. In [30] the authors prefer to follow an argument of Titchmarsh [278, p. 8.85] which appears to be based on the original one of Schottky and which, in many respects, seems the most natural way to arrive at Schottky's theorem.

The main part of the proof of Schottky's theorem in [30] follows very closely the lines of Titchmarsh's proof. However, what Titchmarsh describes as "Schottky's theorem" is not the full form of the theorem as understood by other authors, and they are obliged to supplement his argument in order to reach that full form.

Scottky's theorem can be given a completely elementary proof, based on nothing more than Montel's theorem and general properties of normal families, which works for the cases $n = 1$ and $n > 1$ at the same time.

Theorem 4.5.1 (Schottky Theorem). *Suppose that* $f(z)$ *is holomorphic in* $\Omega \subset \mathbb{C}^n$, $n \geq 1$, *whose range omits* 0 *and* 1 *there. If* $a \in \Omega$ *and* $|f(a)| < M$, *then for every* $\varepsilon > 0$ *sufficiently small we have*

$$|f(z)| < C \text{ for all } z \in \Omega, \quad \delta(z) \geq \varepsilon, \tag{4.1}$$

where $\delta(z)$ *denotes the Euclidean distance from* $z \in \Omega$ *to* $\partial\Omega$. *Here* C *is a positive constant depending only on* $f(a)$ *and* ε.

The proof of Theorem 4.5.1 is based on the following lemma.

Lemma 4.5.2. *Let* $\mathcal{F}$ *be a normal family of holomorphic functions in a domain* $\Omega \subset \mathbb{C}^n$, $n \geq 1$. *Let* M *be a positive number and* $a \in \Omega$. *Assume that for each function* $f \in \mathcal{F}$, *we have*

$$|f(a)| < M.$$

Then the family $\mathcal{F}$ *is uniformly bounded on each compact subset of* Ω.

Proof. Let K be a compact set of points belonging to Ω. Assume that $\mathcal{F}$ is not uniformly bounded on K. Then to each positive integer k, corresponds a function $f_k \in \mathcal{F}$ such that

$$\max_{z\in K} |f_k(z)| > k.$$

The sequence $\{f_k\}$ cannot converge on compact $\{a\} \cup K$ to a holomorphic function or to ∞ in Ω. This contradicts the hypothesis that $\mathcal{F}$ is a normal family. □

Proof of Theorem 4.5.1. Let $\mathbf{G}$ denote the family of all functions holomorphic on Ω which omit the values 0 and 1. According to Montel's theorem [66, Theorem 1.4] the family $\mathbf{G}$ is normal. By Lemma 4.5.2, the family $\mathbf{G}$ is uniformly bounded on $\overline{\Omega}_\varepsilon = \{z \in \Omega,\ \delta(z) \geq \varepsilon\}$; hence there is a positive constant C such that (4.1) holds for each function $f(z)$ of $\mathcal{F}$. This constant C has then the required property. □

Remark 4.5.3. The method of proof in Theorem 4.5.1 does not give any information about what the constant C is. The essence of this theorem is not the exact value of the constant C, but the fact that such a constant exists.

In the case $n = 1$ the Schottky theorem has numerous proofs. We refer to [31] for an exposition (the history, methods and references) of the theorem. Schottky's original theorem [262] did not give an explicit bound for f. Let $K(f(0), r)$ denote the best possible bound in Theorem 4.5.1. Various authors have dealt with the problem of giving an explicit estimate for this bound. See Jenkins [120]; Hempel [113] who gave some bounds which are in some sense the best possible.

It was shown in Murali Rao [243, Theorem 7.17, p. 118] that Shottky's theorem holds with

$$C = \tan(\arctan M + (c(1 - r^2))^{-1}),$$

where c is some constant. The proof is based on the work of Minda and Schober [190].

We shall give an alternative argument based on the classical Montel theorem, and Marty's criterion for normality.

Recall that to every a in the unit disk $D := \{z \in \mathbb{C} : |z| < 1\}$ corresponds an automorphism φ_a of the disk that interchanges a and 0, namely $\varphi_a(w) := (w-a)/(1-\overline{a}w)$. Note for later reference that $\varphi_a(0) = -a$ and $\varphi_a'(0) = 1-|a|^2$. Define $\mathrm{Aut}(D) := \{\varphi_a, a \in D\}$.

The following theorem is the main result of this section.

Theorem 4.5.4 (see [243, Theorem 7.17, p. 118]). *Suppose that* $f(z)$ *is holomorphic in the unit disk* D *and with range omitting* 0 *and* 1. *If* $|f(0)| < M$, *then, for every* $r \in (0, 1)$, *we have*

$$|f(z)| < \tan(\arctan M + 2rL/(1 - r^2)) \text{ for all } |z| < r, r \in [0, 1). \tag{4.2}$$

Proof. Let $\mathbf{G}$ denote the family of all functions holomorphic on the unit disk D which omit the values 0 and 1. By Montel's theorem [292, p. 218] the family $\mathbf{G} \circ \mathrm{Aut}(D)$ is normal. Marty's theorem [292, p. 216] now yields a nonnegative constant L such that, for any $f \circ \varphi_a \in \mathbf{G} \circ \mathrm{Aut}(D)$

$$\frac{|f'(\varphi_a(0))|}{1 + |f(\varphi_a(0))|^2} \le L.$$

The constant L in the inequality above does not depend on $f \circ \varphi_a$. The exact value of L is not known.

Using the chain rule for differentiation and replacing $-a$ by z we obtain

$$\frac{|f'(z)|\,(1 - |z|^2)}{1 + |f(z)|^2} \le L \text{ for all } z \in D.$$

Since f never takes the value 0, the function $t \to |f(tz)|$ is continuously differentiable on $(0, 1)$ for any fixed $z \in D[r] = \{|z| < r\}$ and

$$\frac{d}{dt}\Big(\arctan(|f(tz)|)\Big) \le \frac{2\,|f'(tz)|\,|z|}{1 + |f(tz)|^2} \le \frac{2rL}{1 - r^2}.$$

Integrating this from $t = 0$ to 1, we deduce

$$\Big|\arctan(|f(z)|) - \arctan(|f(0)|)\Big| \le \int_0^1 \Big|\frac{d}{dt}(\arctan(|f(tz)|))\Big| dt \le \frac{2rL}{1 - r^2}$$

and hence

$$\arctan(|f(z)|) \le \arctan(|f(0)|) + \frac{2rL}{1 - r^2} < \arctan M + \frac{2rL}{1 - r^2}$$

which is the theorem with $C = \tan(\arctan M + 2rL/(1 - r^2))$. This completes the proof of Theorem 4.5.4. □

Remark 4.5.5. The same proof works for $n > 1$.

It still does not matter that the numerical upper bound is equal to $\tan(\arctan M + 2rL/(1 - r^2))$, and this estimate is still not the optimal one (by the way, the optimal estimate is not known to this day). The important things are, first, the fact that the estimate depends only on the radius of the disk and on the value of function at its center; and, second, the form of its dependence on $f(0)$ and r.

This type of argument can also be used to prove:

Theorem 4.5.6 (Picard's Little Theorem). *At most one complex number is absent from the range of a nonconstant entire function (= holomorphic in* $\mathbb{C}$*).*

4.5.1 Picard's little theorem

If $f \in H(\mathbb{C})$ and $f(\mathbb{C}) \subset U$ where $\mathbb{C} \setminus U$ contains at least two points, then f is constant. This is a dramatic strengthening of Liouville's theorem, as Liouville's theorem can be thought of as the preceding theorem where the condition on U is replaced by the condition $U = D$. We will actually prove an even stronger result known as Picard's big theorem [185, Theorem 10.14, p. 166].

Proof. Assume, to the contrary, that an entire function f omits two distinct values from its range. Composing with a linear fractional transformation, we may also assume that the omitted values are 0 and 1.

The function $f(2^k\lambda)$, where k is positive integer, is holomorphic in D, $f(2^k\lambda)$ takes the same values in D as does f in $|z| < 2^k$, and $f(2^k\lambda)$ omits the values 0 and 1.

The family $\{f(2^k\varphi_a), \varphi_a \in \mathrm{Aut}(D), k = 1, 2 \ldots\}$, being contained in the family of functions avoiding 0 and 1, is normal in D by Montel's theorem [292, p. 218]. By Marty's theorem [292, p. 216], there exists $L > 0$ such that, for any $f(2^k\varphi_a)$,

$$\frac{|f'(2^k\varphi_a(0))|}{1 + |f(2^k\varphi_a(0))|^2} \leq L.$$

After some manipulations, using the chain rule for differentiation and the equalities $\varphi_a(0) = -a$ and $\varphi_a'(0) = 1 - |a|^2$, we find that

$$\frac{2^k\,|f'(-2^ka)|\,(1 - |a|^2)}{1 + |f(-2^ka)|^2} \leq L\,.$$

Replacing in this inequality -2^ka by z we get

$$\frac{2^k\,|f'(z)|\,(1 - |z/2^k|^2)}{1 + |f(z)|^2} \leq L.$$

Thus

$$\frac{|f'(z)|}{1 + |f(z)|^2} \leq \frac{2^kL}{2^{2k} - |z|^2}.$$

Keeping z fixed, let k tend to infinity to deduce that $f'(z) = 0$ for every point z in the complex plane. But this clearly implies that $f(z)$ is a constant, not a nonconstant entire function. The obtained contradiction proves the theorem. □

An immediate (and easy) corollary of Picard's little theorem is Liouville's theorem:

Corollary 4.5.7. *Every bounded entire function is constant.*

Theorem 4.5.6 is a remarkable generalization of Liouville's theorem. It is simple to find entire functions whose range is the entire $\mathbb{C}$—nonconstant polynomials, for instance. The exponential function is an example of an entire function whose range omits only one value, namely zero. But there does not exist a nonconstant entire function whose range omits two different values.

The fundamental theorem of algebra is a consequence of Liouville's theorem (for the proof see, for example, [243, Corollary 4, p. 98]).

An alternative program for obtaining the above theorems (and not only) in elementary fashion was indicated by Zalcman [292, p. 817].

Theorem 4.5.8 (Picard's big theorem). *Let f be holomorphic in*

$$\Omega = \{z \in \mathbb{C} : 0 < |z - z_0| < \delta\} .$$

If z_0 is an essential singularity of the function f, then $f(z)$ assumes in an arbitrary neighborhood of z all finite complex values with one possible exception.

Proof. Let us assume that $z_0 = 0$. Suppose that there exist two finite numbers a and b and a neighborhood of zero in which $f(z)$ does not assume either the value a or the value b. Then the functions $f_j(z) = f(z/2^j)$, $j = 1, 2, \ldots$, also do not assume the values a and b in that neighborhood. Consequently, by Montel's theorem, the family $\{f(z/2^j)\}$ is normal in the chordal metric on compact subsets of $\mathbb{C} \setminus \{0\}$. Relabeling a subsequence, we may suppose f_{j_k} converges uniformly on compact subsets of $\mathbb{C} \setminus \{0\}$ to a function g holomorphic on $\mathbb{C} \setminus \{0\}$ or to $g \equiv \infty$ by Hurwitz's theorem. If g is holomorphic, then $|g(z)| \leq M < \infty$ on $|z| = 1$ and hence $|f| \leq M + 1$ on $|z| = 1/2^{j_k}$ for $k \geq k_0$. But then, by the maximum principle, $|f| \leq M + 1$ on $1/2^{j_{k+1}} \leq |z| \leq 1/2^{j_k}$ for $k \geq k_0$. Thus $|f| \leq M + 1$ on $0 < |z| < 1/2^{j_{k_0}}$, so that, by Riemanns theorem on removable singularities, f extends to be holomorphic in a neighborhood of 0.

If $g \equiv \infty$, we can apply a similar argument to $1/f(z/2^j)$ to conclude that $1/f$ extends to be holomorphic at 0 and hence f is meromorphic in a neighborhood of 0. □

4.6 K-normal Functions

In the paper [47] an important equivalent formulation of normal function was derived using ideas from invariant geometry.

Let Ω be a bounded domain in $\mathbb{C}^n$. Set

$$Q_f(z) := \sup_{v \in \mathbb{C}^n \setminus \{0\}} \frac{ds(f(z), f_*(z)v)}{F_\Omega^K(z, v)}$$

for all $z \in \Omega$.

Definition 4.6.1. A holomorphic function $f : \Omega \subset \widehat{\mathbb{C}}$ is *normal* if the derivative $Q_f(z)$ is bounded from the Kobayashi metric on Ω (equivalently, the Poincaré-Bergman metric on Ω) to the spherical metric on $\widehat{\mathbb{C}}$.

The equivalence of this definition with the previous one is a sophisticated exercise with Marty's theorem.

Theorem 4.6.2. *Let Ω be a bounded domain in $\mathbb{C}^n$. The following statements are equivalent for $f \in H(\Omega)$:*

(a) $\mathcal{F} = \{f \circ \varphi : \varphi \in H(D, \Omega)\}$ *is normal;*

(b) *there exists a constant $Q > 0$ such that*

$$Q_f(z) := \sup_{v \in \mathbb{C}^n \setminus \{0\}} \frac{ds(f(z), f_*(z)v)}{F_\Omega^K(z, v)} < Q \text{ for all } z \in \Omega. \tag{4.1}$$

Proof. $(a) \Rightarrow (b)$ Suppose $\mathcal{F} = \{f \circ \varphi : \varphi \in H(D, \Omega)\}$ is normal. Marty's characterization of normal families yields a non-negative constant $L > 0$ such that for any $\varphi \in H(D, \Omega)$

$$\frac{(f \circ \varphi)'(0)}{1 + |f \circ \varphi(0)|^2} < L. \tag{4.2}$$

Since $F_\Omega^K(z, v) = \inf r$ over all maps $\varphi \in H(D, \Omega)$ with the normalization $\varphi(0) = z$, $\varphi'(0)r = v$, where $r > 0$, there exists $\psi \in H(D, \Omega)$ such that $\psi(0) = z$, $\psi_*(0)a = v$ and $a/2 < F_\Omega^K(z, v) \leq a$. Therefore, from (4.2),

$$ds(f(z), f_*(\zeta)v) < 2L \cdot F_\Omega^K(\zeta, v) \text{ for all } (\zeta, v) \in \mathbb{C}^{n+1}.$$

Namely, $Q_f \leq 2L$.

$(b) \Rightarrow (a)$ Let μ and ζ be distinct points in Ω. Let $\gamma : [0, 1] \to \Omega$ be a piecewise C^1 curves joining μ to ζ in Ω. Then $f \circ \gamma : [0, 1] \to \mathbb{C}$ is a piecewise C^1 curves joining $f(\mu)$ to $f(\zeta)$ in $\mathbb{C}$ and

$$s(f(\mu), f(\zeta)) \leq \int_0^1 ds(f(\gamma(t)), f_*(\gamma(t))\gamma'(t))dt \leq$$
$$2L \cdot \int_0^1 F_\Omega^K(\gamma(t), \gamma_*(t)1)dt = Q \cdot L_\Omega(\gamma).$$

Hence

$$s(f(\mu), f(\zeta)) \leq Q \cdot L_\Omega(\gamma).$$

Taking the infimum over all piecewise C^1 curves γ satisfying $\gamma(0) = \mu$, $\gamma(1) = \zeta$, gives

$$s(f(\mu), f(\zeta)) \leq Q \cdot \widetilde{K}_\Omega(\mu, \zeta) \text{ for all } \mu, zeta \in \Omega.$$

Since K_Ω is the integrated form of the infinitesimal pseudometric F_Ω^K, i.e., $K_\Omega = \widetilde{K}_\Omega$ (see [7, Corollary 3]), from the previous inequality we obtain

$$s(f(\varphi(z)), f(\varphi(w))) \leq Q \cdot K_\Omega(\varphi(z), \varphi(w))$$

for all $\varphi \in H(D,\Omega)$ and all $z,w \in D$. By the distance decreasing property of the Kobayashi distance,

$$K_\Omega(\varphi(z),\varphi(w)) \le K_D(z,w)$$

for all $\varphi \in H(D,\Omega)$ and all $z,w \in D$. Therefore

$$K_\Omega(\varphi(z),\varphi(w)) \le Q \cdot K_D(z,w)$$

for all $\varphi \in H(D,\Omega)$ and all $z,w \in D$. Since $K_D = d_D$ it follows that the family $\mathcal{F} = \{f\circ\varphi : \varphi \in H(D,\Omega)\}$ is equicontinuous from the Poincaré metric to the spherical metric. By the Ascoli-Arzelà Theorem, the family $\mathcal{F} = \{f\circ\varphi : \varphi \in H(D,\Omega)\}$ is normal on D. □

Theorem 4.6.2 shows that Definition 4.6.1 is equivalent to the definition of K-normal functions given earlier, i.e. one implies the other.

Proposition 4.6.3. *Let $\mathcal{F}$ be the family of all holomorphic functions on Ω with values in $\mathbb{C}\setminus\{0,1\}$. Then, for all $\mu,\zeta \in \Omega$ and all $f \in \mathcal{F}$,*

$$\exp(-K_\Omega(\mu,\zeta)) \le \frac{c+\log|f(\mu)|}{c+\log|f(\zeta)|} \le \exp(K_\Omega(\mu,\zeta)) \tag{4.3}$$

where

$$c = \frac{\Gamma(\frac{1}{4})}{4\pi^2} = 4.3768796\ldots.$$

Furthermore, if there exists continuous $\log f$ *on* Ω *then*

$$\exp(-K_\Omega(\mu,\zeta)) \le \frac{c+|\log f(\mu)|}{c+|\log f(\zeta)|} \le \exp(K_\Omega(\mu,\zeta)) \tag{4.4}$$

In the proof of this proposition, we combine the result of Lai [154] with the definition of the Kobayashi metric and obtain a very elementary proof of Proposition 3 in [45].

Proof. The classical theorem of Landau may be stated in the form that if the function $f(z)$ is holomorphic in the unit disk D and does not take the values 0 and 1, then $f'(0)$ has a bound depending only on $f(0)$. In fact

$$|f'(0)| \le 2\,|f(0)|\,(c+|\log|f(0)||), \tag{4.5}$$

where

$$c = \frac{\Gamma(\frac{1}{4})}{4\pi^2} = 4.3768796\ldots$$

(see, for example, [154]).

For $z \in D$ define $\psi_z(w) = (w+z)/(1+\overline{z}w)$. Since $f\circ\psi_z$ does not take the values 0 and 1, from (4.5) we derive the following inequality

$$(1-|z|^2)\,|(f\circ\psi_z)'(0)| \le 2\,|(f\circ\psi_z)(0)|\,(c+|\log|f\circ\psi_z(0)||)\,. \tag{4.6}$$

Let μ, ζ be a pair of points in Ω. Since $K_\Omega(\mu,\zeta)$ is an inner pseudometric (see [6]), for each $\varepsilon > 0$ there exist an integer $\ell > 1$, $\phi_1, \ldots, \phi_\ell \in H(D,\Omega)$ and $a_1, \ldots, a_\ell \subset (0,1)$ satisfying $\phi_1(0) = \zeta$, $\phi_j(0) = \phi_{j+1}$ for $j = 1, \ldots, L-1$ and $\phi_\ell(0) = \mu$, and

$$\Sigma_{j=1}^{l}\rho(0,a_j) < K_\Omega(\mu,\zeta) + \varepsilon/2.$$

Set $g_j = f \circ \phi$. From inequality (4.6) we obtain

$$\frac{\partial}{\partial t}\left|\log(c + \log|g_j(t)|)\right| \leq \frac{2}{1-t^2} \tag{4.7}$$

If we integrate both sides from $t = 0$ to a_j, the result becomes

$$\log\left\{\frac{c + \log|g_j(a_j)|}{c + \log|g_j(0)|}\right\} \leq \rho(0,a_j).$$

Then

$$\log\left\{\frac{c + \log|f(\mu)|}{c + \log|f(\zeta)|}\right\} \leq K_\Omega(\mu,\zeta) + \varepsilon.$$

Letting $\varepsilon \to 0$, we finally get

$$\frac{c + \log|f(\mu)|}{c + \log|f(\zeta)|} \leq \exp(K_\Omega(\mu,\zeta))$$

so the second inequality in (4.3) is proved. Since μ and ζ play symmetric roles, it is evident that the first inequality in (4.3) also holds.

For obtaining inequalities (4.4), let us notice that there exists continuous $\log(\log g_j)$ on $t \in [0, a_l]$. Since

$$\left|\frac{\partial}{\partial t}\log\log g_j(t)\right| \geq \frac{\partial}{\partial t}\log|\log g_j(t)|$$

we have

$$\left|\frac{g_j'(t)}{g(t)}\right| = \left|\frac{\partial}{\partial t}\log g_j(t)\right| \geq \frac{\partial}{\partial t}|\log g_j(t)| = \frac{\partial}{\partial t}(c + |\log g_j(t)|)\,.$$

From this inequality and inequality (4.7), we obtain

$$\frac{\partial}{\partial t}(c + |\log g_j(t)|) \leq \frac{2}{1-t^2} \text{ for } t \in [0, a_j].$$

Integrating both sides of this inequality as above we obtain the inequality (4.4).

The proof of proposition is now complete. □

By Montel's theorem the family of all holomorphic functions on Ω with values in $\mathbb{C} \setminus \{0,1\}$ is necessary normal. In a similar fashion we may extend the proposition to obtain the following:

Lemma 4.6.4. *Let f be a K-normal function on a domain $\Omega \subset \mathbb{C}^n$. There exist a constant $c > 1$ such that*

$$\log(c\mu(f, x)) \leq \log(c\mu(f, y)) \exp(2K_\Omega(x, y)) \text{ for all } x, y \in \Omega.$$

Here $\mu(f, x) := \max\{1, |f(x)|\}$.

Proof. We observe that $\mathcal{F} \circ H(D, \Omega) \equiv \mathcal{F}$, i.e., $\mathcal{F}$ is an invariant normal family. Hayman [111, p. 165] has obtained stronger versions of the theorems of Schottky [262] and Landau [156]. He has shown that for an invariant normal family in $H(D)$ there is a constant $c > 0$ such that for all $g \in \mathcal{F}$,

$$|g'(0)| < 2\mu_0(\log \mu_0 + c),$$

where $\mu_0 = \max\{1, |g(0)|\}$. For $z \in D$ define ψ_z by $\psi_z(w) = (w + z)/(1 + \overline{z}w)$. Then, for each $g \in \mathcal{F}$, we have

$$(1 - |z|^2) |g'(z)| = |(g \circ \psi_z)'(0)| < 2\mu_z(\log \mu_z + c),$$

where $\mu_z = \max\{1, |g(z)|\}$.

Let $\mu, \zeta \in \Omega$, let $f \in \mathcal{F}$. For each $\varepsilon > 0$ there exist an integer $j > 1$, $\varphi_1, \varphi_2, \ldots, \varphi_j$ holomorphic maps of D into Ω and points $a_1, \ldots, a_j \subset (0, 1)$ satisfying $\varphi_1(0) = \zeta$, $\varphi_i(a_i) = \varphi_{i+1}(0)$ for $i = 1, \ldots, j - 1$ and $\phi_j(a_j) = x$, such that

$$\sum_i k_D(0, a_i) < K_\Omega(\mu, \zeta) + \varepsilon/2.$$

We may assume that $|f(\mu)| > 1$, that $g_i([0, a_i]) \subset \overline{D}$ or $g_i([0, a_i]) \subset \mathbb{C} \setminus D$ where $g_i = f \circ \varphi_i$. Let $I = \{i : g_i([0, a_i]) \subset \mathbb{C} \setminus D\}$. For each $i \in I$

$$\frac{|g_i'(z)|}{|g_i(z)| (\log |g_i(z)| + c)} \leq \frac{2}{1 - |z|^2} \text{ for all } z \in [0, a_i];$$

hence, after integrating above inequality along $[0, a_i]$, we have

$$\log\Big(\frac{\log |g_i(a_i)| + c}{\log |g_i(0)| + c}\Big) \leq 2k_\Omega(0, a_i) \quad (i \in I).$$

If $1 \in I$, then

$$\log\Big(\frac{\log |f(\mu)| + c}{\log |f(\zeta)| + c}\Big) \leq 2K_\Omega(\mu, \zeta) + \varepsilon.$$

If $1 \notin I$, let α the smallest element of I. Then $|g_\alpha(0)| = 1$ and

$$\log\Big(\frac{\log |f(\mu)| + c}{\log |g_\alpha(0)| + c}\Big) \leq 2K_\Omega(\mu, \zeta) + \varepsilon.$$

Letting $\varepsilon \to 0$, we finally get

$$\log\Big(\frac{\log |f(\mu)| + c}{\log |g_\alpha(0)| + c}\Big) \leq 2K_\Omega(\mu, \zeta).$$

The proof is easy to complete. □

Remark 4.6.5. Proposition 4.6.3 holds also for holomorfic functions defined on an infinite dimensional complex Banach manifold with values in $\mathbb{C}\setminus\{0,1\}$; the same proof works. An alternative but equally complicated proof of Proposition 4.6.3 can be found in [45, Proposition 3].

For the sake of convenience we shall now restate the lemma of Zalcman [291].

Lemma 4.6.6 (Zalcman Lemma). *Let $\mathcal{F}$ be a family of analytic functions in D. Then $\mathcal{F}$ is not normal in D if and only if there exist:*

(i) *a number r with $0 < r < 1$;*

(ii) *points z_n satisfying $|z_n| < r$;*

(iii) *functions $f_n \in \mathcal{F}$;*

(iv) *positive numbers $\rho_n \to 0$ as $n \to \infty$; such that*

$$f_n(z_n + \rho_n \xi) \to g(\xi) \text{ as } n \to \infty, \tag{4.8}$$

uniformly on compact subsets of $\mathbb{C}$, where g is a nonconstant entire function in $\mathbb{C}$. The function g may be taken to satisfy the normalization

$$g^{\sharp}(z) < g(0) = 1, \quad z \in \mathbb{C}.$$

As before, we desire a criterion for the non-normality of a holomorphic function. We are now ready to present our main result.

Theorem 4.6.7. *Let Ω be a bounded domain in $\mathbb{C}^n$. The following statements are equivalent for $f \in \mathcal{O}(\Omega)$:*

(a) *f is not a K-normal function;*

(b) *There exist sequences $\{\mu_m\}$, $\{\zeta_m\}$ in Ω, and a positive constant M such that $k_\Omega(\mu_m, \zeta_m) < M$ for all $m \geq 1$*

$$\lim_{m\to\infty} f(\mu_m) = \infty \text{ and } \lim_{m\to\infty} f(\zeta_m) = a \in \mathbb{C}.$$

Proof. $(a) \Rightarrow (b)$ Since f is not normal on Ω it follows that the family

$$\mathcal{F} = \{f \circ \varphi : \varphi \in \mathcal{O}(D, \Omega)\}$$

is not normal in D. Apply Lemma 4.6.6 to $\mathcal{F}$ with r, z_n, ρ_n, $f \circ \varphi_n$, and g as given therein. Since g is a nonconstant entire function in $\mathbb{C}$ it follows that there exists a sequence $\{\xi_m\} \subset \mathbb{C}$ such that $|g(\xi_m)| > m$. In fact, if this were not the case, g is bounded in $\mathbb{C}$. By Liouville's theorem $g \equiv$ constant.

For fixed ξ_m chose n_m large enough so that

(i) $|z_{n_m} + \rho_{n_m}\xi_m| < (1+r)/2$;

(ii) $|f \circ \varphi_{n_m}(z_{n_m} + \rho_{n_m}\xi_m)| > m/2$.

Put $w_{n_m} = z_{n_m} + \rho_{n_m}\xi_m$, $\mu_m = \varphi_{n_m}(w_{n_m})$, and $y_m = \varphi_{n_m}(z_{n_m})$.

By the distance decreasing property of the Kobayashi metric,

$$k_\Omega(\mu_m, \zeta_m) = k_\Omega(\varphi_{n_m}(w_{n_m}), \varphi_{n_m}(z_{n_m})) \leq d_D(w_{n_m}, z_{n_m}).$$

By the triangle inequality of the Poincaré distance,

$$d_D(w_{n_m}, z_{n_m}) \leq d_D(0, z_{n_m}) + d_D(0, w_{n_m}).$$

For any $z \in D$ the Poincaré distance $d_D(0, z)$ is equal to be $\log((1+|z|)/(1-|z|))$. Since $|z_{n_m}| < r$ and $|w_{n_m}| < (1+r)/2$, it follows that

$$d_D(w_{n_m}, z_{n_m}) \leq \log((1+r)/(1-r)) + \log((3+r)/(1-r)).$$

Putting the above together we get $k_\Omega(\mu_m, \zeta_m) \leq M$, where $M = \log((1+r)/(1-r)) + \log((3+r)/(1-r)) < \infty$. By *(ii)*, we have that

$$\lim_{m\to\infty} f(\mu_m) = \lim_{m\to\infty} f \circ \varphi_{n_m}(z_{n_m} + \rho_{n_m}\xi_m) = \infty.$$

By (4.8), we have that

$$\lim_{m\to\infty} f(\zeta_m) = \lim_{m\to\infty} f \circ \varphi_{n_m}(z_{n_m} + \rho_{n_m}0) = g(0) \in \mathbb{C}.$$

(b) $\Rightarrow$ *(a)* Assume, to get a contradiction, that f is a normal function on Ω. By Lemma 2.2,

$$\log(c\mu(1, f(\mu_m))) \leq \log(c\mu(1, f(\zeta_m))) \exp(2k_\Omega(\mu_m, \zeta_m)) \text{ for all } m \geq 1. \quad (4.9)$$

The left hand side of (4.9) tends to infinity as $m \to \infty$ while the right hand side is bounded since $\log(c\mu(1, f(\zeta_m)))$ is convergent and $\exp(2k_\Omega(\mu_m, \zeta_m))$ is bounded by $\exp(2M)$ and we have a contradiction. This contradiction proves that f is not a K-normal function. That concludes the proof. □

Lehto and Virtanen [14, p. 53] remark that the sum of a normal function and a bounded function (which is necessary normal) is a normal function. In the finite dimensional case we have:

Corollary 4.6.8. *Let Ω be a bounded domain in $\mathbb{C}^n$. If $f_1, \ldots, f_l$ are a finite number of K-normal holomorphic function on $\Omega \subset \mathbb{C}^n$ such that each sequence $\{\mu_n\}$ of points in Ω contains a subsequence $\{\mu_{n_m}\}$ on which at most one of f_j $(1 \leq j \leq l)$ is unbounded, then $h := \sum_{j=1}^{l} f_j$ is a K-normal function.*

Proof. Suppose, to the contrary, that h is not a K-normal function. By Theorem 4.6.7, we can find two sequences $\{\mu_n\}$ and $\{\zeta_n\}$ in Ω, and a positive constant M such that $k_\Omega(\mu_m, \zeta_m) < M$ for all $m \geq 1$, $\lim_{m\to\infty} h(\mu_m) = \infty$, and $\lim_{m\to\infty} h(\zeta_m) = a \in \mathbb{C}$. Since $\lim_{m\to\infty} h(\mu_m) = \infty$ then $\{\mu_m\}$ contains a subsequence again denoted by $\{\mu_m\}$ such that at most one of f_j, say f_1, is unbounded on $\{\mu_m\}$. Since $\lim_{m\to\infty} h(\zeta_m) = a \in \mathbb{C}$ then $\{\zeta_m\}$ contains a subsequence $\{\zeta_{m_k}\}$ such that either:

(i) at least two of f_j $(1 \leq j \leq l)$ is unbounded on $\{\zeta_{m_k}\}$;

(ii) or $\lim_{m\to\infty} h(\zeta_m) = a_j \in \mathbb{C}$, $(1 \leq j \leq l)$.

The case (i) is excluded by the assumption of the corollary. Hence

$$\lim_{k\to\infty} f_1(\mu_{m_k}) = \infty, \quad \lim_{k\to\infty} f_1(\zeta_{m_k}) = a_1 \in \mathbb{C},$$

and $k_\Omega(\mu_{m_k}, \zeta_{m_k}) < M$ for all $k \geq 1$. By Theorem 4.6.7 f_1 is not a K-normal function on Ω, a contradiction which proves the corollary. □

In [14] F. Bagemihl and W. Seidel posed the following question: Given a sequence $\{z_j\} \subset D$ converging to some $\zeta \in \partial D$ and a holomorphic mapping $f \in \mathcal{O}(D)$ such that $\lim_{j\to\infty} s(f(z_j), l) = 0$ for some $l \in \mathbb{C}$, under what conditions on f and $\{z_j\}$ can f have the limit l along some continuum in D which is asymptotic at ζ? Theorem 4.6.7 yields the following multidimensional analogue to one of their results.

Corollary 4.6.9. *Let Ω be a bounded domain in $\mathbb{C}^n$. Let $\{\mu_n\}$ and $\{\zeta_n\}$ be two sequences of points in Ω, and let M be a positive constant such that $k_\Omega(\mu_m, \zeta_m) < M$ for all $m \geq 1$. If $f \in \mathcal{O}(\Omega)$ is a K-normal function which omits $\ell \in \widehat{\mathbb{C}}$ in Ω but $\lim_{m\to\infty} s(f(\mu_m), \ell) = 0$ then $\lim_{m\to\infty} s(f(\zeta_m), l) = 0$.*

Proof. Assume first that $\ell \in \mathbb{C}$. Since

$$\frac{1 + |f(\mu) - l|^2}{1 + |f(\mu)|^2} \leq 2(1 + |\ell|^2)$$

it follows from the hypothesis and Theorem 4.6.7 that $g(z) = 1/(f(z) - l)$ is a K-normal holomorphic function on Ω. It is easy to see that $\lim_{m\to\infty} g(\mu_m) = \infty$. Therefore, we have $\lim_{m\to\infty} g(\zeta_m) = \infty$ by Theorem 4.6.7. Hence $\lim_{m\to\infty} s(f(y_m), \ell) = 0$ as desired. If $\ell = \infty$ then the corollary is an immediate consequence of Theorem 4.6.7. Thus the proof is complete. □

Corollary 4.6.10. *Let Ω be a bounded domain in $\mathbb{C}^n$. If $f \in H(\Omega)$ is a K-normal function and if h is not a K-normal holomorphic function on Ω such that each sequence $\{\mu_m\}$ of points in Ω contains a subsequence $\{\mu_{m_k}\}$ on which at most one of f or h is unbounded, then $f + h$ is not a K-normal function.*

Proof. Since h is not K-normal there exists $\{\mu_m\}$ and $\{\zeta_m\}$ such that $k_\Omega(\mu_m, \zeta_m) < M$, $h(\mu_m) \to \infty$ and $h(\zeta_m) \to a$. It follows that $f(\mu_m)$ is bounded and hence it contains a convergent subsequence. By passing to this subsequence we can assume $f(\mu_m) \to \alpha$. From Theorem 4.6.7, as f is K-normal, it follows that $f(\zeta_m) \not\to \infty$. By passing once more to this subsequence we may assume that $f(\zeta_m) \to \beta$. We deduce then that $(h + f)(\mu_m) \to \infty$ and $(h + f)(\zeta_m) \to \alpha + \beta$ and therefore, by Theorem 4.6.7, $f + h$ is not K-normal. Thus the proof is complete. □

The proof goes through word for word when Ω is a complex space, or a complex Banach manifold. See [65, 151].

4.7 P-point Sequences

In this chapter we gives a criterion for a holomorphic function, defined in a strictly pseudoconvex domain in $\mathbb{C}^n$, to be a normal function in terms of the distribution of values of holomorphic functions. The first result of the theory of the distribution of values of holomorphic functions goes back to 1868; in the master's dissertation of Yulian Vasil'evich Sokhotskii a theorem was proved which asserted that "at a pole of infinite order" a function always "must take on all possible values".

A clarification of some terminology is in order.

Definition 4.7.1. A strictly pseudoconvex domain in $\mathbb{C}^n$ with $\mathcal{C}^2$ boundary is a bounded domain Ω in $\mathbb{C}^n$ for which there exist a neighborhood U of $\partial\Omega$ and a real-valued function $\phi \in \mathcal{C}^2(U)$ such that

(a) $\Omega \cap U = \{z \in \Omega : \phi(z) < 0\}$;

(b) ϕ is strictly plurisubharmonic in U, i.e.

$$L_\xi(\phi, v) > 0 \text{ for all } v \in T_\xi^c(S) \setminus \{0\};$$

(c) $\nabla\phi(z) \neq 0$ in U.

Primarily, we will do no more than summarize some basic results on strictly pseudoconvex domains which we are needed in the sequel. All these results can be found in Krantz's book [143]; references to the indicated papers indicate that these ideas are formulated in these papers in precisely this form.

Lemma 4.7.2 ([192, Lemma 2.8]). *Let the strictly pseudoconvex domain $\Omega \subset \mathbb{C}^n$. For any point $\xi \in \partial\Omega$ there exist a neighborhood W such that the domain $U = \Omega \cap W$ is biholomorphically equivalent to a convex domain in $\mathbb{C}^n$; moreover, the closure $\overline{U}$ is biholomorphically equivalent to a closed convex set in $\mathbb{C}^n$.*

The next lemma is a standard consequence of Hopf's lemma [143].

Lemma 4.7.3 ([233, Lemma 3]). *Let Ω_1 and Ω_2 be strictly convex domains in $\mathbb{C}^n$ and $f : \Omega_1 \to \Omega_2$ Then there exist constants $\gamma_1, \gamma_2 > 0$ so that for any point $z \in \Omega_1$ the inequalities*

$$\gamma_1\delta(z, \partial\Omega) < \delta(f(z), \Omega_2) < \gamma_1\delta(z, \partial\Omega)$$

are valid.

We record the following localization result.

Proposition 4.7.4 ([85, Proposition 6]). *Let Ω be a strictly pseudoconvex domain in $\mathbb{C}^n$ with $\mathcal{C}^2$ boundary. Let V be any neighborhood of $\xi \in \partial\Omega$. Then for all vectors $v \in \mathbb{C}^n$*

$$\lim_{z \to \xi} \frac{F_{\Omega \cap V}(z, v)}{F_\Omega(z, v)} = 1.$$

The localization is uniform in the sense that: Suppose Ω is fixed and P is a neighborhood of the (variable) boundary point ξ of uniform size. Then, given $\varepsilon > 0$, there exists a neighborhood Q of ξ of uniform size, such that $F_{\Omega \cap P}(z, v)/F_\Omega(z, v) < 1 + \varepsilon$ for $z \in \Omega \cap Q$ and all $v \in \mathbb{C}^n$.

Lemma 4.7.5 ([234, Lemma 2.2, p. 71]). *Let $\Omega \subset \mathbb{C}^n$ be a strictly pseudoconvex domain, $\rho \in \mathcal{C}^2(D)$ $(\Omega \subset D)$ its defining function, and ξ^0 an arbitrary point of $\partial\Omega$. Then there are a neighborhood $U \ni \xi^0$ and a family of biholomorphic mappings $h^\xi : \mathbb{C}^n \to \mathbb{C}^n$, depending continuously on $\xi \in \partial\Omega \cap U$, satisfying the following conditions:*

1. $h^\xi(\xi) = 0$.

2. The defining function $\rho^\xi = \rho \circ (h^\xi)^{-1}$ of the domain $\Omega^\xi = h^\xi(\Omega)$ has the form

$$\rho^\xi(z) = 2Re(z_n + K^\xi(z)) + H^\xi(z) + \alpha^\xi(z)$$

in a neighborhood of the origin, where

$$\alpha^\xi(z) = o(|z|^2),$$

$$K^\xi(z) = \Sigma^n_{\mu,\nu=1} a_{\mu,\nu} z_\mu z_\nu \quad H^\xi(z) = \Sigma^n_{\mu,\nu=1} a_{\mu,\overline{\nu}} z_\mu \overline{z}_\nu$$

with $K^\xi('z, 0) \equiv 0$ and $H^\xi('z, 0) \equiv |'z|^2$ (here and in the sequel $'z = (z_1, \ldots, z_{n-1})$).

3. The mapping h^ξ takes the real normal n_ξ to $\partial\Omega$ at the point ξ into the real normal $\{'z = y_n = 0\}$ to $\partial\Omega^\xi$ at the origin.

For a strictly pseudoconvex domain in $\mathbb{C}^n$ (cf. [114, 235]) S.I. Pinčuk and G.M. Khenkin independently have obtained a much more refined estimate, which takes into account the position of the vector v with respect to the complex structure of the boundary of the domain.

The next result follows from ideas of Graham [85]. See also [5].

Theorem 4.7.6 (see [114, Lemma 1], [115]). *For each strictly pseudoconvex domain $\Omega \subset \mathbb{C}^n$, there exists positive constants k_1 and k_2 such that for all points $z \in \Omega$, sufficiently close to the boundary $S = \partial\Omega$, and all $v \in \mathbb{C}^n \setminus \{0\}$*

$$k_1 \leq C_\Omega(z, v) \left(\frac{|\pi_T(v)|}{\sqrt{\delta(z, \partial\Omega)}} + \frac{\pi_N(v)}{\delta(z, \partial\Omega)} \right)^{-1} \leq k_2,$$

where $\pi_T(v)$ and $\pi_N(v)$ are the projections of v onto the complex tangent plane and the complex normal to $\partial\Omega$ at the point $\xi = \xi(z)$ closest to z.

In the case of the ball such a different behavior of $C_\Omega(z,v)$ near the boundary in different directions is evident. There is a refinement of Theorem 4.7.6 due to Graham [85]. It can be shown that an analog of Theorem 4.7.6 also holds for the invariant Bergman metric.

For strictly pseudoconvex domains in any dimension, the Bergman, Caratheodory, and Kobayashi metrics are all comparable; such comparability follows from the results of I. Graham [84] and K. Diederich [52].

Let Ω be a bounded domain in $\mathbb{C}^n$. We shall use $c_\Omega(z,\zeta)$ to denote the Carathéodory distance between points $z,\zeta \in \Omega$. Denote by $B_c(z^0,\varepsilon) = \{z \in \Omega : c_\Omega(z^0,z) < \varepsilon\}$ the ball in the Carathéodory metric with center z^0 and radious ε.

It is known that for the ball $B = B(0,R) = \{z \in \mathbb{C}^n : |z| < R\}$ the Caratheodory metric coincides with the Bergman metric and has the form

$$ds^2 = (n+1)\left(\frac{|dz|^2}{R^2 - |z|^2} + \sum_{\mu,\nu=1} \frac{\overline{z}_\mu z_\nu dz_\mu d\overline{z}_\nu}{(R^2 - |z|^2)^2}\right).$$

Let $z^0 = (0,\ldots,0,z_n^0) \in B$. Then at the point z^0 we have

$$ds^2 = (n+1)\left(\frac{|dz|^2}{R^2 - |z_n^0|^2} + \frac{|z_n^0|^2\,|dz_n|^2}{\left(R^2 - |z_n^0|^2\right)^2}\right).$$

Hence it can be readily seen that constants $c_1(R), c_2(R) > 0$ exist so that, in suitable local coordinates,

$$\frac{c_1(R)}{\varepsilon^2}\left\{\sum_{j=1}^{n-1} \frac{|z_j|^2}{r} + \frac{|z_n - z_n^0|^2}{r^2} < 1\right\} \subset$$

$$B_c(z^0,\varepsilon) \subset \frac{c_2(R)}{\varepsilon^2}\left\{\sum_{j=1}^{n-1} \frac{|z_j|^2}{r} + \frac{|z_n - z_n^0|^2}{r^2} < 1\right\}, \quad (4.1)$$

where $r = \delta(z^0,\partial B) = R - |z_n^0|$, $B_c(z^0,\varepsilon) = \{z \in B : c_B(z^0,z) < \varepsilon\}$.

We now have now some grasp on the fundamental facts about the strictly pseudoconvex domains in $\mathbb{C}^n$. It is time to apply our knowledge to the study of normal functions of several complex variables.

The notion of P-point sequence was originally defined by V. I. Gavrilov [82] for meromorphic functions in the unit disk D. Gavrilov has shown that a holomorphic function $f(z)$ in the unit disk D is normal, in the sense of Lehto and Virtanen [164, p. 86], if and only if $f(z)$ does not possess a sequence of P-points.

We shall now state the conditions for $f(z) \in \mathcal{O}(\Omega)$ to be normal. For this purpose, we introduce

Definition 4.7.7. Let $f(z)$ be a holomorphic function in a domain $\Omega \subset \mathbb{C}^n$. A sequence of points $\{p_j\}$ of the domain Ω is called a *sequence of P-points*

for the function $f(z)$ if for each $\varepsilon > 0$ and each subsequence $\{p_k\} \subset \{p_j\}$ the function $f(z)$ assumes every value, with at most one exception, infinitely often in the union of balls $B_c(p_k, \varepsilon) = \{c_\Omega(p_k, z) < \varepsilon : z \in \Omega\}$.

For $f \in \mathcal{O}(\Omega)$, set

$$|\nabla f(z)| = \sup\left\{|f_*(v)| \cdot (F_\Omega^C(z,v))^{-1} : \quad v \in \mathbb{C}^n \setminus \{0\}\right\},$$

where $f_* = df$ is the differential of f at point $z \in \Omega$.

It is obvious that this quantity $|\nabla f(z)|$ is well defined for bounded domains in $\mathbb{C}^n$ and is invariant under biholomorphic maps of Ω.

We now state our main result.

Theorem 4.7.8. *(see [59]) A function $f(z)$ holomorphic in a bounded domain $\Omega \subset \mathbb{C}^n$ has a P-point sequence if and only if*

$$\overline{\lim}_{z \to \partial\Omega} |\nabla f(z)| \left(1 + |f(z)|^2\right)^{-1} = +\infty. \tag{4.2}$$

Proof. *Sufficiency.* Suppose that there is a subsequence $\{p^j\}$, $p^j \in \Omega$, $j = 1, 2, \ldots$, which converges to some point $\xi \in \partial\Omega$ such that

$$\lim_{j \to \infty} \left|\nabla f(p^j)\right| \left(1 + \left|f(p^j)\right|^2\right)^{-1} = +\infty, \tag{4.3}$$

but $\{p^j\}$ is not a P-point sequence of f. Then there exist $\varepsilon > 0$ and a subsequence $\{p^k\} \subset \{p^j\}$ such that $f(z)$ assumes every value, with at most two exceptions, in the union of the balls $\cup_{k=1}^{\infty} B_c(p^k, \varepsilon)$.

Using the fact that the boundary $\partial\Omega$ is a doubly smooth compact hypersurface, we have the number $r > 0$ possessing the following property: the inclusion

$$B_\zeta(r) \subset \Omega,$$

where $B_\zeta(r)$ is a Euclidean ball of radius r, whose boundary contains the point $\zeta \in \partial\Omega$, is valid for any points $\zeta \in \partial\Omega$. Let

$$\Omega_r = \{z \in \Omega : \delta(z, \partial\Omega) < r\}.$$

For any point $z \in \Omega_r$ a ball $B_z' = B(z'(z), r)$ will be found contained in Ω_r and such a one that $\delta(z, \partial\Omega) = \delta(z, \partial B_z)$.

By Lemma 4.7.2, for any point $\zeta \in \partial\Omega$, there exist a neighborhood W such that the domain $U = \Omega \cap W$ is biholomorphically equivalent to a convex domain in $\mathbb{C}^n$; by Lemma 4.7.4 F_U^C is equivalent to F_Ω^C. Since our considerations are in neighborhood of ξ, without loss of generality we assume that the boundary of domain Ω in a neighborhood of ξ is convex. It follows that for any point $\zeta \in \partial\Omega$ sufficiently close to ξ a ball $B(\zeta'', R)$ exists with a center at some point $\zeta'' \in \mathbb{C}^n$, so that

$$\Omega \subset B(\zeta'', R),$$

$\zeta \in \partial B(\zeta'', R)$. From this condition it follows that regardless of the nature of the point $w \in \Omega$ a ball $B''_w = B(\zeta''(w), R)$ will be also found containing the domain Ω and a ball such that $\delta(w, \partial\Omega) = \delta(w, \partial B''_w)$.

Denote by $\xi(z)$ the point on $\partial\Omega$ such that

$$|z - \xi(z)| = \inf \{\omega \in \partial\Omega : |z - \omega|\} = \delta(z, \partial\Omega).$$

Let us take any point $z^0 \in \Omega_r$. After a possible translation and unitary linear transformation h_{z^0} in the ambient space $\mathbb{C}^n$, we may assume that the point $\xi(z^0)$ is the origin, that the outward unit normal at $\xi(z^0)$ is in the negative y_n direction, (here $z_n = x_n + iy_n$) and thus $N^c_\xi(z^0) = \{('0, z_n)\}$ (here and later $'z = (z_1, \dots, z_{n-1})$), while $T^c_\xi(z^0) = \{('z, 0)\}$. It is easy to see that $h_{z^0}(z^0) = ('0, -\delta(z_0, \partial\Omega))$. Then

$$B'_0 = \left\{ z : \sum_{\mu=1}^{n-1} |z_\mu|^2 + |z_n - z'_n|^2 < r^2 \right\} \subset \Omega \subset$$

$$B''_0 = \left\{ z : \sum_{\mu=1}^{n-1} |z_\mu|^2 + |z_n - z''_n|^2 < R^2 \right\}$$

and $\delta(z^0, \partial\Omega) = \delta(z^0, \partial B'_0) = \delta(z^0, \partial B''_0)$.

Let

$$b_{B'_0}(\varepsilon) = \left\{ z \in B'_0 : c_{B'_0}(z^0, z) < \varepsilon \right\}, \quad b_{B''_0}(\varepsilon) = \left\{ z \in B''_0 : c_{B''_0}(z^0, z) < \varepsilon \right\}.$$

Then, by virtue of (4.1), for adequately small $\varepsilon > 0$, we have

$$G_1 = \left\{ z : \frac{c_1(r)}{\varepsilon^2} \left(\sum_{j=1}^{n-1} \frac{|z_j|^2}{r} + \frac{|z_n + \delta(z^0)|^2}{r^2} < 1 \right) \right\} \subset b_{B'_0}(\varepsilon) \subset b_\Omega(\varepsilon)$$

$$\subset b_{B''_0}(\varepsilon) \subset \left\{ z : \frac{c_2(R)}{\varepsilon^2} \left(\sum_{j=1}^{n-1} \frac{|z_j|^2}{r} + \frac{|z_n + \delta(z^0)|^2}{r^2} < 1 \right) \right\} = G_2,$$

where $r = \delta(z^0, \partial\Omega)$.

Denote by $h_j = h_{p^j}$ and $\delta_j = \delta(p^j, \partial\Omega)$. The idea is to set

$$g_j(w) = f \circ h_j^{-1}(\sqrt{\delta_j}\,'w, \delta_j w_n), \quad j = 1, 2, \dots$$

All function g_j, beginning with some m_0, are defined and holomorphic in the ball

$$B_\varepsilon('0, -1) = \left\{ w \in \mathbb{C}^n : |'w|^2 + |w_n + 1|^2 < \varepsilon^2 / c_1(r) \right\}.$$

According to our initial assumption this means that for each $m \in \mathbf{N}$ and each $\varepsilon > 0$ the family $G = \{g_k\}_{k \geq m}$ omits two complex numbers on $B_\varepsilon('0, -1)$, say 0 and 1. By Montel's criterion G is normal in $B_\varepsilon('0, -1)$. Since the singleton

$\{('0,-1)\}$ is a compact subset in $B_\varepsilon('0,-1)$, by Marty's normality criterion there is a positive constant K such that

$$\sum_{\mu,\nu=1}^{n} \frac{\partial^2 \log(1+|g_k|^2)}{\partial w_\mu \partial \overline{w}_\nu}\Big|_{('0,-1)} v_\mu \overline{v}_\nu \le K\,|v|^2 \tag{4.4}$$

for all $k \ge m_0$, $v \in \mathbb{C}^n$. Since this holds for arbitrary v in $\mathbb{C}^n$, we may replace v by $('v/\delta_k, v_n/\delta_k^2)$ and rewrite (4.4) as

$$\sum_{\mu,\nu=1}^{n} \frac{\partial^2 \log(1+|f\circ h_k^{-1}|^2)}{\partial w_\mu \partial \overline{w}_\nu}\Big|_{('0,-\delta_k)} v_\mu \overline{v}_\nu \le K \left\{ \frac{|'v|^2}{\delta_k} + \frac{|v_n|^2}{\delta_k^2} \right\}$$

for all $k \ge m_0$, $v \in \mathbb{C}^n$. Using this inequality and Theorem 4.7.6, we find a positive constant K_1 such that

$$\sum_{\mu,\nu=1}^{n} \frac{\partial^2 \log(1+|f\circ h_k^{-1}|^2)}{\partial w_\mu \partial \overline{w}_\nu}\Big|_{('0,-\delta_k)} v_\mu \overline{v}_\nu \le K_1 \Big(F^C_{h_k^{-1}(\Omega)}(('0,-\delta_k),v) \Big)^2$$

for all $k \ge m_0$, $v \in \mathbb{C}^n$. Since the infinitesimal Caratheodory metric is invariant under biholomorphic mappings, we have

$$\sum_{\mu,\nu=1}^{n} \frac{\partial^2 \log(1+|f|^2)}{\partial w_\mu \partial \overline{w}_\nu}\Big|_{p^k} v_\mu \overline{v}_\nu \le K_1 \Big(F^c_\Omega(p^k,v) \Big)^2 \tag{4.5}$$

for all $k \ge m_0$, $v \in \mathbb{C}^n$. It follows readily from (4.5) that

$$|f_*(v)|\,(F^C_\Omega\left(p^k,v)\right)^{-1}\left(1+|f(p^k)|^2\right)^{-1} \le K_1$$

for all $k \ge m_0$ and all $v \in \mathbb{C}^n \setminus \{0\}$. Taking the supremum over all $v \in \mathbb{C}^n \setminus \{0\}$ we obtain

$$|\nabla f(p^k)| \left(1+|f(p^k)|^2\right)^{-1} \le K_1$$

for all $k \ge m_0$. This is a clear contradiction to (4.3).

Necessity. Let $\{z^k\}$ be a P-point sequence of f, then the family $\{g^k\}$ assumes every value, with at most one exception, infinitely often in any ball $B_\varepsilon('0,-1)$, $\varepsilon > 0$. It follows from the Hurwitz Theorem 2.0.14 and the definition of a normal family of functions that $\{g^k\}$ cannot be normal on $B_\varepsilon('0,-1)$. It follows from Marty's normality criterion 3.2.5 that there exist two sequences $\{\widetilde{w}^k\}$, $\widetilde{w}^k \in B_\varepsilon('0,-1)$, $k = 1,2,\ldots$, and $\{\widetilde{v}^k\}$, $\widetilde{v}^k \in \mathbb{C}^n$, $k = 1,2,\ldots$, such that

$$\sum_{\mu,\nu=1}^{n} \frac{\partial^2 \log(1+|g_k|^2)}{\partial \widetilde{w}_\mu \partial \overline{\widetilde{w}}_\nu}\Big|_{\widetilde{w}^k} \widetilde{v}^k_\mu \overline{\widetilde{v}}^k_\nu \ge k\,|\widetilde{v}^k|^2$$

for all $k \ge 1$. It follows that

$$\sum_{\mu,\nu=1}^{n} \frac{\partial^2 \log(1+|f\circ h_k^{-1}|^2)}{\partial w_\mu \partial \overline{w}_\nu}\Big|_{w^k} v_\mu \overline{v}_\nu \ge k \left\{ \frac{|'v|^2}{\delta_k} + \frac{|v_n|^2}{\delta_k^2} \right\}$$

where $w_k = (\sqrt{\delta_k}\,'\widetilde{w}, \delta_k\,'\widetilde{w}_n^k)$, $v_k = (\sqrt{\delta_k}\,'\widetilde{v}, \delta_k\,'\widetilde{v}_n^k)$. By Theorem 4.7.6,

$$\sum_{\mu,\nu=1}^{n} \frac{\partial^2 \log(1+|f|^2)}{\partial w_\mu \partial \overline{w}_\nu}\Big|_{p^k} v_\mu \overline{v}_\nu \geq \frac{k}{k_1}\left(F_\Omega^c(p^k, v)\right)^2.$$

Here $p^k = h_k^{-1}(w^k)$, $v^k = (h_k^{-1})_*(v^k)$, where $(h_k^{-1})_* = dh_k^{-1}$ is the differential of h_k^{-1} at point $w^k \in \Omega$. From the above inequality we have

$$\lim_{j\to\infty} |\nabla f(p^j)| \left(1 + |f(p^j)|^2\right)^{-1} = +\infty,$$

which completes the proof □

This theorem was motivated by a theorem of V.I. Gavrilov [82] about P-point sequences.

The argument used here goes back to an idea which can be found in [234]. It follows from the definition of a normal function that Theorem 4.7.8 gives a criterion for a holomorphic function defined in a strictly pseudoconvex domain to be a normal function.

We deal with a proven theorem in which one gets sufficient conditions for two limiting values of a function which is bounded and holomorphic in the unit disk D, corresponding to two different sequences of points of D, to be equal [267]. In this paper we prove a generalization of this theorem of Seidel to the case of bounded holomorphic functions in strictly pseudoconvex domains of the space $\mathbb{C}^n$.

We shall prove the following theorem:

Theorem 4.7.9. *Let Ω be a strictly pseudoconvex domain in $\mathbb{C}^n$, $(n \geq 1)$, $A \in \mathbb{C}$ and let the function $f \in H^\infty(\Omega)$ which omits the value A in domain Ω. Let us assume that $\{a_m\}$ and $\{b_m\}$ are two sequences of points of Ω, such that*

$$\lim_{m\to\infty} a_m = \xi \in \partial\Omega \text{ and } \lim_{m\to\infty} b_m = \xi.$$

If the Carathéodory distance $c_\Omega(a_m, b_m)$ is less than a constant ε, independent of m

$$c_\Omega(a_m, b_m) < \varepsilon, \quad m = 1, 2, \ldots,$$

and

$$\lim_{m\to\infty} f(a_m) = A,$$

then also

$$\lim_{m\to\infty} f(b_m) = A.$$

If Ω is the unit disk D, then, as is known, the distance in the Carathéodory metric c_D coincides with the distance in the metric of the Poincaré disk D, so if $n = 1$ our theorem is precisely Seidel's theorem of [267, Theorem 3, p. 10].

The proof of our theorem is given for the case $n > 1$; the only difference between the one-dimensional case and the multidimensional one is the richer geometry of balls with respect to the standard invariant Caratheodory metric of multidimensional domains. It will be clear from the proof that the geometric conditions $c_\Omega(a_m, b_m) < \varepsilon$, $m = 1, 2, \ldots$, mean that for all sufficiently large numbers m, the points b_m lie inside ellipsoids with centers at the points a_m, whose semiaxes in the normal and tangential directions are equal to

$$-k_1 \varepsilon \rho(a_m) \quad \text{and} \quad k_1 \varepsilon \sqrt{-\rho(a_m)}$$

respectively.

Here and everywhere below k_ℓ, $\ell = 1, \ldots$, will denote certain positive constants.

Proof of Theorem 4.7.9. Let $g(z) = f(z) - A$. We can without loss of generality assume that the upper bound M of $|g(z)|$ in Ω is 1. It follows from the definition of the infinitesimal form of the Carathéodory metric that, for all $z \in \Omega$, $v \in \mathbb{C}^n$,

$$|dg(v)| \leq F_\Omega^C(z, v). \tag{4.6}$$

From this, using the definition of the Levi form, we get

$$L_z(\log(1 + |g|^2), v) \leq (F_\Omega^C(z, v))^2 .$$

The function $g(z) \neq 0$ and is holomorphic in the domain Ω, so the function $\log |g(z)|^2$ is pluriharmonic there—since the Levi form vanishes for any pluriharmonic function,

We now calculate that

$$\begin{aligned} L_z(\log(1 + |1/g|^2), v) = \\ L_z(\log(1 + |g|^2), v) - 2L_z(\log |g|, v) \equiv \\ L_z(\log(1 + |g|^2), v). \end{aligned}$$

From this identity and (4.6), we get that the function $h(z) = 1/g(z)$ satisfies

$$L_z(\log(1 + |h|^2), v) \leq (F_\Omega^C(z, v))^2 \tag{4.7}$$

for all $z \in \Omega$, $v \in \mathbb{C}^n$. Without loss of generality we can assume that the defining function ρ of the domain Ω is such that

$$\frac{\partial \rho(\xi)}{\partial z_\mu} = 0, \quad \mu = 1, 2, \ldots, n-1; \quad \frac{\partial \rho(\xi)}{\partial z_n} = 1. \tag{4.8}$$

For each point $\zeta \in \partial\Omega$, belonging to some neighborhood U of the point ξ, we make the linear transformation:

$$h^\zeta = \begin{cases} w_\mu = \frac{\partial \rho(\xi)}{\partial z_n}(z_\mu - \zeta_\mu) - \frac{\partial \rho(\xi)}{\partial z_\mu}(z_n - \zeta_n), & \mu = 1, 2, \ldots, n-1; \\ w_n = \sum_{\mu=1}^n \frac{\partial \rho(\xi)}{\partial z_n}(z_\mu - \zeta_\mu). \end{cases}$$

If the neighborhood U of the point ξ is sufficiently small then, in view of (4.8), the transformation h^ζ is nondegenerate. In coordinates w the complex tangent plane $T^c_\zeta(\partial\Omega)$ to $\partial\Omega$ at the point ζ has the form $\{w_n = 0\}$, and the complex normal $N^c_\zeta(\partial\Omega)$ to $\partial\Omega$ at the point ζ coincides with the set $\{w \in \mathbb{C}^n : {}'w = 0\}$.

By the hypothesis of the theorem, the domain Ω is strictly pseudoconvex so by Theorem 4.7.6 for all $z \in \partial\Omega$, sufficiently close to $\partial\Omega$,

$$F^C_\Omega(z, v) \le k_2\left(\frac{|\pi_T(v)|}{\sqrt{\delta(z, \partial\Omega)}} + \frac{\pi_N(v)}{\delta(z, \partial\Omega)}\right), \tag{4.9}$$

where $\pi_T(v)$ and $\pi_N(v)$ are the projections of v onto the complex tangent plane and the complex normal to $\partial\Omega$ at the boundary point $\xi = \xi(z)$ closest to z.

We let $\xi_m = \xi(a_m)$, $\delta_m = \delta(a_m)$, and $h^m = h^{\xi_m}$. Under the map h^m the ball in the Carathéodory metric $B^C_\varepsilon(a_m) = \{z \in \Omega, c_\Omega(a_m, z) < \varepsilon\}$ is mapped into a set contained in the ellipsoid

$$E(k_3\varepsilon\sqrt{\delta_m}, k_3\varepsilon\delta_m) = \left\{\frac{|'w|^2}{\sqrt{\delta_m}} + \frac{|w_n - \delta_m|^2}{\delta_m^2} \le (k_3\varepsilon)^2\right\} \subset \Omega^m,$$

where $\Omega^m = h^m(\Omega)$.

The maps h^ζ, $\zeta \in U$, are nondegenerate and depend continuously on the parameter ζ, so it follows from (4.9) that for all sufficiently large numbers m and all w sufficiently close to $\partial\Omega^m$,

$$F^C_{\Omega^m}(w, h^m_* v) \le k_4\left(\frac{|'h^m_* v|}{\sqrt{\delta(w)}} + \frac{(h^m_* v)_n}{\delta(w)}\right), \tag{4.10}$$

where $h^m_* = dh^m$ is the differential of the map.

Making use of the invariance of the Levi form under holomorphic maps, and the invariance of the Carathéodory metric with respect to biholomorphic maps, from (4.7) and (4.10) we get that for all sufficiently large numbers m,

$$L_w(\log(1 + \left|g \circ (h^m)^{-1}\right|^2), v) \le k_5\left(\frac{|'v|}{\sqrt{\delta(w)}} + \frac{(v)_n}{\delta(w)}\right).$$

Under the map

$$'\widetilde{w} \to \frac{'w}{\sqrt{\delta_m}}, \quad \widetilde{w}_n \to \frac{w_n}{\sqrt{\delta_m}}$$

the preceding inequality is transformed to the inequality

$$L_{\widetilde{w}}(\log(1 + |F_m|^2), v) \le k_6 |v|^2, \tag{4.11}$$

where $v \in \mathbb{C}^n$ and

$$F_m(\widetilde{w}) = g \circ (h^m)^{-1}(\sqrt{\delta_m}\,'\widetilde{w}, \delta_m\widetilde{w}_n),$$

and the ellipsoid $E(k_3\varepsilon\sqrt{\delta_m}, k_3\varepsilon\delta_m)$ goes into the ball

$$B_{k_7\varepsilon}('0,1) = \left\{\widetilde{w} \in \mathbb{C}^n : |'\widetilde{w}|^2 + |\widetilde{w}_n - 1|^2 < (k_7\varepsilon)^2\right\}.$$

From (4.11) and the Marty normality criterion we conclude that the family of functions $\{F_m\}$ is normal in the ball $B_{k_7\varepsilon}('0,1)$. Since $F_m('0,1) = (f(a_m) - A) \to \infty$ as $m \to \infty$, the functions F_m converge uniformly on compact subsets of $B_{k_7\varepsilon}('0,1)$ to infinity identically, so it is easy to see that $\lim_{m\to\infty} f(b_m) = A$. □

4.8 Lohwater-Pommerenke's Theorem in $\mathbb{C}^n$

In the one-dimensional case there are many criteria known for a meromorphic function to be normal, and Lohwater and Pommerenke [176] add a very valuable criterion to this list: a nonconstant function f meromorphic in the unit disk $D \subset \mathbb{C}$ is normal if and only if there do not exist sequences $\{z_n\}$ and $\{\rho_n\}$ with $z_n \in D$, $\rho_n > 0$, $\rho_n \to 0$, such that

$$\lim_{n\to\infty} f(z_n + \rho_n t) = g(t)$$

locally uniformly in $\mathbb{C}$, where $g(t)$ is a nonconstant meromorphic function in $\mathbb{C}$.

The following criterion is essentially a reformulation of (4.10).

Theorem 4.8.1. *Let Ω be a bounded domain in $\mathbb{C}^n$. Let F be either the infinitesimal Carathéodory or Kobayashi metrics on Ω. If $f : \Omega \to \mathbb{C}$ is a F-normal holomorphic function, then for every choice of sequence $\{p_j\}$ in Ω and $\{r_j\}$, $r_j > 0$, with $\lim_{j\to\infty} r_j/\delta_j = 0$, where δ_j is the Euclidean distance function from p_j to $\partial\Omega$, the sequence $\{f(p_j + r_j\zeta)\}$ converges locally uniformly to a constant function in $\mathbb{C}^n$.*

Proof. Set $R_j = \delta_j/r_j$. It is clear that $R_j \to \infty$. Without restriction we can assume that $R_j > j$. Let $\zeta \in \mathbb{C}^n$. Then, for all $|\zeta| < j$, we have

$$|(p_j + r_j\zeta) - p_j| = r_j|\zeta| < \delta_j$$

so that $p_j + r_j\zeta \in B(p_j, \delta_j) \subset \Omega$. Hence $g_j(\zeta) := f(p_j + r_j\zeta)$ is a holomorphic function on the ball $B(0,j) = \{\zeta \in \mathbb{C}^n : |\zeta| < j\}$.

Suppose that f is normal in Ω. By definition, "f is normal" means that there is a positive constant C such that

$$L_z(\log(1 + |f|^2), v) \leq C \cdot (F_\Omega(z,v))^2$$

for all $(z,v) \in \Omega \times \mathbb{C}^n$.

By the distance-decreasing property, we have

$$F_\Omega(z,v) \le F_{B(p_j,\delta_j)}(z,v)$$

for all $(z,v) \in B(p_j,\delta_j) \times \mathbb{C}^n$.

Recall that

$$F_{B(p_j,\delta_j)}(z,v) = \frac{[(\delta_j^2 - |z-p_j|^2)|v|^2 + |(z-p_j,v)|^2]^{1/2}}{\delta_j^2 - |z-p_j|^2}.$$

As a result, we may use Schwarz's inequality to obtain that

$$F_{B(p_j,\delta_j)}(z,v) \le \frac{\delta_j |v|}{\delta_j^2 - |z-p_j|^2}.$$

From the above two inequalities, we obtain the inequality

$$F_\Omega(z,v) \le \frac{\delta_j |v|}{\delta_j^2 - |z-p_j|^2}$$

for all $(z,v) \in B(p_j,\delta_j) \times \mathbb{C}^n$.

Therefore

$$\sqrt{L_{p_j+r_j\zeta}(\log(1+|f_j|^2),v)} \le \frac{\sqrt{C}\delta_j |v|}{\delta_j^2 - |r_j\zeta|^2}$$

for all $(\zeta,v) \in B(0,j) \times \mathbb{C}^n$. Taking the supremum on both sides over $|v|=1$, we have

$$f^\sharp(p_j + r_j\zeta) \le \frac{\sqrt{C}\delta_j}{\delta_j^2 - |r_j\zeta|^2} \tag{4.1}$$

See Chapter 3 equality (3.2) for the definition of $f^\sharp$.

By the invariance of the Levi form under biholomorphic mappings, we have

$$L_\zeta(\log(1+|g_j|^2),v) = L_{p_j+r_j\zeta}(\log(1+|f_j|^2), r_j v).$$

Hence, if in equality (3.2) of Chapter 3 we replace f by g_j, we obtain

$$g_j^\sharp(\zeta) = r_j f^\sharp(p_j + r_j\zeta). \tag{4.2}$$

By (4.1) and (4.2), we have

$$g_j^\sharp(\zeta) \le \frac{\sqrt{C} r_j \delta_j}{\delta_j^2 - |r_j\zeta|^2} \le \frac{\sqrt{C}/j}{1-(1/j)^2|\zeta|^2}$$

for all j sufficiently large and all ζ, $|\zeta| < j$.

For every $m \in \mathbb{N}$ the sequence $\{g_j\}_{j>m}$ is normal in $|\zeta| < m$ by Marty's criterion (Theorem [68, Theorem 2.1]). The Cantor diagonalization process now yields a subsequence $\{g_k = g_{j_k}\}$ which converges uniformly on every ball $|\zeta| < R$. The limit function g is holomorphic and satisfies $g^\sharp(\zeta) = 0$ which yields: $dg(\zeta) = 0$ for all $\zeta \in \mathbb{C}^n$, i.e. $g(\zeta) \equiv$ constant in $\mathbb{C}^n$. □

Remark 4.8.2. In a sense, the wheel has come full circle. Theorem 4.8.1 shows that the concept of normal holomorphic function of several complex variables (see Definition 4.2.7) is equivalent to the privious one (see Definition 4.2.1) and allows one to attribute to one function the properties related to some class of functions, namely with a normal family of holomorphic functions defined on the unit ball in $\mathbb{C}^n$.

Theorem 4.8.1 may be rephrased as follows:

Corollary 4.8.3. *Let Ω be a bounded domain in $\mathbb{C}^n$. Let F be either the infinitesimal Carathéodory or Kobayashi metric on Ω. If $f : \Omega \to \mathbb{C}$ is a holomorphic function, and there exist sequences $\{p_j\}$ in Ω and $\{r_j\}$, $r_j > 0$, with $\lim_{j\to\infty} r_j/\delta_j = 0$, where δ_j is the Euclidean distance function from p_j to $\partial\Omega$, such that $\{f(p_j + r_j\zeta)\}$ converges locally uniformly to a nonconstant holomorphic function in $\mathbb{C}^n$, then f is not F-normal.*

The proofs of Theorem 4.8.1 and Corollary 4.8.3 given in [101] follow closely the method of [176]. Unfortunately several inaccuracies occur in [101], e.g., in [101, Theorem 2] concerning the selection of convergent subsequences of normal families.

Theorem 4.8.4. *Let Ω be a bounded domain in $\mathbb{C}^n$. Let F be either the infinitesimal Carathéodory or Kobayashi metrics on Ω. A nonconstant function f holomorphic on Ω is not F-normal if there exist sequences $z_j \in \Omega$, $\rho_j = 1/f^\sharp(z_j) \to 0$, such that the sequence*

$$g_j(\zeta) = f(z_j + \rho_j\zeta)$$

converges locally uniformly in $\mathbb{C}^n$ to a nonconstant entire function g satisfying $g^\sharp(\zeta) \le g^\sharp(0) = 1$.

Proof. Definition 4.2.7 imply that there is a positive constant C such that

$$L_z(\log(1 + |f|^2), v) \le C \cdot (F_\Omega(z, v))^2$$

for all $(z, v) \in \Omega \times \mathbb{C}^n$.

Let $\{p_j\}$ be an arbitrary sequence of points in Ω. Then $B(p_j, \delta_j) \subset \Omega$. By the distance-decreasing property,

$$F_\Omega(z, v) \le F_{B(p_j,\delta_j)}(z, v)$$

for all $(z, v) \in B(p_j, \delta_j) \times \mathbb{C}^n$. As in the proof of Theorem 4.8.1 this yields

$$f^\sharp(p_j + \delta_j\zeta) \le \frac{\sqrt{C}\delta_j}{\delta_j^2 - |\delta_j\zeta|^2}. \tag{4.3}$$

Set $g_j(\zeta) := f(p_j + \delta_j\zeta)$. By the invariance of the Levi form under biholomorphic mappings, we have

$$L_\zeta(\log(1 + |g_j|^2), v) = L_{p_j+\delta_j\zeta}(\log(1 + |f_j|^2), \delta_j v)$$

and so

$$g_j^\sharp(\zeta) = \delta_j f^\sharp(p_j + \delta_j \zeta).$$

It follows from (4.3) that

$$g_j^\sharp(\zeta) \leq \frac{\sqrt{C}}{1 - |\zeta|^2} \tag{4.4}$$

for all j and all $\zeta \in B(0,1)$. By Marty's criterion (Theorem [68, Theorem 2.1]), the family $\{g_j(\zeta)\}$ is normal in the unit ball $B(0,1)$.

So if f is not an F-normal function in Ω, then there exists a sequence $\{p_j\}$ in Ω such that $\{g_j(\zeta) := f(p_j + \delta_j\zeta)\}$ is not a normal sequence at a point, say, ζ_0, $\zeta_0 \in B(0,1)$. It follows from Zalcman's lemma [68, Theorem 3.1] that there exist $\zeta_j \to \zeta_0$, $\rho_j = 1/g_j^\sharp(\zeta_j) \to 0$, such that the sequence

$$g_j(\zeta) = f(p_j + \delta_j(\zeta_j + \rho_j \zeta))$$

converges locally uniformly in $\mathbb{C}^n$ to a nonconstant entire function g satisfying $g^\sharp(\zeta) \leq g^\sharp(0) = 1$. A simple calculation shows that $\delta_j \rho_j = 1/f^\sharp(p_j + \delta_j \zeta_j)$ and therefore

$$g_j(\zeta) = f(p_j + \delta_j \zeta_j + \zeta / f^\sharp(p_j + \delta_j \zeta_j))$$

converges locally uniformly in $\mathbb{C}^n$ to a nonconstant entire function g satisfying $g^\sharp(\zeta) \leq g^\sharp(0) = 1$. It follows that $z_j = p_j + \delta_j \zeta_j$, $\rho_j = 1/f^\sharp(p_j + \delta_j \zeta_j)$ satisfy our conditions. This completes the proof of Theorem 4.8.4. □

Remark 4.8.5. Lohwater and Pommerenke [176, Theorem 1] originally stated their theorem with no restriction on the speed at which $\rho_n \to 0$. In proving their theorem they assert, "if f is normal and $f(z_n + \rho_n \zeta) \to g(\zeta)$ locally uniformly, then $\rho_n/(1 - |z_n|) \to 0$". The statement in quotes is false as one can see from the example $f(z) = z$, $z_n = 1 - n^{-3}$, $\rho_n = n^{-2}$, $g(\zeta) \equiv 1$.

In general we can not even assert that $\lim_{j\to\infty} f^\sharp(z_j)\delta_j$ exists. This is the reason that the true generalization of Lohwater-Pommerenke's theorem to the higher-dimensional setting most likely breaks down. See also [101] and [33, Remark 6.1, p. 68].

4.9 The Scaling Method

One of the most useful and most profound techniques in the study of automorphism groups of domains is the scaling method (also called the flattening method by some geometers). This is a rather profound application of normal families, and we take the opportunity to describe it here.

This description will be rather informal, and not detailed. But it will give the reader an idea of how the method works. And we shall provide references for further reading. We particularly recommend [93].

Let D be the open unit disk in the complex plane $\mathbb{C}$. Choose a sequence a_j in D satisfying the conditions

$$0 < a_j < a_{j+1} < 1, \ \forall j = 1, 2, \ldots ,$$

and

$$\lim_{j \to \infty} a_j = 1.$$

Define the sequence of dilations

$$L_j(z) = \frac{1}{1 - a_j}(z - 1) .$$

We write $\lambda_j = 1 - a_j$. Then

$$\begin{aligned} L_j(D) &= \{\zeta \in \mathbb{C} : (1 + \lambda_j \zeta)(1 + \lambda_j \bar{\zeta}) < 1\} \\ &= \{\zeta \in \mathbb{C} : 2 \operatorname{Re}\zeta < -\lambda_j |\zeta|^2\}. \end{aligned}$$

Hence the sequence of sets $L_j(D)$ *converges* to the left half-plane $H = \{\zeta \in \mathbb{C} : \operatorname{Re}\zeta < 0\}$ in the sense that

$$L_j(D) \subset L_{j+1}(D), \ \forall j = 1, 2, \ldots,$$

and

$$\bigcup_{j=1}^{\infty} L_j(D) = H.$$

Now we combine this simple observation with the fact that there exists the sequence of maps

$$\varphi_j(z) \equiv \frac{z + a_j}{1 + a_j z}$$

that are automorphisms of D satisfying $\varphi_j(0) = a_j$. Consider the sequence of composite maps

$$\sigma_j \equiv L_j \circ \varphi_j : D \to \mathbb{C}.$$

A simple computation yields that

$$\begin{aligned} L_j \circ \varphi_j(z) &= \frac{1}{1 - a_j}\left(\frac{z + a_j}{1 + a_j z} - 1\right) \\ &= \frac{z - 1}{1 + a_j z}. \end{aligned}$$

Hence we see that the sequence of holomorphic mappings $L_j \circ \varphi_j$ converges uniformly on compact subsets of D to the mapping

$$\widehat{\sigma}(z) = \frac{z - 1}{z + 1}$$

that is a biholomorphic mapping from the open unit disk D onto the left half plane H.

What we have done here is to produce the classical Cayley map using scaling.

Of course scaling is most interesting, and most effective, when it is applied in the function theory of several complex variables. Let us now formulate one of the most famous theorems that can be proved with scaling.

Definition 4.9.1. Let $\Omega \subseteq \mathbb{C}^n$ be a domain in space. Let $\xi \in \partial\Omega$. We say that ξ is a *boundary orbit accumulation point* if there exist biholomorphic self maps $\varphi_j : \Omega \to \Omega$ and a point $\zeta \in \Omega$ such that $\varphi_j(\zeta) \to \xi$ as $j \to \infty$.

The famous theorem of Bun Wong and Rosay ([250, 286]) says this:

Theorem 4.9.2. *Let $\Omega \subseteq \mathbb{C}^n$ be a bounded domain. Suppose that $\xi \in \partial\Omega$ has a smooth neighborhood in the boundary and is strongly pseudoconvex. If ξ is a boundary orbit accumulation point, then Ω is biholomorphic to the unit ball in $\mathbb{C}^n$.*

This result has been generalized by Greene/Krantz and others to finite type boundary orbit accumulation points. These studies have led to the following conjecture of Greene/Krantz:

> **CONJECTURE:** Let $\Omega \subseteq \mathbb{C}^n$ be a bounded domain and let $P \in \partial\Omega$ have a smooth boundary neighborhood and be a boundary orbit accumulation point. Then P is of finite type in the sense of Catlin/D'Angelo/Kohn (see [50]).

4.10 Asymptotic Values of Holomorphic Functions

After the appearance of the paper by Lehto and Virtanen [164], one of the main directions of research was simply to test various properties of bounded holomorphic functions in order to establish the same properties for normal functions, or to construct examples of normal functions that do not have this property.

The most important work related to normal functions was published by MacLane [181], who considered the general question of the asymptotic values of holomorphic functions. It should be noted here that there is a natural division in the study of normal functions into normal meromorphic functions (which may not have asymptotic values) and normal holomorphic functions; the study of the latter is more fruitful, since a holomorphic function always has at least one asymptotic value.

The purpose of this section is to list some results on the asymptotic values of functions $f(z)$ holomorphic in the unit disk D and give example showing that some of these results do not hold in the multidimensional case.

It should be observed that all the results are for holomorphic functions. The arguments break down in various places for meromorphic functions. In particular, nothing in the nature of Theorem 4.10.4 can hold for meromorphic functions, for, as was shown in [183], there exist meromorphic functions without any asymptotic values for which Nevanlinna original characteristic $T(r)$ grows arbitrarily slowly.

Definition 4.10.1. We shall say that f has the *asymptotic value* a at ζ, $|\zeta| = 1$, if and only if there is an arc $\Gamma \subset D$ tending to ζ such that $f \to a$ on Γ.

Let $f(z)$ be holomorphic and nonconstant in D. For any complex number a ($a = \infty$ permitted) we define the set A_a as follows. The point ζ belongs to A_a if and only if $|\zeta| = 1$ and f has the asymptotic value a at ζ. See (4.10.1). If S is any set on the Riemann sphere, we set

$$A(S) = \bigcup_{a \in S} A_a, \qquad A(S) = \emptyset \text{ if } S = \emptyset.$$

In particular we set

$$A^* = \bigcup_{a \neq \infty} A_a, \qquad A = A^* \cup A_\infty.$$

Definition 4.10.2. The function $f(z)$ belongs to the class $\mathcal{N}$ if and only if f is holomorphic, nonconstant, in D, and normal.

Definition 4.10.3. The function $f(z)$ belongs to the class $\mathcal{A}$ if and only if f is holomorphic and nonconstant in D and A is dense in ∂D.

Section 7 of MacLane's book [181] is devoted to deriving some sufficient conditions for $f \in \mathcal{A}$. The fundamental condition, Theorem 4.10.4, is

$$(I) \qquad \int_0^1 (1-r)\log^+ \left|f(re^{i\theta})\right| dr < \infty \qquad (\theta \in \Theta),$$

where $\Theta \subset [0, 2\pi]$ is dense in $[0, 2\pi]$.

Set

$$m(r) = \frac{1}{2\pi}\int_0^{2\pi} \log^+ \left|f(re^{i\theta})\right| d\theta \qquad (0 \leq r < 1).$$

We shall say that $f(z)$, holomorphic in D, satisfies the condition (II) if

$$(II) \qquad \int_0^1 (1-r)m(r)dr < \infty.$$

A more restrictive condition is

$$(III) \qquad \int_0^1 (1-r)\log^+ M(r)dr < \infty,$$

where $M(r)$ is the maximum modulus of f on the circle of radius r.

Theorem 4.10.4. *Let $f(z)$ be holomorphic, nonconstant, in D and satisfy (I). Then $f \in \mathcal{A}$.*

Remark 4.10.5. The hypothesis (I) in Theorem 4.10.4 may be replaced by either (II) or (III).

It is well known that the characteristic function, $T(r)$, of $f(z)$ is given by the formula

$$T(r) = S(r) + O(1)$$

where

$$S(r) = \int_0^r \frac{A(t)}{t} dt \text{ and } A(r) = \int_0^r \int_0^{2\pi} \frac{\left|f'(re^{i\theta})\right|^2}{\left(1 + |f(re^{i\theta})|^2\right)^2} r dr d\theta.$$

Theorem 4.10.6. *We have $\mathcal{N} \subset \mathcal{A}$ and the inclusion is proper. Also, if $f \in \mathcal{N}$, then:*

Given $\zeta, |\zeta| = 1, f$ has at most one asymptotic value at ζ. If f has the asymptotic value a at ζ then f has the angular limit a at ζ.

Remark 4.10.7. Lehto and Virtanen gave an example [164, p. 58] of a normal function without any asymptotic values. But that function is *meromorphic*, not holomorphic, and Theorem 4.10.4 shows that such must be the case.

Proof. We assume that the nonconstant holomorphic function f satisfies (4.2) and prove that $f \in \mathcal{A}$.

Then from (4.2) we have

$$A(t) \leq C^2 \int_0^t \frac{2r dr}{\left(1 - r^2\right)^2} = \frac{C^2 t^2}{1 - t^2} \leq \frac{C^2 t}{1 - t}$$

and hence

$$S(r) \leq C^2 \int_0^r \frac{dt}{1 - t} = C^2 \log \frac{1}{1 - r}$$

Since $f(z)$ is holomorphic,

$$T(r) = S(r) + O(1) < C^2 \log \frac{1}{1 - r} + O(1) \qquad (0 \leq r < 1) \tag{4.1}$$

where $T(r)$ is the Nevanlinna characteristic function of f. It is immediately clear from (4.1) that $T(r)$ satisfies condition (II), and hence $f \in \mathcal{A}$.

Next, the assertion in 4.10.6 follows immediately from Theorem 4.10.9 below. It is then clear that the inclusion $\mathcal{N} \subset \mathcal{A}$ is proper. For in [182] it was shown that there exist holomorphic functions of arbitrarily slow growth (and hence some that satisfy condition (III)) without any radial limits. ☐

Asymptotic values always imply angular limits, but we cannot say anything about the existence of asymptotic values. In fact, there exist normal functions which possess no asymptotic values at all.

Lehto and Virtanen [164, Theorem 1, p.49-52] prove the following theorem.

Theorem 4.10.8. *Let $f(z)$ be meromorphic in D. Suppose that $f(z)$ has the asymptotic value zero along a Jordan curve lying in D and terminating at a point ξ in ∂D and that $f(z)$ does not possess the angular limit zero at the point ξ. Then there exists a Jordan curve L in D with end-point at ξ, on which $f(z)$ tends to zero, and a sequence of points $\{z_n\}$, in D, which converges to ξ and at which $f(z_n) = a$ for a certain value $a \neq 0$, such that the points z_n have a bounded hyperbolic distance from L less than M, where M is any fixed positive number.*

As an immediate consequence of the above theorem, we have

Theorem 4.10.9. *Let $f(z)$ be meromorphic and normal in ∂D and let $f(z)$ have an asymptotic value α at a point ξ on ∂D along a Jordan curve lying in D. Then, $f(z)$ possesses the angular limit α at the point ξ.*

Proof. [4] Without loss of generality, we may suppose that $\alpha = 0$. Contrary to the assertion, suppose that $f(z)$ does not possess the angular limit zero at ξ. Consider now the asymptotic path L and the sequence of points $\{z_n\}$ in Theorem 4.10.8. Denote by $z' = S_n(z)$ a function which maps D conformally onto itself and satisfies the condition $S_n(0) = z_n$. We denote by K the hyperbolic circle with center at $z = 0$ and of hyperbolic radius $M + 1$. By hypothesis, the family $\{f(S_n(z))\}$ is normal in D; hence, we can select a subsequence $\{f(S_{n_k}(z))\}$ which converges uniformly on every compact subset of D to a meromorphic function $\varphi(z)$ in D. Since the images of the arcs of L mapped by the inverse functions $z = S_{n_k}^{-1}(z')$ into the interior of K has at least one accumulation continuum γ, it follows that $\varphi(z) = 0$ on γ and therefore $\varphi(z)$ vanishes identically. However, this is a contradiction, because $f(S_{n_k}(0)) = a \neq 0$. □

Now the following question arises. Is it clear that there is a holomorphic function on the disk/ball that is not normal? Example 4.10.10 shows that a holomorphic function might fail to be normal.

Example 4.10.10. Let B be the unit ball in $\mathbb{C}^2$. The function

$$f(z_1) = \frac{\exp \frac{z_1+1}{z_1-1}}{z_1 - 1}$$

is holomorphic in the unit disk $D = \{z_1 \in \mathbb{C} : |z_1| < 1\}$ and tends to ∞ when $z_1 \to 1$ along the path $|z_1 - 1/2| = 1/2$ and tends to 0 when $z_1 \to 1$ along the radius. By Noshiro-Lehto-Virtanen's theorem about asymptotic values of normal functions (Theorem 4.10.9) $f(z_1)$ is not a normal holomorphic function in D.

Since the normality of the function remains a hereditary property under projections this example also shows that $f(z_1)$ considered as a holomorphic functions in B is not normal in B (use Corollary 4.2.16).

[4] An entirely similar argument was used in [226]. See also related results of Seidel [266].

Let $p(r)$ be a function on $[0,1)$ such that $0 < p(r) \uparrow \infty$. It was shown in [182, Theorem 2, p. 23] that there exists a function $f(z)$ holomorphic in D for which $f(|z|) < p(|z|)$, $0 \le |z| < 1$ and which has no radial limit at every point $\xi \in \partial D$.

Lehto and Virtanen also proved results which indicate that, in some sense, the class of normal functions is the largest class for which the Lindelöf phenomenon holds.

For example, if f is meromorphic in D and has a nontangential limit at ξ, then in each Stolz angle $\Gamma_\alpha(\xi) = \{z \in D : |z - \xi| < \alpha(1 - |z|)\}$ we have

$$\frac{|f'(z)|}{1 + |f(z)|^2} = O\Big(\frac{1}{1 - |z|}\Big)$$

as $|z| \to 1$. Since the hyperbolic (Poincaré) metric on D is

$$\rho(z) = \frac{1}{1 - |z|^2} = O\Big(\frac{1}{1 - |z|}\Big),$$

this is a local version of (4.2).

Some of the basic theorems and concepts of complex analysis generalize to several complex variables, but many others do not.

Normal holomorphic functions in the plane have many remarkable properties that are not shared by their higher-dimensional analogues. Attempts to generalize these special results even to two-dimensional complex space seem doomed to failure, although some open questions of this sort still remain.

In fact, Lehto–Virtanen's theorem is false in dimensions higher than two. The following simple example (probably due to Čhirka [48] but discovered independently by several others, see Cima and Krantz [47]) shows that Theorem 4.10.9 is not true in several complex variables, and even bounded holomorphic functions (which are normal, since they omits two different values in D) can have asymptotic limits along curves other than radial ones.

Example 4.10.11. Let $B \subseteq \mathbb{C}^2$ be the unit ball. Let $f : B \to \mathbb{C}$ be given by

$$f(z_1, z_2) = \frac{z_2^2}{1 - z_1}.$$

The function $f(z)$ is holomorphic in B and is bounded there, since

$$|f(z)| \le \frac{1 - |z_1|^2}{1 - |z_1|} \le 2.$$

Then f has radial limit 0 at $\xi = (1, 0) \in \partial B$. However f does not have asymptotic limit 0 at ξ — indeed, f possesses limit $1/\lambda$ along the surfaces $S = \{1 - z_1 = \lambda z_2^2\}$, $\lambda \in \mathbb{C}$ and $|\lambda| \ge 1/2$ which belong to B and are tangent to ∂B at the point ξ. Indeed $f|_S \equiv 1/\lambda$. It is straightforward to compute that the set of all asymptotic values of f at ξ contains a closed disk of radius 1.

For domains of the form $\Omega_m = \left\{(z_1, z_2) \in \Omega : |z_1|^2 + |z_2|^{2m} < 1\right\}$, the functions $f^m : \Omega_m \to \mathbb{C}$ given by

$$f^m(z_1, z_2) = \frac{z_2^{2m}}{1 - z_1}$$

provide similar examples. Indeed Range's estimates [242] show that the surface $S = \left\{1 - z_1 = \lambda z_2^{2m}\right\}$ approaches $\xi = (1, 0)$, however $f^m|_S \equiv 1/\lambda$ while f^m has radial limit 0 at ξ.

This example demonstrates the failure of the Noshiro-Lehto-Virtanen theorem (Theorem 4.10.9) in $\mathbb{C}^2$, even for bounded holomorphic functions.

Definition 4.10.12. If f is any function with domain the ball B, and $0 < r < 1$, then f_r denotes the dilated function defined for $|z| < 1/r$ by $f_r(z) = f(rz)$.

A function $f \in \mathcal{O}(B)$ is in the Hardy space $H^p(B)$ $(0 < p < \infty)$ provided that

$$\sup_{0<r<1} \int_{\partial B} |f_r|^p \, d\sigma < \infty. \tag{4.2}$$

In the latter case, we write $\|f\|_p$ for the pth root of the left side of (4.2). This is evidently a norm when $p \geq 1$, and $\|f - g\|_p^p$ defines a metric when $0 < p < 1$.

Clearly, $H^{p_1} \subset H^{p_2}$ if $p_1 > p_2$. Note that $H^\infty \subset H^1$.

It may be of interest to compare the class of normal holomorphic functions with Hardy space H^p.

It follows from the definition that bounded functions are always normal. Hence Hardy space H^p has nonempty intersection with the class $\mathcal{N}$ for every $0 < p < \infty$.

The first question that we need to consider is: Is there an unbounded normal function on B which is in the Hardy space $H^1(B)$? The answer to this question is given by the following example.

Example 4.10.13. Let B be the unit ball in $\mathbb{C}^2$. Set

$$f(z) = (1 - z_1)^{-1/2},$$

then $f \in \mathcal{O}(B)$.

We now show that this function is in the Hardy space $H^1(B)$. Let σ denote the usual rotationally invariant measure on ∂B and let m denote area measure on D. Then, for fixed $r < 1$, integrating over the slices $z_1 =$ constant (see

Proposition 1.47 of [257, Proposition 1.4.7, p.15]) we obtain

$$\begin{aligned}\int_{\partial B}|f(rz)|d\sigma(z) &= \frac{1}{2\pi}\int_D dm(z_1)\int_{-\pi}^{\pi}\left|f\left(rz_1, re^{i\phi}\sqrt{1-|z_1|^2}\right)\right|d\phi \\ &= \frac{1}{2\pi}\int_D dm(z_1)\int_{-\pi}^{\pi}\frac{1}{\sqrt{|1-rz_1|}}d\phi \\ &\le \int_D \frac{1}{\sqrt{1-|z_1|}}dm(z_1) \\ &\le \int_0^1 \frac{1}{\sqrt{1-\rho}}\rho d\rho \\ &= \frac{-2\rho-4}{3}\sqrt{1-\rho}\Big|_0^1 = \frac{4}{3} < \infty.\end{aligned}$$

(Here and below $d\phi$ denote the differential of function ϕ defined on $[-\pi,\pi]$.) The latter integral is bounded (independently of r), and therefore $f \in H^1(B)$.

The explicit calculation (which we perform below) shows that the Levi form of $\log(1+|f|^2)$ at z is

$$L_z(\log(1+|f|^2), v) = \frac{|v_1|^2}{(1+|1-z_1|)^2} \le 1\cdot|v|^2 \text{ for all } z\in B \text{ and all } v \in \mathbb{C}^2.$$

From the definition of the normal function we see that $f(z)$ is a normal function in B.

Hence the intersection of the Hardy space $H^1(B)$ with the class of normal holomorphic functions in B contains the unbounded normal function f defined above.

Remark 4.10.14. It is not known whether the Hardy space $H^1(D)$ coincide with the class of normal holomorphic functions in the unit disk D.

A fundamental fact about a normal holomorphic function f of one complex variable is that f can posses only one asymptotic value at the boundary point ξ. Theorem 4.10.9 implies that no normal function in the unit disk can have more than one asymptotic value at any boundary point. This is false if the unit disk is replaced by the unit ball; the function f mentioned in Example 4.10.11 has every c with $|c| < 1$ as an asymptotic value at $(1,0)$, even though $|f| \le 2$.

Example 4.10.11 shows that there is a major difference between the boundary behavior of the normal holomorphic functions of one and several variables. When $n \ge 2$, even bounded holomorphic functions (which is definitely normal) can posses infinitely many asymptotic values at a fixed boundary point, which is impossible in one complex variable by Theorem 4.10.9.

But we will see in the next section that we still have uniqueness if we restrict ourselves to special ξ-curves.

4.11 Lindelöf Theorem in $\mathbb{C}^n$

Lehto and Virtanen [164] studied boundary behavior of normal meromorphic functions by applying Montel's theory of normal families and the theory of hyperbolic measure.

The primary motivation for this investigation is an important theorem of Lindelöf (see [37, Vol 2, p. 42]).

Theorem 4.11.1. *(Lindelöf). Let f be a bounded analytic function in the open unit disk, D, of the complex plane with associated boundary function $f(e^{i\theta})$. If*

$$\textit{ess} \lim f(e^{i\theta}) = 0 \textit{ as } \theta \to 0^+ (\textit{ or } 0^-)$$

then $f(z)$ tends to zero in every "Stolz angle" at 1.

For smooth domains in $\mathbb{C}^n$ the analogous theorem for bounded holomorphic functions holds, and one can even supplant cones with the wider class of so-called hypoadmissible approach regions, allowing tangential approach to the boundary in certain directions [48].

In [60] we restrict to the class of strongly pseudoconvex domains in $\mathbb{C}^n$ but extend the class of functions from bounded holomorphic functions to normal holomorphic functions; in this setting we prove that if f has a limit along one nontangential path terminating at a boundary point ξ then f has a limit at ξ within any hypoadmissible region.

The following terminology describes the limiting behavior involved in this circle of ideas.

Let Ω be a bounded domain in $\mathbb{C}^n$ and $\delta(z, \partial\Omega)$ the Euclidean distance from the point z to the boundary of the domain.

Let $T_\xi = T_\xi(\partial\Omega)$ be the tangent plane to the boundary of domain Ω at the point ξ and ν_ξ be the unit outward normal. T_ξ contains a unique complex hyperplane T_ξ^c passing through ξ; this T_ξ^c is called the *complex tangent plane* to $\partial\Omega$ at the point ξ, while the vectors parallel to it are called *complex-valued tangent vectors* at this point. The distance up to T_ξ^c in $\mathbb{C}^n$ equals $|\langle z - \xi, \nu_\xi \rangle|$ where $\langle a, b \rangle = a_1\bar{b}_1 + \cdots + a_n\bar{b}_n$ is the Hermitian scalar product in $\mathbb{C}^n$, whose real part coincides with the usual scalar product in $\mathbf{R}^{2n} = \mathbb{C}^n$.

By $\delta_\xi(z)$ we denote the minimum of the distances from point z to $\partial\Omega$ and to the tangent plane T_ξ; then the nontangential limit value at a point ξ is the limit value over the sets

$$C_\alpha(\xi) = \{z \in \Omega : |z - \xi| < (1 + \alpha)\delta_\xi(z)\}, \quad \alpha > 0.$$

Of course, the term "nontangential" refers to the fact that the boundary curves of $C_\alpha(\xi)$ have a corner at ξ, with angle less than π.

A function f is said to have a *nontangential limit* at ζ if $\lim_{z\to\zeta} f(z)$ exists in each nontangential region $\Gamma(\zeta,\alpha)$.

A set $A \subset \Omega$ is said to be *asymptotic* if A has limit points belonging to $\partial\Omega$, *asymptotic* at the point ξ if $A \cap \partial\Omega = \{\xi\}$, and nontangentially asymptotic at the point ξ if $A \subset C_\alpha(\xi)$ for some $\alpha > 0$. An *asymptotic curve* at the point ξ is a curve $\{\gamma(t) : 0 < t < 1\} \subset \Omega$ such that $\lim_{t\to 1}\gamma(t) = \xi$.

The Lindelöf principle, even for normal functions, is not really a theorem about automorphisms of domains or about normal families (as in [164]). Indeed it is in the nature of a Schwarz lemma, i.e., it exploits metric inequalities. The results below depend in an essential way on the one-variable Lehto-Virtanen result, but the proof of that result consists primarily of a careful study of the hyperbolic metric. The portions of the Lehto-Virtanen proof which involve normal families arguments may be entirely recast in metric terms.

Lemma 4.11.2. *Let $\Omega \subset\subset \mathbb{C}^n$ have $\mathcal{C}^2$ boundary. Let $f : \Omega \to \widehat{\mathbb{C}}$ be K-normal and $\xi \in \partial\Omega$. Let $\ell \in \widehat{\mathbb{C}}$ and suppose that f has radial limit ℓ at ξ. Then f has non-tangential limit ℓ at ξ.*

Proof. Let $\alpha > 1$. Let $\widetilde{\Gamma}_\alpha(P) = \Gamma_\alpha(P) \cap \mathbb{C}\nu_\xi$. By Proposition 1.6, $f|_{\tilde{\Gamma}_\alpha(\xi)\cap B(\xi,\epsilon)}$ is normal in the classical one variable sense for some small $\epsilon > 0$. So, by the classical theorem of Lehto-Virtanen,

$$\lim_{\tilde{\Gamma}_\alpha(P)\ni z\to\xi} f(z) = \ell\,.$$

Let $\mathcal{T}_p$ denote the Hermitian orthogonal complement to $\mathbb{C}\nu_\xi$. By elementary geometry there is a $C_0 > 0$ so that if $z \in \tilde{\Gamma}_\alpha(\xi)$, $|z-\xi| < C_0$, and $\pi(z) \in \mathcal{T}_P$, $|\pi(z)| < C_0\sqrt{\delta_\Omega(z)}$, then $z + \pi(z) \in \Omega$. Let $w \in \Gamma_\alpha(P)$, $|w-\xi| < C_0/2$. Let w' be the orthogonal projection of w into $\tilde{\Gamma}_\alpha(P)$. Let q be the complex line determined by w, w'. Then $q \subseteq \mathcal{T}_P + w'$. Hence $q \cap B\left(w', C_0\sqrt{\delta_\Omega(w')}\right)$ is a complex disk contained in Ω. If ξ is a Euclidean unit vector parallel to q then the map

$$\begin{aligned}\varphi :&D \to \Omega\\ &\zeta_1 \mapsto w' + C_0\sqrt{\delta_\Omega(w')}\xi\cdot\zeta\end{aligned}$$

is a holomorphic parametrization of this disk. Notice that $\varphi(0) = w'$. Say that $\varphi(\beta) = w$. Then $F_\Omega^K(w,w') \le F_D^K(0,\beta)$. But, since $\|w-w'\| \le \alpha\delta_\Omega(w')$, it follows that $|0-\beta| \le C\sqrt{\delta_\Omega(w')}$. So $F_\Omega^K(w,w') \le C\sqrt{\delta_\Omega(w')}$. As a result,

$$\lim_{\Gamma_\alpha(\xi)\ni w\to\xi} f(w) = \lim_{\tilde{\Gamma}_\alpha(\xi)\ni w'\to\xi} f(w') = \ell\,.$$

□

Remark 4.11.3. The same proof shows that the hypothesis "radial limit" may be replaced by "limit along some non-tangential curve γ".

At this point we can give the answer to the following question: Is there a function $f \in H^1(B)$ which is not normal? Lemma 4.11.2 allows us to answer this question.

Example 4.11.4 (cf. [257, 8.4.7. Example, 173]). Define f in

$$B = \{z \in \mathbb{C}^2 : |z| < 1\}$$

by

$$f(z) = \frac{z_2}{1 - z_1}.$$

Take a point $(a, b) \in \partial B$, $a \neq 1$, and consider the rectilinear path

$$\Gamma(t) = (t + (1-t)a, (1-t)b)$$

from (a, b) to $(1, 0)$. On this path,

$$f(\Gamma(t)) \equiv \frac{b}{1-a}.$$

All rectilinear limits of f exist therefore at $(1, 0)$, but they are not equal. In fact, the set of limiting values is all of $\mathbb{C}$. Since $f(z_1, 0) = 0$, the radial limit is 0 at $(1, 0)$.

Then f is holomorphic in B and has radial limit zero at $(1, 0)$ since $f(z_1, 0) = 0$ for all $z_1 \in D$. Nevertheless f does not have non-tangential limit at $(1, 0)$. By Lemma 4.11.2, f cannot be normal in B.

We now estimate the $H^1(B)$ norm of f. Let σ denote the usual rotationally invariant measure on ∂B and let m denote area measure on D. Then, for fixed $r < 1$, integrating over the slices $z =$ constant (see Proposition 1.47 of [257, Proposition 1.4.7, p.15]) we obtain

$$\begin{aligned}
\int_{\partial B} |f(rz)| d\sigma(z) &= \frac{1}{2\pi} \int_D dm(z_1) \int_{-\pi}^{\pi} \left| f\left(rz_1, re^{i\phi}\sqrt{1 - |z_1|^2}\right) \right| d\phi \\
&= \frac{1}{2\pi} \int_D dm(z_1) \int_{-\pi}^{\pi} \left| \frac{re^{i\phi}\sqrt{1 - |z_1|^2}}{1 - rz_1} \right| d\phi \\
&\leq \frac{1}{2\pi} \int_D dm(z_1) \int_{-\pi}^{\pi} \frac{\sqrt{1 - |z_1|^2}}{1 - |z_1|} d\phi \\
&\leq \int_D \frac{\sqrt{1 - |z_1|^2}}{1 - |z_1|} dm(z_1) \\
&\leq 2 \int_0^1 \frac{1}{\sqrt{1-\rho}} \rho d\rho \\
&\leq 2 \int_0^1 \frac{1}{\sqrt{1-\rho}} \rho d\rho \\
&= 2 \left\{ \frac{-2\rho - 4}{3} \sqrt{1-\rho} \right\} \Big|_0^1 = \frac{8}{3}.
\end{aligned}$$

The latter integral is bounded (independently of r), and therefore $f \in H^1(B)$.

By Lindelöf's theorem, no normal function can behave in this way. It follows that the class $\mathcal{N}(B)$ is different from the Hardy space $H^p(B)$, $p < 1$.

In Lemma 4.3.2 we saw that if $f(z)$ is a normal function on a ball B, and $g(z)$ is a bounded analytic function on B, then the sum $f(z)+g(z)$ is a normal function; the power $[f(z)]^j$ for any integer j is also normal.

A natural question arises: To what extent is the class of normal functions closed with respect to various arithmetic operations?

Lappan [162] showed that the sum of two normal functions in the unit disk need not be normal, and that if $f(z)$ is a normal meromorphic function with an infinite set of poles, then there exists a Blaschke product $B(z)$ such that the product $f(z)B(z)$ is no longer a normal function. The condition that the functions in question omit three values does not help. Lappan [163] also showed that these conclusions remain in force for holomorphic functions: the sum of two normal analytic functions in D may not have Fatou points.

Having Example 4.11.4 at our disposal it is now quite obvious that the class of normal functions is not closed under the operations of addition or multiplication.

Example 4.11.5. Let B be the unit ball in $\mathbb{C}^2$. It is easy to check that $-2/(1-z_1)$ is a normal function on B. The sum of two normal holomorphic functions in B need not be normal. Indeed, the sum of two normal functions $(z_2+2)/(1-z_1)$ (which is normal by Lemma 4.3.3) and $-2/(1-z_1)$ is equal to $z_2/(1-z_1)$ which is not normal in B (see Example 4.11.4).

The product of two normal holomorphic functions z_2 and $1/(1-z_1)$ is equal to $z_2/(1-z_1)$ which is not normal (see Example 4.11.4).

The product of a bounded normal holomorphic function z_2 and an unbounded non-normal holomorphic function $z_2/(1-z_1)$ is equal to $z_2^2/(1-z_1)$ which is a bounded holomorphic function in B, since $|f(z)| \leq |z_2^2|/(1-|z_1|) \leq 2$, and is thus definitely normal.

For many of the basic function theory results in a bounded domains in $\mathbb{C}^n$, approach within a *hypoadmissible region* replaces nontangential approach. Suppose that its boundary $\partial\Omega$ is a (compact) real hypersurface of class $\mathcal{C}^2$ in $\mathbb{C}^n$. Then there exists a $\mathcal{C}^2$-smooth real function ρ in a neighborhood U of $\partial\Omega$ such that

$$\Omega \cap U = \{z \in U : \rho(z) < 0\},$$

where $\rho \in \mathcal{C}^2(U)$ with gradient

$$\nabla\rho = \Big(\frac{\partial\rho}{\partial z_1}, \dots, \frac{\partial\rho}{\partial z_n}\Big) \neq 0$$

on $\partial\Omega = S$; then at a point $\xi \in S$ the unit vector $\nu = \overline{\nabla\rho}/|\nabla\rho|$ is in the direction of the normal, while $\tau = i\nu$ belongs to the tangent plane $T_\xi S$ and is orthogonal to the complex tangent plane $T_\xi^c(S)$. We call such a function ρ a *global defining function.*

For each fixed $\alpha > 0$, $k > 0$, and ε, $0 < \varepsilon < 1/2$, we define a hypoadmissible region $A_{\varepsilon,\alpha}$, with vertex ξ, by the equation

$$A_{\varepsilon,\alpha}(\xi) = \left\{ z \in \Omega : |(z - \xi, \nu)| < (1 + \alpha)\delta_\xi(z), |z - \xi| < k{\delta_\xi}^{1/2+\varepsilon}(z) \right\}.$$

The domain $A_{\varepsilon,\alpha}(\xi)$ is tangent to $T_\xi(S)$ in directions of the complex tangent plane $T_\xi^c(S)$ and gives an acute angle in the orthogonal direction τ.

We shall say that f has a *hypoadmissible limit* at ζ if $\lim_{z \to \zeta, z \in A_{\varepsilon,\alpha}} f(z)$ exists for all $\alpha > 0$.

For each fixed $\alpha > 0$ and $k > 0$ we define a *admissible region* $A_\alpha(\xi)$, with vertex ξ, by the equation

$$A_\alpha(\xi) = \left\{ z \in \Omega : |(z - \xi, \nu)| < (1 + \alpha)\delta_\xi(z), |z - \xi| < k{\delta_\xi}^{1/2}(z) \right\}.$$

We shall say that f has a *admissible limit* at ζ if $\lim_{z \to \zeta, z \in A_\alpha} f(z)$ exists, for all $\alpha > 0$.

The notion of hypoadmissible limit arises because the full analogue of Lindelof's theorem, with admissible limits replacing nontangential limits, does not hold for $n > 1$.

Remark 4.11.6. It is well known that for arbitrary tangent directions the assertion of the full analogue of Lindelöf's theorem becomes false: the function $f(z) = z_2^2/(1 - z_1)$ is holomorphic in the unit ball $B \subset \mathbb{C}^2$ and is bounded (and hence normal) there, since $|f(z)| \leq |z_2^2|/(1 - |z_1|) \leq 2$, but as z tends to the boundary point $\xi = (1, 0)$ it has different limits along the surfaces $1 - z_1 = \lambda z_2$, tangent to ∂B at the point ξ.

The theorem is also false for domains with nonsmooth boundaries: the function $f(z) = z_1/z_2$ is holomorphic and bounded in the domain $\{z \in \mathbb{C}^n : |z_1| < |z_2|\}$, but its limits as $z \to (0, 0)$ along different lines $z_1 = \lambda z_2$ are different.

The following theorem, a version of Cirka's theorem, provides an adequate substitute.

Theorem 4.11.7. *If the function f is normal in a strictly pseudoconvex domain $\Omega \subset \mathbb{C}^n$ and it has a limit along some nontangential asymptotic curve at $\xi \in \partial\Omega$, then it has the same limit as $z \to \xi$ through the points of $A_{\varepsilon,\alpha}(\xi)$.*

For strictly pseudoconvex domains in any dimension, the Bergman, Caratheodory, and Kobayashi metrics are all comparable; such comparability follows from the results of I. Graham [84] and K. Diederich [52]. Therefore we can just talk about normal functions without specifying the infinitesimal metric used in Definition 4.2.7. Nevertheless, suppose for the moment that f is C-normal in Ω. The proof of Theorem 4.11.7 hinges on the classical one-dimensional Lindelöf-Lehto-Virtanen's theorem [164, Theorem 2, p. 53] and the following lemma:

Lemma 4.11.8. *Let Ω be a bounded domain in $\mathbb{C}^n$ and let $\{z_j\}$, $\{z'_j\}$ be two sequences of points in Ω such that*

$$\lim_{j\to\infty} z_j = \xi, \quad \lim_{j\to\infty} C^i_\Omega(z_j, z'_j) = 0.$$

If f is a C-normal function in Ω and $\lim_{j\to\infty} f(z_j) = f^*(\xi)$, then $\lim_{j\to\infty} f(z'_j) = f^*(\xi)$.

Proof. Since f is by assumption normal in Ω from the definition of the distance C^i_Ω, we deduce the inequality

$$\frac{1}{\sqrt{c}} \inf \int_0^1 \frac{|g'(\ell)|}{1+|g(\ell)|^2} d\ell \le C^i_\Omega(z_j, z'_j), \tag{4.1}$$

where the infimum is taken over all $g(\ell) = f(\gamma(\ell))$, $\ell \in [0,1]$, and γ is a smooth curve in Ω joining z_j with z'_j. The length of the curve $g(\ell)$, $g(0) = f(z_j)$, $g(1) = f(z'_j)$, in the spherical metric $(1+|z|^2)^{-2} dz \wedge d\bar{z}$ is equal to

$$\int_0^1 \frac{|g'(\ell)|}{1+|g(\ell)|^2} d\ell \,.$$

By the hypothesis, $C^i_\Omega(z_j, z'_j) \to 0$ as $j \to \infty$. The inequality (4.1) implies that the spherical distance between the points $f(z_j)$ and $f(z'_j)$ tends to zero as $j \to \infty$ and therefore the assertion of the lemma follows. □

Proof of Theorem 4.11.7. After a possible translation and unitary linear transformation in the ambient space $\mathbb{C}^n$ we may assume that the point ξ is the origin, that the outward unit normal at 0 is in the negative y_1 direction, (here $z_1 = x_1 + iy_1$) and thus $N^c_\xi = \{(z_1, 0, \ldots, 0)\}$, while $T^c_\xi = \{(0, z_2, \ldots, z_n)\}$.

Denote by π_ξ the orthogonal projection of $\mathbb{C}^n$ to N_ξ; that is if $z = (z_1, \ldots, z_n) \in \mathbb{C}^n$, then $\pi_\xi(z) = (z_1, 0, \ldots, 0)$.

Let $\{z(t) : 0 < t \le 1\} \subset \Omega$ be a nontangential asymptotic curve at a point ξ along which f has a limit $f(\xi)$. From the definition of the distance C^i_Ω we have

$$C^i_\Omega(z(t), \pi_\xi(z(t))) \le \int_0^1 F^C_\Omega(\gamma(\ell), \dot\gamma(\ell)) d\ell, \tag{4.2}$$

where

$$\gamma(\ell) = (z_1(t), \ell z_2(t), \ldots, \ell z_n(t)), \quad \ell \in [0,1].$$

By Theorem 4.7.6, since Ω is strictly pseudoconvex and the vector $\dot\gamma(\ell) = (0, z_2(t), \ldots, z_n(t))$ is parallel to T^c_ξ, we see that

$$F^C_\Omega(\gamma(\ell), \dot\gamma(\ell)) \le C \cdot \frac{|\dot\gamma(\ell)|}{\sqrt{\delta(\gamma(\ell))}}. \tag{4.3}$$

The curve $\{\gamma(t) : 0 < t < 1\} \subset \Omega$ belongs to some $C_\alpha(\xi)$, therefore

$$|\dot\gamma(\ell)| \le |z(t)| < \alpha\delta_\xi(z(t)) \le \alpha\delta(z(t)).$$

Since $\delta(z)$ is a continuous function in Ω there exist two constants c_1, c_2 such that $c_1 \leq \delta(\gamma(\ell))/\delta(z) \leq c_2$ for all $\ell \in [0,1]$. From (4.2) and (4.3) we obtain

$$C_\Omega^i(z(t), \pi_\xi(z(t))) \leq C \cdot \sqrt{\delta(z(t))} \to 0$$

as $t \to 0$.

Since the boundary of Ω belongs to the class C^2 then there is a ball B_R with center in some point $w(\xi) = (iR, 0, \ldots, 0)$ and radius R such that $B_R \subset \Omega$ and $\partial B_R \cap \partial\Omega = \{\xi\}$.

From the monotonicity property of the F_Ω^C we have

$$B_R \subset \Omega \Rightarrow F_\Omega^C(z,v) \leq F_{B_R}^C(z,v) \text{ for all } z \in B_R, v \in \mathbb{C}^n.$$

The above inequality and inequality (4.10) give

$$L_z(\log(1+|f|^2), v) \leq C(F_{B_R}^C(z,v))^2 \quad \text{for all } z \in B_R, v \in \mathbb{C}^n.$$

Hence

$$\frac{|f'(z_1, 0, \ldots, 0)|}{1+|f(z_1, 0, \ldots, 0)|^2} \leq \frac{\sqrt{C}R}{R^2 - |z_1 - iR|^2} \quad \text{for all } z_1 \in |z_1 - iR| < R.$$

Thus $f(z_1, 0, \ldots, 0)$ is a normal function in the classical one-variable sense (see [164, Theorem 3, p. 56]) which has limit $f(\xi)$ along the nontangential asymptotic curve $\pi(z(t))$ at the point ξ. By the classical Lindelöf-Lehto-Virtanen's theorem [164, Theorem 2, p. 53], $f(\pi_\xi)$ has the angular limit $f(\xi)$ at the point ξ.

The same proof shows that

$$C_\Omega^i(z, \pi_\xi(z)) \leq C \cdot \delta^\varepsilon(z) \to 0$$

as $z \to \xi, z \in A_\alpha(\xi)$. By Lemma 4.11.8 and the fact that the function $f(\pi(z))$ has the angular limit $f(\xi)$ at the point ξ, we obtain that f has an admissible limit at ξ. □

Proposition 4.11.9. *Let Ω be a strictly pseudoconvex domain in $\mathbb{C}^n$, $n > 1$, and assume that the normal function f has the limit of $f(z)$ over a sequence $\{z_m\} \subset A_{\varepsilon,\alpha}(\xi)$ so that $\lim_{m\to\infty} z_m = \xi$ exists and is such that $\lim_{m\to\infty} c_\Omega(z_m, z_{m+1}) = 0$. Then*

$$\lim_{z\to\xi, z\in A_{\varepsilon,\alpha}(\xi)} f(z) = f^*(\xi).$$

Proof. It is easy to see that the polygonal arc L with nodes at the points z_m belongs to some approach region $A_{\varepsilon,\beta}(\xi)$, $\beta > \alpha$.

Since f is a normal function for all points z belonging to the segment $[z_m, z_{m+1}]$, the spherical distance between $f(z)$ and $f(z_m)$ is

$$s(f(z), f(z_m)) \leq const \cdot c_\Omega(z, z_m). \tag{4.4}$$

From the properties of the Caratheodory metric it follows that $c_\Omega(z, z_m) \leq c_\Omega(z_m, z_{m+1}) + c_\Omega(z, z_{m+1})$. Hence it follows from (4.4) that the function f has the limit $f^*(\xi)$ along L. Now we find ourselves under the conditions of Theorem 4.11.7, according to which

$$\lim_{z \to \xi, z \in A_{\varepsilon,\alpha}(\xi)} f(z) = f^*(\xi).$$

□

Lindelöf's theorem for normal functions in $\mathbb{C}^n$ was proved by the first author in [60]. See also J. A. Cima, and S. G. Krantz [47].

4.12 Lindelöf Principle in $\mathbb{C}^n$

The most classical formulation of the Lindelöf principle on the disk (see [175]) is as follows:

Theorem: Let f be a bounded, holomorphic function on the unit disk $D \subset \mathbb{C}$. Suppose that the radial limit

$$\lim_{r \to 1^-} f(re^{i\theta}) \equiv \lambda \in \mathbb{C} \qquad (*)$$

of f exists at the boundary point $e^{i\theta}$. Then in fact f has nontangential limit λ at $e^{i\theta}$.

Thus we have a sort of tauberian theorem: for *bounded* holomorphic functions, radial convergence implies nontangential convergence. It is of interest to have a result of this nature in several complex variables. Pioneering work on the Lindelöf principle in several complex variables was done by Cirka [48].

The paper [47] established a Lindelöf principle for holomorphic functions of several variables. That result was new and optimal in the following senses:

(i) It was proved for *normal functions* (see [164]), in a sense the most natural function space for which to consider a Lindelöf principle.

(ii) It was formulated in terms of the Kobayashi metric, thus providing an "optimal" result in terms of the intrinsic Levi geometry of the domain.

But the result of [47] has certain drawbacks. Notable among these is that, whereas the most natural mode of boundary convergence in the several variable setting is admissible convergence (see [139, 140, 143, 270]), that used in the results of [47] is hypoadmissible converge—a strictly weaker concept. In addition, when the radial curve, as in $(*)$, is replaced by a fairly arbitrary curve—especially by a curve with a significant complex tangential component—then a rather unsatisfying result obtains.

In this section we introduce new techniques that address the shortcomings of [47] and produce a sharp version of the Lindelöf principle. In common with [47], we shall be able to prove our result not only for bounded holomorphic functions but also for normal functions. We refer the reader also to [2] and [3] for modern work on the Lindelöf principle in several variables. Also the paper [147] must be mentioned in this context.

We begin by proving the following proposition and discussing it.

Proposition 4.12.1. *Let f be a bounded, holomorphic function on the unit ball $B \subset \mathbb{C}^2$. Let*

$$T = \{(s + i0, t + i0) \in \mathbb{C}^2 : s, t \in \mathbb{R}, 0 < s < 1, 0 < |t| < \sqrt{1 - 2s}\}.$$

Let $1 = (1 + i0, 0 + i0)$. Suppose that

$$\lim_{T \ni z \to 1} f(z) \equiv \lambda \in \mathbb{C}$$

exists. Then, for any $\alpha > 1$,

$$\lim_{\mathcal{A}_\alpha(1) \ni z \to 1} f(z) = \lambda .$$

Proof. The proof follows classical lines. We may assume that $\lambda = 0$. Define, for $j = 1, 2, \ldots,$

$$\begin{aligned} \Omega_j &= \Big\{(z_1, z_2) \in \mathbb{C}^2 : 1 - 2^{-j+1}/(2\alpha) \le \operatorname{Re} z_1 < 1 - 2^{-j-1}/(2\alpha), \\ &\qquad |\operatorname{Im} z_1| < 2^{-j+1}, \ |z_2| < \sqrt{2^{-j+1}}/\sqrt{\alpha}\Big\}. \end{aligned}$$

Observe that, for all $1 \le j_0 \in N$, the map

$$\varphi_j(z_1, z_2) = \left(2^{j-j_0}(z_1 - 1) + 1, \sqrt{2^{j-j_0}}z_2\right)$$

sends Ω_j biholomorphically onto Ω_{j_0}. Also, following the paradigm of the classical Lindelöf principle on the disk, we may see that

$$\bigcup_{j=1}^{\infty} \Omega_j = \bigcup_{j=1}^{\infty} \varphi_j^{-1}(\Omega_{j_0}) \supseteq \mathcal{A}_\alpha(1) .$$

Furthermore, we see that

$$T \cap \Omega_j = \left\{(s, t) : 1 - 2^{-j+1}/(2\alpha) \le s < 1 - 2^{-j-1}/(2\alpha) \ , 0 < t < \sqrt{1 - 2s}\right\}.$$

Of course φ_j maps $T \cap \Omega_j$ onto

$$T \cap \Omega_{j_0} = \left\{(s, t) : 1 - 2^{-j_0+1}/(2\alpha) \le s < 1 - 2^{-j_0-1}/(2\alpha) \ , 0 < t < \sqrt{1 - 2s}\right\}.$$

Now consider the function

$$g_j \equiv f \circ \varphi_j^{-1} : \Omega_{j_0} \to \mathbb{C}\,.$$

Being uniformly bounded, the g_j form a normal family. Let g_0 be a subsequential limit function.

It follows from our hypotheses that g_0 vanishes on $T \cap \Omega_{j_0}$. But $T \cap \Omega_{j_0}$, being two-dimensional and totally real, is a set of determinacy for holomorphic functions. Hence $g_0 \equiv 0$.

Unravelling the logic, we find that if $K \subset \Omega_{j_0}$ is a compact set such that

$$\bigcup_{j=1}^{\infty} \varphi_j^{-1}(K) \supseteq \mathcal{A}_\alpha(1)\,,$$

then $g_j \to 0$ uniformly on K. It follows that f itself has admissible limit 0 on K. □

The proof that we have just presented is misleadingly simple. For it is an artifact of the special geometry of T, and the way that it is imbedded in space, that $\varphi_j^{-1}(T \cap \Omega_j) = T \cap \Omega_{j_0}$ for each j. For a very general sort of Lindelöf principle, we would like to replace the flat T with a rather arbitrary, two-dimensional, totally real surface. In that situation, the sets

$$\varphi_j^{-1}(T \cap \Omega_j)$$

could be pairwise disjoint. Thus additional arguments will be required. It is worth noting that this problem arises even on the disk in the complex plane—in the situation where the hypothesis of the Lindelöf principle is the existence of a limit along a somewhat arbitrary curve rather than along a radius.

With these thoughts in mind, we now formulate a more sophisticated version of the Lindelöf principle on the ball:

Proposition 4.12.2. *Let f be a bounded, holomorphic function on the unit ball $B \subset \mathbb{C}^2$. Let*

$$T = \{(s + i0, t + i0) : s, t \in \mathbb{R}, 0 < s < 1, 0 < |t| < \sqrt{1-2s}\}\,.$$

Suppose that $\rho : T \to \mathbb{R}^2$ is a C^1 function, with bounded first and second derivatives, such that (writing $\rho(s,t) = (\rho_1(s,t), \rho_2(s,t))$)

$$\mathcal{T} = \{(s + i\rho_1(s,t), t + i\rho_2(s,t)) : (s,t) \in T\}$$

is a two-dimensional, totally real manifold in $B \subseteq \mathbb{C}^2$.

Let $\mathbf{1} = (1 + i0, 0 + i0)$. Suppose that

$$\lim_{\mathcal{T} \ni z \to \mathbf{1}} f(z) \equiv \lambda \in \mathbb{C}$$

exists. Then, for any $\alpha > 1$,

$$\lim_{\mathcal{A}_\alpha(\mathbf{1}) \ni z \to \mathbf{1}} f(z) = \lambda\,.$$

This result is more nearly like the general Lindelöf principle that one may find in [164, Theorem 2, p. 53]. It hypothesizes the existence of a limit along a fairly "arbitrary" surface terminating at the boundary point $1 \in \partial B$, and it concludes the existence of an admissible limit. It still leaves open the question of obtaining a result on more general domains, and also the question of treating normal functions. We save those two topics for subsequent sections.

Proof. Let Ω_j, ϕ_j, g_j be as in the proof of the preceding proposition. As before, we may assume that $\lambda = 0$ and we may obtain a subsequential limit function g_0.

But it is important for us now to note that the sets

$$\mathcal{T}_j \cap \Omega_{j_0} \equiv \varphi_j(\mathcal{T} \cap \Omega_j)$$

are all graphs over $T \cap \Omega_{j_0}$ of functions τ_j, and each τ_j has, by design, bounded derivatives (with the bound uniform in j). Thus we may extract (using the Ascoli-Arzelà theorem) a subsequence τ_{j_k} that converges uniformly, along with its first derivatives, on compacta to some τ_0. We pass to a corresponding subsequence of the g_js, and continue to call the limit function g_0. Now let $\mathcal{T}_0$ be the graph of τ_0.

It follows that $\mathcal{T}_0$ is a totally real, two-dimensional manifold. And certainly g_0 vanishes on $\mathcal{T}_0$. Thus $g_0 \equiv 0$. Arguing as in the proof of Proposition 4.12.1, we conclude that f has admissible limit 0 at 1. □

We continue, for the moment, to work on the unit ball $B \subset \mathbb{C}^n$. Recall (see [47]) that a *normal* function has at least two equivalent definitions. Here we let $\widehat{\mathbb{C}}$ denote the Riemann sphere.

Definition 4.12.3. Let $f : B \subset \widehat{\mathbb{C}}$ be holomorphic. We say that f is *normal* if, whenever $\{\varphi_j\}$ are biholomorphic self-maps of B then $\{f \circ \varphi_j\}$ is a normal family.

In the paper [47] an important equivalent formulation was derived using ideas from invariant geometry.

Definition 4.12.4. A holomorphic function $f : B \subset \widehat{\mathbb{C}}$ is *normal* if the derivative ∇f is bounded from the Kobayashi metric on B (equivalently, the Poincaré-Bergman metric on B) to the spherical metric on $\widehat{\mathbb{C}}$.

The equivalence of these two definitions is a sophisticated exercise with Marty's theorem. Now we have

Proposition 4.12.5. *Let $B \subset \widehat{\mathbb{C}}$ be the unit ball. Let $f : B \subset \widehat{\mathbb{C}}$ be holomorphic and normal. Let*

$$T = \{(s + i0, t + i0) : s, t \in \mathbb{R}, 0 < s < 1, 0 < |t| < \sqrt{1 - 2s}\}.$$

Suppose that ρ *on* $T \subset \mathbb{R}^2$ *is a* C^1 *function, with bounded first and second derivatives, such that (writing* $\rho(s,t) = (\rho_1(s,t), \rho_2(s,t))$*)*

$$\mathcal{T} = \{(s + i\rho_1(s,t), t + i\rho_2(s,t)) : (s,t) \in T\}$$

is a two-dimensional, totally real manifold in $B \subseteq \mathbb{C}^2$.

Let $e_1 = (1 + i0, 0 + i0)$. *Suppose that*

$$\lim_{\mathcal{T} \ni z \to e_1} f(z) \equiv \lambda \in \mathbb{C}$$

exists. Then, for any $\alpha > 1$,

$$\lim_{\mathcal{A}_\alpha(e_1) \ni z \to e_1} f(z) = \lambda\,.$$

Proof. The key here is to let

$$\Phi : B \to B$$

be the automorphism

$$\Phi(z_1, z_2) = \left(\frac{z_1 + 1/4}{1 + (1/4)z_1}, \frac{\sqrt{1 - 1/16}\,z_2}{1 + (1/4)z_1} \right).$$

Then we define Ω_{j_0} as before but now we let

$$\Omega_j = \Phi^{j - j_0}(\Omega_{j_0})\,.$$

Here Φ^ℓ is the mapping Φ composed with itself ℓ times, $\ell > 0$.

Now the proof goes through just as that for Proposition 4.12.2. We merely must note that now we are examining the mappings $g_j \equiv f \circ \Phi^{-(j - j_0)}$. These are compositions of f with automorphisms. By the definition of "normal function", we may be sure that the g_j form a normal family. The proof is completed then as before. □

It should be stressed that it is misleading, indeed essentially incorrect, to think of normal functions on an arbitrary domain in terms of automorphisms of the domain. For most domains in $\mathbb{C}$ or $\mathbb{C}^n$ have only the identity as an automorphism (see [91] for a discussion of this notion). One of the main motivations for the development in [47] of normal functions using the Kobayashi metric was to address this difficulty. Thus, if we wish to prove a Lindelöf principle on general domains, we certainly cannot use the ideas in Section 3.

Instead we examine the invariant Kobayashi metric. Let us begin by looking at a strongly pseudoconvex domain Ω with C^2 boundary. Let $P \in \partial\Omega$. By normalizing coordinates, we may assume as usual that $P = e_1 = (1, 0)$, that $\operatorname{Re} z_1$ is the real normal direction, and that $\operatorname{Im} z_1$ is the complex normal direction. Thus $z_2, \ldots, z_n$ are the complex tangential directions at P. For simplicity, we assume as above that the dimension $n = 2$. Following the paradigm in [84] or [74] (see also [143]), we may assume that there is an internally tangent ball at $P \in \partial\Omega$. There is no loss of generality to assume that this ball is the unit ball. Thus we may define regions Ω_j just as on the unit ball above. And the maps φ_j are defined as before.

Theorem 4.12.6. *Let* $\Omega \subset \mathbb{C}^2$ *be a domain with* C^2 *boundary. Let* $f : \Omega \to \widehat{\mathbb{C}}$ *be holomorphic and normal. Let*

$$T = \{(s + i0, t + i0) : s, t \in \mathbb{R}, 0 < s < 1, 0 < |t| < \sqrt{1 - 2s}\}.$$

Suppose that $\rho : T \to \mathbb{R}^2$ *is a* C^1 *function, with bounded first and second derivatives, such that (writing* $\rho(s,t) = (\rho_1(s,t), \rho_2(s,t))$*)*

$$\mathcal{T} = \{(s + i\rho_1(s,t), t + i\rho_2(s,t)) : (s,t) \in T\}$$

is a two-dimensional, totally real manifold in $B \subseteq \mathbb{C}^2$.

Let $e_1 = (1 + i0, 0 + i0)$. *Suppose that*

$$\lim_{\mathcal{T} \ni z \to e_1} f(z) \equiv \lambda \in \mathbb{C}$$

exists. Then, for any $\alpha > 1$,

$$\lim_{\mathcal{A}_\alpha(e_1) \ni z \to e_1} f(z) = \lambda .$$

Proof. It is propitious to consider a holomorphic mapping $\psi : D \to \Omega_0$, where $D \subset \mathbb{C}^n$ is the unit disk. If $\psi(0) = p \in \Omega_0$, then we may take ψ to be an extremal function for the Kobayashi metric at the point p. Now look at $\mu_j : D \to \widehat{\mathbb{C}}$ given by $f \circ \varphi_j^{-1} \circ \psi$.

Then we calculate that

$$|\mu_j'(0)| \le \left|\nabla f(p) \left[(\varphi_j^{-1}(p) \circ \psi)'(0)\right]\right| . \qquad (*)$$

Of course the expression in brackets on the right is nothing other than the reciprocal of the Kobayashi metric for Ω_j at $\varphi_j^{-1}(p)$. The first expression, as we know from the definition of "normal function", is bounded from the Kobayashi metric on Ω (which is smaller than the Kobayashi metric on Ω_j) to the spherical metric on $\widehat{\mathbb{C}}$. In sum, the expression $(*)$ is bounded on compact subsets of D. And this bound is certainly independent of j.

So we may extract a normally convergent subsequence of μ_j'. Call the limit function μ_0. Arguing now as in the proof of Proposition 4.12.2, we see that there is a corresponding sub-subsequence τ_{j_k} convergin to τ_0 and a limiting totally real, two-dimensional manifold $\mathcal{T}_0$ in Ω_0 which is the graph of τ_0. Thus we find that μ_{j_k}' converges on $\mathcal{T}_0$. Putting together the convergence of the derivatives together with the convergence of the functions on $\mathcal{T}_0$, we see that the functions themselves converge uniformly on compact subsets of Ω_0. By the usual logic, we find that f has admissible limit λ at P. □

4.13 Admissible Limits of Normal Functions in $\mathbb{C}^n$

The existence almost everywhere of admissible limits of bounded holomorphic functions of several complex variables was established by Korányi [141] and Stein [270]; the complex geometric nature of this phenomenon was studied by Čhirka [48]. A natural generalization of the class of bounded holomorphic functions is the class of normal holomorphic functions. However, it is known that normal functions can have bad boundary behavior almost everywhere.

Montel [200] used normal families in a simple but ingenious way to investigate boundary behavior of holomorphic functions in angular domains. We used his ideas in studying boundary behavior in the case of functions of several complex variables.

The basic goal of the present section is to find conditions under which normal functions of several complex variables have admissible limits almost everywhere.

We have the following theorem:

Theorem 4.13.1. *Let Ω be a domain in $\mathbb{C}^n$, $(n > 1)$ with C^2-smooth boundary and let the function f be normal in the domain Ω. If*

$$\lim_{z \to \xi, z \in A_\beta(\xi)} f(z) \text{ exist for some } \beta > 0,$$

then f has admissible limit at the point ξ.

For bounded holomorphic functions of several complex variables this result was first proved by Čhirka [48]. One should note that although Theorem 4.13.1 contains the result of Čhirka we do not give a new proof of it but one essentially based on the old proof.

Proof of Theorem 4.13.1. First let us assume that f has finite limit when z tends to the point ξ remaining inside the domain $A_\beta(\xi)$.

We shall give a proof by contradiction. Let us first assume that one can find a number α, $\alpha > \beta$, and a sequence of points $\{a^m\} \in A_\alpha(\xi)$, $a^m \to \xi$ as $m \to \infty$ such that $f(a^m) \to \infty$ as $m \to \infty$.

We choose coordinates $z_1, \ldots, z_n$ in $\mathbb{C}^n$, so that $\xi = (0, \ldots, 0)$ and $\nu_0 = ('0, \ldots, 1)$. Here and later, $'z = (z_1, \ldots, z_{n-1})$.

Let

$$P_c(z) = \left\{\omega \in \mathbb{C}^n : |\omega_\mu - z_\mu| < c\sqrt{\delta(z)}, \mu = 1, \ldots, n-1; |\omega_n - z_n| < c\delta(z)\right\}$$

be a polydisk in $\mathbb{C}^n$, with center at the point z. One can choose the number c so that for all $z \in A_\alpha(\xi)$ sufficiently close to zero, one has $P_c(z) \subset A_\alpha(\xi)$. (cf. [241, Lemmas 5.2, 7.2]).

Let $\pi(z) = ('0, z_n)$ be the projection of the point $z \in \mathbb{C}^n$ to the complex normal to $\partial\Omega$ at the point 0. For points $z \in A_\alpha(0)$ sufficiently close to zero, the ratio $\delta(z)/\delta(\pi(z))$ remains between two positive constants depending only on α and Ω (cf. [241, Lemma 7.2]); hence, for all sufficiently large m, the segment $[a^m, \pi(a^m)]$ belongs to $A_\alpha(0)$ and it can be covered by a finite number of polydisks $P_c(a^{1,m}) \subset A_\alpha(0)$ such that

$$(a)\quad P_c(a^{i,m}) \cap P_c(a^{i+1,m}) \neq \emptyset, \quad i = 1, \dots, N-1,$$

$$(b)\quad a^{i,m} = a^m, \quad a^{N,m} = \pi(a^m), \quad a^{i,m} \in [a^m, \pi(a^m)], \quad i = 1, \dots, N.$$

Under the map

$$'\tilde{\omega} \to \frac{'\omega - 'z}{\alpha\sqrt{\delta(z)}}, \quad \tilde{\omega}_n \to \frac{\omega_n - z_n}{\alpha\delta(z)},$$

the polydisk $P_c(z)$ goes into the unit polydisk

$$P = \{\tilde{\omega} \in \mathbb{C}^n : |\tilde{\omega}_\mu| < 1, \mu = 1, \dots, n\}.$$

Since the Kobayashi metric is invariant with respect to biholomorphic transformations, we conclude from Definition 4.2.7 and Marty's criterion that the family of holomorphic functions

$$\left\{ g^{i,m}(\tilde{\omega}) = f\left(c\sqrt{\delta(a^{i,m})}'\tilde{\omega}, c\delta(a^{i,m})\tilde{\omega}_n\right) \right\}$$

is normal in the polydisk P. By definition of a normal family, from the sequence $\{g^{i,m}\}$ one can choose a subsequence converging uniformly inside P (i.e., converging uniformly on compact subsets of P) either to a holomorphic function, or identically to infinity. From (a) we conclude that the sequence $\{g^{2,m}\}$ also converges uniformly inside P identically to infinity. After a finite number of steps, considering (b) we get that $g^{N,m}(0) = f(\pi(a^m)) \to \infty$ as $m \to \infty$. But $f(\pi(z))$ is a normal function of one complex variable, so from the statement about the finiteness of the limit with respect to $D_\beta(0)$ and the theorem of Lehto and Virtanen [164], $f(\pi(z))$ has finite angular limit at zero. The contradiction obtained shows that f, for any $\alpha > 0$, is bounded in the domain $A_\alpha(0)$ (each time by a constant). Applying the theorem of [48] to the bounded function $f|_{D_{2\alpha}}(0)$, analogously to Theorem 1, we conclude that the limit of f with respect to $A_\alpha(0)$ exists; in view of the arbitrariness of α, we see that f has admissible limit at zero.

Let us now assume that f has infinite limit when z tends to zero while remaining inside the $D_\beta(0)$. From this assumption and the theorem of Lehto and Virtanen [164] we conclude that $f(\pi(z)) \to \infty$ as $z \to 0, z \in A_\alpha(0)$. We fix any $\alpha > 0$ and any sequence $\{a^m\} \subset A_\alpha(0)$, as $m \to \infty$. In condition (b) we let $a^{1,m} = \pi(a^m)$, and $a^{N,m} = a^m$. Just as above, the family $\{g^{i,m}\}$ is normal in the polydisk P. Since $g^{N,m}(0) = f(\pi(a^m)) \to \infty$ as $m \to \infty$, the sequence $\{g^{1,m}\}$ converges uniformly inside P identically to infinity. From (a) and (b), just as above we conclude that $g^{N,m}(0) = f(a^m) \to \infty$ as $m \to \infty$. In view of the arbitrariness of α, f has admissible limit equal to infinity at zero. □

Let

$$ds^2_\Omega(z,v) = \sum_{\mu,\nu=1}^{n} g_{\mu,\nu}(z)v_\mu \overline{v}_\nu$$

be the Bergman form of the domain Ω (cf., e.g., [238]).

The square of the modulus of the gradient of a holomorphic function f in the Bergman metric of the domain Ω is defined as follows:

$$|\nabla_\Omega f(z)|^2 = \sum_{\mu,\nu=1}^{n} g^{\mu,\nu}(z)\frac{\partial f(z)}{z_\mu}\overline{\frac{\partial f(z)}{z_\nu}}.$$

Here $(g^{\mu,\nu}(z))$ is a matrix inverse to $(g_{\mu,\nu}(z))$. This quantity is invariant with respect to biholomorphic transformations since

$$|\nabla_\Omega f(z)|^2 = \sup_{v\in\mathbb{C}^n\setminus\{0\}} \frac{|df(z)|^2}{ds^2_\Omega(z,v)}. \tag{4.1}$$

The volume element $d_\Omega(z)$ corresponding to the Bergman metric is

$$d_\Omega(z) = \det(g_{\mu,\nu}(z))d\omega,$$

where $d\omega$ is the Euclidean volume element in $\mathbb{C}^n \approx \mathbb{R}^{2n}$.

Stein [270] proved that for any function f holomorphic in a domain $\Omega \subset \mathbb{C}^n$, $n > 1$, and almost all points $\xi \in \partial\Omega$ the following conditions are equivalent:

$$(a) \quad \sup_{z\in A_\alpha(\xi)} |f(z)| < +\infty;$$

$$(b) \quad \lim_{z\to\xi, z\in A_\alpha(\xi)} f(z) \text{ exists };$$

$$(c) \quad \int_{A_\alpha(\xi)} |\nabla_\Omega f(z)|^2 \, d\Omega(z) < +\infty.$$

In strictly pseudoconvex domains of the space $\mathbb{C}^n$ the Kobayashi and Berman metrics are equivalent (cf., e.g., [238, p. 113]), so it follows from Definition 4.2.7 and (4.1) that the function f is normal in the domain Ω if and only if

$$\frac{|\nabla_\Omega f(z)|}{1+|f(z)|^2} = O(1)\,.$$

For normal functions we can give a global and a weaker condition replacing (c).

Theorem 4.13.2. *Let Ω be a strictly pseudoconvex domain in $\mathbb{C}^n$, $n > 1$, and let f be a normal function in the domain Ω. If*

$$\int_\Omega \frac{|\nabla_\Omega f(z)|^2}{1+|f(z)|^2} d_\Omega(z) < +\infty, \tag{4.2}$$

then f has admissible limits almost everywhere on $\partial\Omega$.

Proof. We choose coordinates $z_1, \dots, z_n$ in $\mathbb{C}^n$, so that $0 \in \partial\Omega$ and $\nu_0 = ('0, 1)$. Since $\partial\Omega$ is of class C^2, one can find a neighborhood U of the point 0 such that for all points $\xi \in \partial\Omega \cap U$ the vectors ν_ξ make an acute angle with the vector ν_0. It follows from (4.2) and Fubini's theorem that for almost all points $\xi \in \partial\Omega \cap U$

$$\int_0^1 \frac{|\nabla_\Omega f(\ell_\xi(t))|^2}{1 + |f(\ell_\xi(t))|^2} \det(g_{\mu,\nu}(\ell_\xi(t)))dt < +\infty, \tag{4.3}$$

where $\ell_\xi(t) = \{\xi - t\nu_0, t \in [0, \delta)\}$ and $\delta > 0$ is sufficiently small that $\ell_\xi(t) \in \partial\Omega$ for all $\xi \in \partial\Omega \cap U$ and $0 < t \leq \delta$. It is proved in [270] that for all $z \in \partial A_\alpha(0)$, sufficiently close to $\partial\Omega$,

$$|\nabla_\Omega f(z)| \approx \sum_{\mu=1}^{n} \left|\frac{\partial f(z)}{\partial z_\mu}\right| \sqrt{\delta(z)} + \left|\frac{\partial f(z)}{\partial z_n}\right| \delta(z)$$

$$d\Omega(z) \approx \delta(z)^{-n-1} d\omega$$

(the expression $A \approx B$ means that the ratio A/B is between two positive constants).

Hence from (4.3) and the Cauchy-Bunyakovskii inequality we get that for the same

$$\int_0^1 \frac{|f'(\ell_\xi(t))|}{\sqrt{1 + |f(\ell_\xi(t))|^2}} dt < +\infty,$$

where the apostrophe means derivative with respect to t.

Further, since

$$\log(1 + |f(\ell_\xi(t))|^2)' \leq \frac{2\,|f'(\ell_\xi(t))|}{\sqrt{1 + |f(\ell_\xi(t))|^2}}$$

it follows from (4.3) and the preceding inequality that for almost all points $\xi \in \partial\Omega \cap U$

$$\int_0^\delta \log(1 + |f(\ell_\xi(t))|^2)' dt < +\infty.$$

Consequently, the function $|f(\ell_\xi(t))|$ has finite limit as $t \to 0$. From this and (4.3) we conclude that for almost all $\xi \in \partial\Omega \cap U$,

$$\left|\int_0^\delta f'(\ell_\xi(t))dt\right| \leq \int_0^\delta |f'(\ell_\xi(t))|\, dt < +\infty\,,$$

and hence $f(\ell_\xi(t))$ tends to a finite limit as $t \to 0$. Since the vector ν_ξ makes an acute angle with the vector ν_0 it follows from the proof of Theorem 4.13.1 that for such ξ the function f is bounded in the domains $A_\alpha(\xi)$. Let $\Omega' \subset \Omega \cap U$ be a domain with C^2-smooth boundary such that $\partial\Omega \supset \partial\Omega' \cap V$ for some neighborhood $V \ni 0$. Then the admissible domains $A_\alpha(\xi)$ coincide for Ω and

Ω' in a neighborhood $W \ni 0$. According to the equivalence of properties (a) and (b) for the domain Ω' indicated above, we thus get that f has admissible limits almost everywhere on $\partial\Omega \cap W$. Since the assertion of the theorem is local, the theorem is proved. □

One sufficient condition that guarantees existence of an admissible limit for an arbitrary function f which is holomorphic in a domain $\Omega \subset \mathbb{C}^n$ follows from the argument Stein used to prove Theorem 12 in [270]: if $f \in \mathcal{O}(\Omega)$ has a limit along normal at $\xi \in \partial\Omega$ and

$$\int_{A_\alpha(\xi)} |\nabla_\Omega f(z)| \, d\Omega(z) < +\infty$$

for all $\alpha > 0$, then f has an admissible limit at ξ.

Another sufficient condition for existence of an admissible limit is given by the following

Proposition 4.13.3. *Let Ω be a domain in $\mathbb{C}^n$, $n > 1$, with C^2-smooth boundary. If a function f holomorphic in Ω has a finite limit along the normal at $\xi \in \partial\Omega$ and Re f has an admissible limit at ξ, then f has an admissible limit at ξ.*

Proof. Without loss of generality, set $\xi = 0$, $\nu_0 = ('0, 1)$. Here and henceforth $z = (z_1, \ldots, z_n)$, $'z = (z_1, \ldots, z_{n-1})$.

We also assume that f has the limit zero along ν_0; for, in the event that f has a finite limit a along ν_0, it suffices to consider the function $f - a$. The function e^f has the limit unity along ν_0 and is bounded in any domain $A_\alpha(0)$; note that $|e^f| < e^{\Re f}$ and, by hypothesis, Re f has an admissible limit at zero. The one-dimensional Lindelöf theorem now implies that $e^{f('0,z_n)}$ and, consequently, $f('0, z_n)$ have angular limits at 0. We write $g(\mu) = f(\mu' z, z_n)$, $\mu \in \mathbb{C}$. For every domain $A_\alpha(0)$, we can find constants $c > 0$ and $\alpha' > \alpha$ such that, for all $z \in A_\alpha(0)$, the following inclusion holds:

$$\left\{w \in \mathbb{C}^n : |z_n - w_n| < c\Re z_n, |'z - 'w| < c\sqrt{\Re z_n}\right\} \subset A_{\alpha'}(0)$$

(cf. [270]). Therefore the function $u(\mu) = \Re g(\mu)$ is harmonic in the closed disk $\overline{U} = \{\mu \in \mathbb{C}; |1 - \mu| \leq c/\sqrt{\alpha}\}$ and, consequently, can be represented by the Poisson formula

$$u(1 + re^{i\phi}) = \frac{1}{2\pi} \int_0^{2\pi} u(1 + Re^{i\phi}) \frac{R^2 - r^2}{R^2 + r^2 - 2Rr\cos(\theta - \phi)} d\theta,$$

with $R = c/2\sqrt{\alpha}$ and $\mu = x + iy = re^{i\phi}$. Differentiating the above function with respect to r, and setting $r = 0$, we obtain

$$\frac{\partial u(1)}{\partial r} = \frac{1}{2\pi} \int_0^{2\pi} u(1 + Re^{i\phi}) \cos(\theta - \phi) \, d\theta \,. \tag{4.4}$$

Denote by $M(z)$ the least upper bound of $\operatorname{Re} f$ on the set

$$A_\alpha(0) \cap \{w \in \mathbb{C} : w_n = z_n\}.$$

Since $\operatorname{Re} f$ has the admissible limit zero at 0, it follows that $M(z) \to 0$ as $z \to 0$, $z \in A_\alpha(0)$. On the other hand, comparing (4.4) with the equality

$$\frac{\partial u}{\partial r} = \frac{\partial u}{\partial x}\cos\phi + \frac{\partial u}{\partial y}\sin\phi,$$

we see that

$$\frac{\partial u(1)}{\partial x} = \frac{1}{2\pi}\int_0^{2\pi} u(1+Re^{i\phi})\cos\theta\, d\theta, \quad \frac{\partial u(1)}{\partial y} = \frac{1}{2\pi}\int_0^{2\pi} u(1+Re^{i\phi})\sin\theta\, d\theta.$$

Therefore

$$\left|\frac{\partial u(1)}{\partial x}\right| \le \frac{2M(z)}{R}, \quad \left|\frac{\partial u(1)}{\partial y}\right| \le \frac{2M(z)}{R}.$$

In virtue of Cauchy-Riemann equations, $g'(\mu) = \frac{\partial u(\mu)}{\partial x} - i\frac{\partial u(\mu)}{\partial x}$; and so, at every point $z \in A_\alpha(0)$, we have

$$\left|\frac{\partial f(\mu' z, z_n)}{\partial \mu}\right| \le \frac{2\sqrt{2}}{R} M(z).$$

In order to estimate $f(z)$ we employ the formula

$$f(z) = \int_0^1 \frac{df(t'z, z_n)}{dt}\, dt + f('0, z_n),$$

which implies

$$|f(z)| \le \frac{2\sqrt{2}}{R} M(z) + |f('0, z_n)|.$$

Since $M(z)$ and $f('0, z_n)$ vanish as $z \to 0$, $z \in A_\alpha(0)$, we conclude that $f(z)$ tends to zero as $z \to 0$, $z \in A_\alpha(0)$. □

Remark 4.13.4. If the inequality $|z - \xi| < \alpha\Psi(\delta(z))$ with $\Psi(\delta(z)) = o(\sqrt{\delta})$ is substituted for the second inequality in the definition of the domain $A_\alpha(\xi)$, then the resulting domain is called *weakly admissible* and the limit of the function over all such domains, whenever it exists, is called *weakly admissible.* The multidimensional Lindelöf theorem established by Čhirka in [48] for bounded holomorphic functions remains valid for pluriharmonic functions. In the proof, the Schwarz lemma should be replaced by its weakened version which ensues immediately from the Poisson formula. Thus, if a bounded function $f \in \mathcal{O}(\Omega)$ is such that $\operatorname{Re} f$ has the limit a along the normal ν_ξ and $\operatorname{Im} f$ has the limit b along normal ν_ξ, then f has a weakly admissible limit at ξ that is equal to $a + ib$.

5

A Geometric Approach to the Theory of Normal Families

Preamble: It is an important insight that the theory of normal families can be rendered in geometric language. We develop that idea in this chapter.

An important tool for us will be the Kobayashi/Royden metric. This is a metric that was devised in the 1960s to generalize the invariant metric that Poincaré created on the unit disk in the early twentieth century. It is fairly simple to define, but it is a powerful tool.

Of course the Ascoli-Arzelà theorem will be a part of our considerations, and we will realize that result as a compactness statement.

Hung-Hsi Wu's idea of taut manifold will be discussed. It is a key idea in the modern theory of normal families. Norberto Kerzman's result that all smoothly bounded pseudoconvex domains are taut will be treated.

Finally, we relate this circle of ideas to the big Picard theorem. That is an attractive way to wrap up our geometric discussion.

5.1 Introduction

In what follows, a *domain* is a connected, open set. The focus of this section is on normal families of mappings from one complete hyperbolic complex manifold to another. Since there are a number of different definitions of normal family, we begin by specifying the definition that we will use here. Our source for this definition is [287]. In what follows we will let M, N denote complex manifolds.

Definition 5.1.1. We let $\mathcal{C}(M, N)$ denote the collection of continuous mappings from M to N. And let $\mathcal{H}(M, N)$ denote the collection of holomorphic mappings from M to N.

Definition 5.1.2. We say that a sequence of mappings $f_j : M \to N$ *converges normally* if, for each compact set $K \subseteq M$, the sequence $f_j|_K$ converges uniformly.

DOI: 10.1201/9781032669861-5

We say that the sequence of mappings f_j is *compactly divergent* if, given any compact $K \subseteq M$ and any compact $L \subseteq N$, there is an index T_0 so that $f_j(K) \cap L = \emptyset$ for $j > T_0$.

Definition 5.1.3. We call a collection $\mathcal{F} \subseteq \mathcal{C}(M, N)$ *normal* if every sequence in $\mathcal{F}$ contains either a subsequence that converges normally or a subsequence that is compactly divergent.

It is a pleasure to acknowledge [20, 287] as the inspiration for some of this work.

5.2 History

Most people think of Montel as the progenitor of the theory of normal families. But in point of fact the basic idea has a considerable history prior to Montel's efforts.

In the early 1860s Weierstrass contributed a key idea about normal convergence of holomorphic functions (see [245]). In 1894 Stieltjes [273] showed that uniform convergence was the natural topology to consider for holomorphic functions. Osgood, in 1901, weakened the hypotheses needed in order to derive uniform convergence of a sequence of holomorphic functions (see [230]). These results were sharpened further by Porter [239, 240] and Vitali [282, 283]. Arzelà and Ascoli proved in [12] and [11] what we now know as the Ascoli–Arzelà theorem.

It was not until 1907 that Montel [198] proved a version of what we now call *Montel's theorem.* In his 1912 paper [200] Montel introduced the term "normal family". Montel's definitive monograph [199] established his name in the firmament.

5.3 The Kobayashi/Royden Pseudometric and Related Ideas

The Kobayashi/Royden pseudometric is one of many invariant metrics that can be defined on a complex manifold. We choose to emphasize this particular one because it is convenient for the study of normal families.

Definition 5.3.1. Let M be a complex manifold. Let $P \in M$. Let $\mathcal{T}_P M$ be the (complex) tangent space to M at P and $\mathcal{T}^* M$ the tangent bundle. Let D be the unit disk in the complex plane. Let e_1 be the vector $1 + i0 = \langle 1, 0\rangle$ in the plane. The infinitesimal form of the *Kobayashi/Royden pseudometric* is given by $F_K^M : \mathcal{T}^*M \to \mathbb{R}$, where

$$\begin{aligned} F_K^M(P,\xi) & \\ \equiv \quad & \inf\left\{\alpha : \alpha > 0 \text{ and } \exists f \in \mathcal{O}(D,M) \text{ with } f(0) = P, f'(0)(e_1) = \xi/\alpha\right\} \\ = \quad & \inf\left\{\frac{|\xi|}{|f'(0)(e_1)|} : f \in \mathcal{H}(D,M), f(0) = P, f'(0)(e_1) \right. \\ & \left. \text{is a constant multiple of } \xi\right\} \end{aligned}$$

We say that M is *Kobayashi hyperbolic* if the integrated Kobayashi distance d_K is actually a metric on M. It is worth noting that a planar domain is hyperbolic if and only if its covering space is the disk. Identifying hyperbolic domains or manifolds in higher dimensions is notably more difficult.

Any domain in the plane that is covered by the disk is complete hyperbolic. Any bounded, strongly pseudoconvex domain in $\mathbb{C}^n$ with C^2 boundary is complete hyperbolic (see [84]). Since any pseudoconvex domain can be exhausted by smoothly bouneded, strongly pseudoconvex domains, this is a useful result. Also a smoothly bounded, finite type domain in $\mathbb{C}^2$ is complete hyperbolic (see [42]). It is straightforward to check that if Ω is a smoothly bounded domain and each boundary point is a peak point, then the domain is complete hyperbolic. The work [75] is relevant here.

A key property of the Kobayashi/Royden pseudometric, and one that we shall use decisively in what follows, is that holomorphic mappings are distance non-increasing in the Kobayashi/Royden pseudometric. See [143, Ch. 11] for details. We sometimes instead use the phrase "distance decreasing".

We will use the notation $B_K^M(P,r)$ to denote an open ball in the Kobayashi/Royden pseudometric of M with center P and radius r. Likewise, $\overline{B}_k^M(P,r)$ denotes the closed ball.

5.4 The Ascoli-Arzelà Theorem and Relative Compactness

We now recall the Ascoli-Arzelà theorem in a form that will be most useful for us. The proof is straightforward and we omit it (or see [258]).

Theorem 5.4.1 (Ascoli-Arzelà). *Let M and N be complex manifolds and let $\mathcal{F} \subseteq C(M,N)$. We have that $\mathcal{F}$ is relatively compact in $C(M,N)$ if and only if*

(a) $\mathcal{F}$ is equicontinuous on M;

(b) for each $z \in M$, the set

$$\mathcal{F}(z) \equiv \{f(z) : f \in \mathcal{F}\}$$

is relatively compact in N.

Now our first basic result about the circle of ideas connected to normal families is this:

Proposition 5.4.2. *Suppose that M and N are complete hyperbolic and that $\mathcal{F} \subseteq \mathcal{H}(M, N)$. Then $\mathcal{F}$ is relatively compact in $\mathcal{C}(M, N)$ if and only if there is a point $z_0 \in M$ such that $\mathcal{F}(z_0) \equiv \{f(z_0) : f \in \mathcal{F}\}$ is relatively compact in N.*

Proof. Of course we equip each of M and N with the Kobayashi/Royden pseudometric. Then, by the distance non-increasing property of the Kobayashi/Royden pseudometric, any holomorphic function $f : M \to N$ is Lipschitz with Lipschitz constant equal to 1. Therefore, condition (**a**) of the Ascoli-Arzelà theorem holds. Because $\mathcal{F}(z_0)$ is relatively compact in N, there is a point $w_0 \in N$ and a constant $R > 0$ so that $\mathcal{F}(z_0) \subseteq \overline{B}_K^N(w_0, R)$. Let z be any point of M and set $r = d_K^M(z_0, z)$. Then, for all $f \in \mathcal{F}$, we know that

$$d_K^N(f(z_0), f(z)) \leq d_K^M(z_0, z)\,.$$

As a result,

$$d_K^N(w_0, f(z)) \leq d_K^N(w_0, f(z_0)) + d_K^N(f(z_0), f(z)) \leq R + r\,.$$

So

$$\mathcal{F}(z) \subseteq B_K^N(w_0, R + r)\,.$$

Because N is complete hyperbolic, we can be sure that closed hyperbolic disks are compact. Therefore, $\mathcal{F}(z)$ is relatively compact in N. □

An easy corollary is this:

Corollary 5.4.3. *Suppose that N and P are complete hyperbolic. Assume that $\overline{N} \subseteq P$. Let M be another complete hyperbolic manifold. Then $\mathcal{H}(M, N)$ is relatively compact in $\mathcal{C}(M, P)$.*

Proof. Certainly $\mathcal{H}(M, N) \subseteq \mathcal{H}(M, P)$. Hence the elements of $\mathcal{H}(M, N)$ satisfy a Lipschitz condition with norm 1. Because $\overline{N} \subseteq P$, we know that every orbit $\mathcal{H}(M, N)(z_0)$ is relatively compact in P. As a result, $\mathcal{H}(M, N)$ is relatively compact in $\mathcal{H}(M, P)$. □

Taking into account the definition of normal family given in Definition 5.1.3, we can now formulate the following result:

Theorem 5.4.4. *Suppose that M and N are complete hyperbolic and that $\mathcal{F} \subseteq \mathcal{H}(M, N)$. Then $\mathcal{F}$ is a normal family.*

Proof. Either there is a point $z_0 \in M$ such that $\mathcal{F}(z_0) \equiv \{f(z_0) : f \in \mathcal{F}\}$ is relatively compact in N or there is not. In the first instance, the proposition applies. In the second instance there is a complactly divergent subsequence. □

Example 5.4.5. Let M be the unit disk in the complex plane and let N also be the unit disk. Then both M and N are complete hyperbolic. And certainly Montel's original theorem tells us that any family $\mathcal{F} \subseteq \mathcal{H}(M, N)$ is normal.

Now let M be as in the last paragraph and N be the upper half plane. Then certainly both M and N are complete hyperbolic. Consider the family of mappings

$$f_j(\zeta) = ij + j\zeta$$

from M to N. Then $f_j(0)$ tends to infinity, and we see that the family $\{f_j\}$ is compactly divergent.

Of course the extended complex plane (the Riemann sphere $\widehat{\mathbb{C}}$) is not hyperbolic. It is instead more convenient to use the spherical metric on $\widehat{\mathbb{C}}$. It is given by

$$\sigma(z) = \frac{2}{1 + |z|^2} .$$

It has corresponding spherical distance function

$$s(z, w) = 2 \tan^{-1} \frac{|z - w|}{1 + \overline{w}z} .$$

The next result is a version of Marty's theorem adapted to our present context.

Theorem 5.4.6. *Let M be complete hyperbolic and $\widehat{\mathbb{C}}$ the Riemann sphere equipped with the spherical metric. Let $\mathcal{F}$ be a family of holomorphic functions from M to $\widehat{\mathbb{C}}$. Assume that, for each $f \in \mathcal{F}$, $|\nabla f|$ is bounded from the Kobayashi/Royden metric to the spherical metric. Then $\mathcal{F}$ is a normal family.*

Proof. This result is immediate from calculus and the mean value theorem. □

5.5 Some More Sophisticated Normal Families Results

Now we have the following result which is in the nature of an Ascoli-Arzelà theorem for Lipschitz spaces. As usual we assume that each of M and N is a complete hyperbolic complex manifold.

Theorem 5.5.1. *Let $\mathcal{F} \subseteq \mathcal{C}(M, N)$ have the property that each $f \in \mathcal{F}$ is Lipschitz with Lipschitz constant equal to 1. Then $\mathcal{F}$ is relatively compact in $\mathcal{C}(M, N)$ if and only if there is a $z_0 \in M$ such that $\mathcal{F}(z_0) \equiv \{f(z_0) : f \in \mathcal{F}\}$ is relatively compact in N.*

Proof. The necessity is immediate from the Ascoli-Arzelà theorem proved in the last section.

For the sufficiency we shall also use the Ascoli-Arzelà theorem. The uniform Lipschitz condition on elements of $\mathcal{F}$ certainly tells us that $\mathcal{F}$ is equicontinuous on M. Fix a point $z \in M$. The uniform Lipschitz condition applied to the set $\{z, z_0\}$, along with the fact that $\mathcal{F}(z_0)$ is relatively compact, tells us immediately that $\mathcal{F}(z)$ is relatively compact in N because the metric is complete. Therefore, Condition (**b**) of the Ascoli-Arzelà theorem holds. □

Again we have a version of Theorem 5.4.4 in this new context. This takes into account the possibility of compact divergence.

Now for holomorphic mappings we will have a stronger result. It will be proved using the theorem that we just treated.

Theorem 5.5.2. *Let $\mathcal{F} \subseteq \mathcal{H}(M, N)$. Then $\mathcal{F}$ is relatively compact in $\mathcal{C}(M, N)$ if and only if there exists a point $z_0 \in M$ such that $\mathcal{F}(z_0) \equiv \{f(z_0) : f \in \mathcal{F}\}$ is relatively compact in Ω.*

Proof. The sufficiency follows immediately from the preceding theorem.

For the necessity it suffices to show that if $\overline{\mathcal{F}}$ is compact in $\mathcal{C}(M, N)$, then it is locally uniformly Lipschitz. So suppose that $\overline{\mathcal{F}}$ is compact in $\mathcal{C}(M, N)$. Fix a compact set $K \subset M$. We then see that the continuous functional

$$f \mapsto |f|_K$$

attains a maximum value (call it L) on $\overline{\mathcal{F}}$. That establishes the relative compactness of $\mathcal{F}(z_0)$. □

As an immediate consequence we have

Corollary 5.5.3. *Let M be complete hyperbolic as usual. Then $\mathcal{F} \subseteq \mathcal{H}(M, \mathbb{C})$ is relatively compact in $\mathcal{C}(M, \mathbb{C})$ if and only if $\mathcal{F}$ is locally uniformly Lipschitz and there exists a point $z_0 \in M$ such that $\mathcal{F}(z_0) \equiv \{f(z_0) : f \in \mathcal{F}\}$ is relatively compact in Ω.*

Corollary 5.5.4 (Royden)**.** *Let M be complete hyperbolic. Equip the extended complex plane $\widehat{\mathbb{C}}$ with the spherical metric. Then $\mathcal{F} \subseteq \mathcal{H}(M, \widehat{\mathbb{C}})$ is relatively compact in $\widehat{\mathbb{C}}(M, \widehat{\mathbb{C}})$ if and only if it is locally uniformly Lipschitz.*

Corollary 5.5.5 (Marty)**.** *Let M be complete hyperbolic. Then $\mathcal{F} \subseteq \mathcal{H}(M, \widehat{\mathbb{C}})$ is relatively compact in $\mathcal{C}(M, \widehat{\mathbb{C}})$ if and only if $\mathcal{F}$ is locally uniformly Lipschitz.*

5.6 Some Examples

Example 5.6.1. Let M be the unit disk in the complex plane and let N be the unit disk in the complex plane. Then the Kobayashi/Royden pseudometric is a true metric on both these manifolds, and it is complete. So our theorems apply, and they yield a version of the usual classical Montel theorem.

Example 5.6.2. Let $M = N = \mathbb{C} \setminus \overline{D}(0,1)$. By an inversion we may map each of M and N to a punctured disk. So each of the domains is complete hyperbolic, and our theorems apply.

Example 5.6.3. Let $N = \widehat{\mathbb{C}} \setminus \{-1, 1\}$. Let $P \in N$. If $f : D \to N$ is a holomorphic function that is a candidate for the Kobayashi/Royden pseudometric at P (with $f(0) = P$ of course), then of course we are endeavoring to maximize $|f'(0)|$ so we may as well suppose that f is univalent. So the image of f is a simply connected region that does not contain -1 and 1. Hence it will not contain a continuous curve X that connects -1 to 1. Since $\widehat{\mathbb{C}} \setminus X$ is conformally equivalent to the disk no matter what the choice of X, we may as well assume that X is simply the interval $\{(x + i0) : -1 \leq x \leq 1\}$. And of course it is obvious that $\widehat{\mathbb{C}} \setminus [-1, 1]$ is conformally equivalent to the disk. Thus N is complete hyperbolic. Let M be the unit disk. Then our theorems apply to this M and N. So we have recovered the generalized Montel theorem for functions that omit two complex values.

Note that the traditional way to handle this example is with the elliptic modular function (see [4, 95]). Our approach of using the Kobayashi/Royden metric is more direct and more accessible.

5.7 Taut Mappings

The paper [287] introduces the important notion of taut mapping.

Definition 5.7.1. A complex manifold M is *taut* if the collection of holomorphic mappings from the unit disk D to M is a normal family.

The paper [127] proves that any smoothly bounded, pseudoconvex domain in $\mathbb{C}^n$ is taut. The ideas in this paper show that any complete hyperbolic manifold is taut.

5.8 Classical Definition of Normal Holomorphic Mapping

The concept of normal meromorphic function allows one to attribute to one function the properties related to some class of functions, namely with a normal family of meromorphic functions. The idea of such a transfer is simple; it was first introduced by Yosida [288] in 1934 and was discussed by Noshiro [226] in 1939.

Noshiro's definition reads as follows: a meromorphic function $f : D \to \widehat{\mathbb{C}}$ belongs to the class (A) if and only if $\{f \circ g : g \in \mathrm{Aut}(D)\}$ is a normal family ([226]). He proved some interesting results about the class (A).

Noshiro's definition only applied to the unit disk but was later extended by Lehto and Virtanen in their seminal paper of 1957 [164]: A meromorphic function $f(z)$ is called *normal* in a simply connected domain Ω, if the family $\{f(S(z))\}$ is normal, where $z' = S(z)$ denotes an arbitrary conformal mapping of Ω onto itself. They showed that the notion of a normal meromorphic function is closely related to some of the most important problems of the boundary behavior of meromorphic functions.

It is well-known (Marty [187]) that a family $\mathcal{F}$ of meromorphic functions is normal in a domain $\Omega \subset \mathbb{C}$ if and only if there exists a positive constant M such that

$$\frac{|f'(z)|}{1+|f(z)|^2} \leq M \text{ for all } f \in \mathcal{F}$$

on every compact set in Ω.

Using this criterion, we get

Theorem 5.8.1 (Noshiro [226], Lehto-Virtanen [164]). *A nonconstant function $f(z)$ is a normal function in the unit disk $D = \{z \in \mathbb{C}\colon |z| < 1\}$ if and only if*

$$\frac{|f'(z)|}{1+|f(z)|^2} \leq C\frac{1}{1-|z|^2} \tag{5.1}$$

is satisfied at every point of D where C is a fixed finite constant.

We see at once that inequality (5.1) can be rewritten in the following form

$$L_z(\log 1 + |f|^2, 1) \leq C(F_K^D(z,1))^2. \tag{5.2}$$

In this section, we define the concept of normal holomorphic mappings similar to normal meromorphic functions (cf. [164]).

Let M and N be complex manifolds. We denote the set of holomorphic mappings from M into N by $\mathcal{O}(M,N)$. We say that a subset $\mathcal{F}$ of $\mathcal{O}(M,N)$ is a normal family if $\mathcal{F}$ is relatively compact in $\mathcal{O}(M,N)$ in the sense of the compact-open topology.

Definition 5.8.2. Let Ω be a homogeneous, bounded domain in $\mathbb{C}^n$ and let N be a complex manifold. We say that a holomorphic mapping $f : D \to N$ is *normal* if the family

$$\mathcal{F} = \{f \circ g; g \in \mathrm{Aut}(\Omega)\}$$

is normal, where $\mathrm{Aut}(\Omega)$ denotes the holomorphic automorphism group of Ω.

Definition 5.8.3. We say that a subset $\mathcal{F}$ of $\mathcal{O}(\Omega, N)$ is $\mathrm{Aut}(\Omega)$ invariant if $f \circ g \in \mathcal{F}$ for every $f \in \mathcal{F}$ and every $g \in \mathrm{Aut}(\Omega)$.

Since Riemann's sphere with the spherical metric is a compact Hermitian manifold and the Kobayasi metric coincides with the Poincarè metric in the unit disk, we find, taking into account (5.2), the following theorem to be a natural result:

Theorem 5.8.4. *Let Ω be a bounded homogeneous domain in $\mathbb{C}^n$ and (N, ds_N^2) be a Hermitian complex manifold. If a subset $\mathcal{F}$ of $\mathcal{O}(\Omega, N)$ is an $Aut(\Omega)$-invariant normal family, then there exists a constant C such that $f^*ds_N^2 \leq C \cdot F_\Omega^2$ for every $f \in \mathcal{F}$.*

For the proof we need the following proposition.

Proposition 5.8.5. *Let F_Ω be the Kobayashi metric of a homogeneous bounded domain Ω in $\mathbb{C}^n$ and (N, ds_N^2) be a compact Hermitian manifold. If a holomorphic mapping $f : \Omega \to N$ satisfies*

$$f^*ds_N^2 \leq C \cdot F_\Omega^2$$

for a finite constant C, then f is a normal holomorphic mapping.

Proof. As F_Ω^2 is $\mathrm{Aut}(\Omega)$-invariant,

$$(f \circ g)^*ds_N^2 = g^*f^*ds_N^2 \leq C \cdot g^*F_\Omega^2 = C \cdot F_\Omega^2$$

for every $g \in \mathrm{Aut}(\Omega)$. Hence $\mathcal{F} = \{f \circ g : g \in \mathrm{Aut}(\Omega)\}$ is equi-continuous. As N is compact, $\mathcal{F} \subset \mathcal{O}(\Omega, N)$ is normal according to the Ascoli-Arzelà theorem. □

Proof of Theorem 5.8.4. Let

$$C(z) = \sup_{\|\mu\|=1, f\in\mathcal{F}} f^*ds_N^2/F_\Omega^2(z, \mu)$$

for each $z \in \Omega$, and $\mu \in T_zC^n$, where $\| \cdot \|$ is the length by a flat metric of C^n. We first prove that $C(0) < \infty$ for a point $0 \in \Omega$.

Suppose that $C(0) = \infty$. Then there exist sequences $\{f_j\} \subset \mathcal{F}$ and $\{\mu_j \in T_0\mathbb{C}^n, \|\mu_j\| = 1\}$ of holomorphic tangent vectors such that

(i) $f_n^*ds_N^2\,(0, \mu_n) \geq n^2 \cdot ds_\Omega^2\,(0, \mu_n)$,

where we may assume that $\{\mu_n\}$ converges to $\mu \in T_0C^n$. From (i) we see

(ii) $\|f_{n*}\mu_n\|_N \geq n \cdot \alpha$,

where $\| \cdot \|_N$ is the length measured by ds_N^2 and α is a positive constant. Since $\mathcal{F}$ is relatively compact in $\mathcal{O}(\Omega, N)$, some subsequence $\{f_{j_k}\}$ of $\{f_j\}$ exists such that $f_{j_k} \to f \in \mathcal{O}(\Omega, N)$ and $f_{j_k}(0) \to p \in N$ as $k \to \infty$. Consequently $f_{j_k*}\mu_{j_k} \to f_*\mu$ and $\|f_{j_k*}\mu_{j_k}\|_N \to \|f_*\mu\|_N < \infty$ as $k \to \infty$. This is contradictory to (ii) and hence $C(0) < \infty$.

We secondly prove that $C(z)$ is constant on Ω. From definition of $C(z)$

$$f^*ds_N^2(z) \leq C(z) \cdot F_\Omega^2(z)$$

for every $f \in \mathcal{F}$. As F_Ω^2 is $\mathrm{Aut}(\Omega)$-invariant,

$$(f \circ g)^*ds_N^2(z) \leq C(g(z)) \cdot F_\Omega^2(z)$$

for every $g \in \mathrm{Aut}(\Omega)$. Hence $C(g(z)) \geq C(z)$. Similarly $C(z) \geq C(g(z))$ and so $C(z) = C(g(z))$. As Ω is homogeneous, $C(z)$ is constant on Ω and the theorem is proved. □

Corollary 5.8.6. *Let Ω and N be as above. If $f : \Omega \to N$ is a normal holomorphic mapping, then there exists a finite constant C such that $f^*ds_N^2 \leq C \cdot F_\Omega^2$.*

5.9 Examples

In this section we give some examples of normal holomorphic mappings.

Example 5.9.1. Let $D = \{|z| < 1\} \subset \mathbb{C}$ be the unit disk.

(i) We choose q arbitrarily distinct points $p_1, \ldots, p_q$ in the Riemann sphere $\widehat{\mathbb{C}}$. Let $\mathcal{F}$ be a family of holomorphic mappings from D into $\widehat{\mathbb{C}}$ which satisfies the following condition (C).

(C) Every f in $\mathcal{F}$ takes each value $p_i \in \widehat{\mathbb{C}}$ $(i = 1, \ldots, q)$ with multiplicity $\geq m_i$ or f omits a point p_i and $\{m_i\}_{i=1}^q$ satisfies

$$-2 + \sum_{i=1}^{q} \left(1 - \frac{1}{m_i}\right) > 0\,,$$

where we set $m_i = \infty$ in the case f omits p_i.

Then $\mathcal{F}$ is normal (Montel-Valiron, cf. Theorem 8.3 in H. Fujimoto [79]). Particularly, if a holomorphic map $f : D \to \widehat{\mathbb{C}}$ satisfies the condition (C), then f is a normal holomorphic mapping, that is, a normal meromorphic function (cf. [164]).

(ii) Let $T = \mathbb{C}/L$ be a complex torus and take a point p of T. We consider the family $\mathcal{F}$ of all holomorphic mappings from D into T which omit a value $p \in T$. Then $\mathcal{F}$ is normal. Particularly, every f in $\mathcal{F}$ is a normal holomorphic mapping.

(iii) Let V be a compact Riemann surface of genus ≥ 2. Then $\mathcal{O}(D, V)$ is normal and hence every holomorphic mapping $f : D \to V$ is normal.

(iv) Let N be a paracompact connected complex analytic space and M be a hyperbolically embedded subspace of N. Then $\mathcal{O}(D, M)$ is relatively compact in $\mathcal{O}(D, N)$ (See [129]). Particularly a holomorphic mapping from D into N is normal if the image of f is contained in M. For such examples see M. L. Green [90] and [89].

Example 5.9.2. **(i)** Let $ds_D^2 = dzd\bar{z}/\left(1 - |z|^2\right)^2$ be the Bergman metric of the unit disk D. We take homogeneous coordinates (z_0, z_1) of $\widehat{\mathbb{C}}$ and let $w = z_1/z_0$ for $z_0 \neq 0$. Then the Fubini-Study metric of $\widehat{\mathbb{C}}$ is given by

$$ds_{P^1}^2 = \frac{1}{(1 + |w|^2)^2} dwd\bar{w}.$$

Consequently a meromorphic function $f : D \to \widehat{\mathbb{C}}$ is normal if and only if

$$\frac{|f'(z)|}{1 + |f(z)|^2} \leq C \cdot \frac{1}{1 - |z|^2} \quad (z \in D)$$

for some finite constant C (cf. [164]).

(ii) Let $T = \mathbb{C}^n/L$ be an n-dimensional complex torus. Then $f : D \to T$ is normal if and only if a lifting $\tilde{f} = (f_1, \ldots, f_n)$ of f to C^n satisfies

$$\sum_{i=1}^{n} |f_i'(z)|^2 \leq C \cdot \frac{1}{\left(1 - |z|^2\right)^2} \tag{5.1}$$

for some finite constant C. A holomorphic function satisfying condition (5.1) is classically called a Bloch function.

5.10 The Estimate for Characteristic Functions

In this section, we consider the case that Ω is the unit ball

$$B = \left\{ \sum_{i=1}^{n} |z_i|^2 < 1 \right\} \subset \mathbb{C}^n.$$

The Kobayashi metric of B is given by

$$F_B^2 = \sum_{i,j=1}^{n} \frac{1}{\left(1 - |z|^2\right)^2} \left[\left(1 - |z|^2\right) \delta_{ij} + z_i \bar{z}_j \right] dz_i d\bar{z}_j$$

where $|z|^2 = \sum_{i=1}^{n} |z_i|^2$. Let (N, ds_N^2) be a Hermitian manifold and let ω_N, ω_B be the $(1,1)$-form associated with ds_N^2, F_B^2 respectively. We calculate the characteristic function of a normal holomorphic mapping $f : B \to N$.

Lemma 5.10.1. *Let $f : B \to N$ be a holomorphic mapping such that $f^*ds_N^2 \leq C \cdot F_B^2$, where C is a constant. Set $\varphi = \sqrt{-1}/2\sum_{i=1}^n dz_i \wedge d\bar{z}_i$ and $B(r) = \{|z| < r\}$. Then*

$$\int_{B(r)} f^*\omega_N \wedge \varphi^{n-1} \leq C \cdot \int_{B(r)} \omega_B \wedge \varphi^{n-1}.$$

Proof. We set

$$f^*\omega_N = \frac{\sqrt{-1}}{2} \sum_{i,j=1}^n g_{ij} dz_i \wedge d\bar{z}_j,$$

$$\omega_B = \frac{\sqrt{-1}}{2} \sum_{i,j=1}^n h_{i\bar{j}} dz_i \wedge d\bar{z}_j.$$

Then

$$f^*\omega_N \wedge \varphi^{n-1} = (n-1)! \cdot \left(\sum_{i=1}^n g_{i\bar{i}}\right) \prod_{j=1}^n \frac{\sqrt{-1}}{2} dz_j \wedge d\bar{z}_j$$
$$= \frac{1}{n}\left(\sum_{i=1}^n g_{i\bar{i}}\right) \varphi^n$$

Since a matrix $\left(g_{k\bar{j}}\right) - C \cdot \left(h_{k\bar{j}}\right)$ is negative semi-definite,

$$g_{k\bar{k}} \leq C \cdot h_{k\bar{k}} \quad (k = 1, \ldots, n).$$

Hence

$$\int_{B(r)} f^*\omega_N \wedge \varphi^{n-1} = \frac{1}{n} \int_{B(r)} \left(\sum_{i=1}^n g_{i\bar{i}}\right) \varphi^n \leq C \cdot \frac{1}{n} \int_{B(r)} \left(\sum_{i=1}^n h_{i\bar{i}}\right) \varphi^n$$
$$= C \cdot \int_{B(r)} \omega_{Ln} \wedge \varphi^{n-1}$$

□

Lemma 5.10.2. *We have that*

$$\frac{1}{\pi^n} \int_0^r \frac{dt}{t^{2n-1}} \left(\int_{B(r)} \omega_{B_n} \wedge \varphi^{n-1}\right) \leq \frac{1}{2} \log \frac{1}{1-r^2} + (n-1) \cdot C_1 \quad (0 < r < 1),$$

where $C_1 = 2\log 2$.

Proof. We know that the area of the boundary $\partial B(r)$ of $B(r)$ is $2\pi^n \cdot r^{2n-1}/(n-1)!$ and

$$\omega_B \wedge \varphi^{n-1} = \frac{n!}{(1-|z|^2)^2}\left(1 - \frac{n-1}{n}|z|^2\right) \cdot \frac{\varphi^n}{n!},$$

$$\int_{B(t)} \omega_B \wedge \varphi^{n-1} = \int_0^t \frac{n!}{(1-u^2)^2}\left(1 - \frac{n-1}{n}u^2\right) \frac{2\pi^n}{(n-1)!} u^{2n-1} du$$
$$= 2\pi^n \int_0^t \frac{u^{2n-1}}{(1-u^2)^2} du + 2\pi^n(n-1) \int_0^t \frac{u^{2n-1}}{1-u^2} du\,.$$

Using the estimates

$$\int_0^t \frac{u^{2n-1}}{(1-u^2)^2}du \le t^{2n-2}\int_0^t \frac{udu}{(1-u^2)^2}$$

and

$$\int_0^t \frac{u^{2n-1}}{1-u^2}du \le t^{2n-1}\int_0^t \frac{du}{1-u^2},$$

we see that

$$\int_{B(t)} \omega_B \wedge \varphi^{n-1} \le \pi^n \frac{t^{2n}}{1-t^2} + (n-1)\pi^n \cdot t^{2n-1}\left[\log\frac{1}{1-t} + \log(1+t)\right].$$

Hence

$$\begin{aligned}
&\frac{1}{\pi^n}\int_0^r \frac{dt}{t^{2n-1}}\left(\int_{B(t)} \omega_B \wedge \varphi^{n-1}\right)\\
&\le \int_0^r \frac{tdt}{1-t^2} + (n-1)\int_0^r \left[\log\frac{1}{1-t} + \log(1+t)\right]dt\\
&= \frac{1}{2}\log\frac{1}{1-r^2} + (n-1)C_1,
\end{aligned}$$

where

$$C_1 = \int_0^1 \left[\log\frac{1}{1-t} + \log(1+t)\right]dt = 2\log 2.$$

□

Definition 5.10.3. Let B, $\left(N, ds_N^2\right)$ be as above and $f : B \to N$ be a holomorphic mapping. We define the *characteristic function* of f by

$$T(f,r) = \frac{1}{\pi^n}\int_0^r \frac{dt}{t^{2n-1}}\left(\int_{B(t)} f^*\omega_B \wedge \varphi^{n-1}\right).$$

Proposition 5.10.4. *Let $\mathcal{F}$ be an Aut(B)-invariant normal family of holomorphic mapping from B to $\left(N, ds_N^2\right)$. Then there exists a constant $C(r)$ depending only r such that*

$$T(f,r) \le C(r) \quad (0 \le r < 1)$$

for every $f \in \mathcal{F}$.

Proof. Since there exists a constant C such that $f^*ds_N^2 \le C \cdot F_B^2$ for every $f \in \mathcal{F}$, according to Theorem 5.8.4, the result is clear by Lemma 5.10.2. In fact we can take

$$C(r) = \frac{C}{2}\log\frac{1}{1-r^2} + (n-1)C_1 \cdot C.$$

□

Corollary 5.10.5. *If a holomorphic mapping $f : B \to N$ is normal, then*

$$T(f \circ g, r) = O\left(\log \frac{1}{1-r}\right) \quad (r \to 1)$$

for every $g \in Aut(B)$.

Remark 5.10.6. The property of the characteristic function for a normal holomorphic mapping is stronger than the property

$$T(f, r) = O\left(\log \frac{1}{1-r}\right) \quad (r \to 1).$$

Remark 5.10.7. If N is the projective space P^n, the inverse of the above proposition holds. This is clear by Theorem 5.10 in H. Fujimoto [79], because B is a homogeneous domain.

5.11 Normal Mappings

In view of Corollary 5.8.6 the definition of normal mappings is extended in the obvious way to arbitrary complex manifolds.

Definition 5.11.1. Let M be a complex manifold and let (N, ds_N^2) be a compact Hermitian complex manifold. Let F_M be the Kobayashi metrics on M. A holomorphic map $f\colon M \to N$ is called normal if there exists a positive constant C such that

$$ds_N^2 f(z, v) \leq C(F_M(z, v))^2 \tag{5.1}$$

for all $(z, v) \in T_z(M)$.

In view of everything else we have said in this book, that is just the right definition.

In the paper [47], Cima and Krantz show that a more natural, geometrically invariant definition of "normal function" is this. Let $\Omega \subseteq \mathbb{C}^n$ be a complete hyperbolic domains. Let $\widehat{\mathbb{C}}$ be as usual the Riemann sphere. Then a function $f : \Omega \to \widehat{\mathbb{C}}$ is normal if and only if the derivative ∇f is bounded from the Kobayashi metric on Ω to the spherical metric on $\widehat{\mathbb{C}}$.

This idea fits naturally with the discussions we have had in earlier sections, because a function as described in the previous paragraph is, by the mean value theorem, Lipschitz of order 1 in the two given metrics. In fact, according to Marty's classical theorem, a family of functions $\mathcal{F} = \{f_\alpha\}_{\alpha \in A}$ of holomorphic functions from Ω to $\widehat{\mathbb{C}}$ is normal if and only if the family $\{\nabla f_\alpha\}$ is bounded from the Kobayashi metric to the spherical metric.

Recall that the Lindelöf principle, in its simplest classical formulation, is this.

Proposition 5.11.2. *Let $f : D \to \mathbb{C}$ be a bounded, holomorphic function. Suppose that $\lim_{r \to 1^-} f(r)$ exists and equals ℓ. Then the nontangential limit of f at 1 exists and equals ℓ.*

Lehto and Virtanen taught us, in the context of the last proposition, **(i)** how to treat a broader class of functions and **(ii)** how to treat a broader class of boundary curves in the hypothesis. It is particularly interesting that this phenomenon of boundary behavior is related to the idea of normal families. And even more interesting that it can be interpreted in the language of invariant metrics.

To put these matters into perspective, we record now a version of the Lindelöf principle in this new context. In particular, what is new here is that we have a Lindelöf principle for mappings of manifolds rather than for function on domains.

Proposition 5.11.3. *Let M, N be complete hyperbolic manifolds as usual. Let $\gamma : [0, 1) \to M$ be a continuous curve with $\lim_{t \to 1^-} f(\gamma(t)) = \ell$. Here ℓ could be a point of N or it could be a "limit point of N". If $\widetilde{\gamma} : [0, 1) \to M$ is another curve with $\lim_{t \to 1^-} F_K^M(\gamma(t), \widetilde{\gamma}(t)) = 0$, then $\lim_{t \to 1^-} f(\widetilde{\gamma}(t)) = \ell$.*

Proof. This result is immediate from the distance non-increasing property of the metrics. □

5.12 A Generalization of the Big Picard Theorem

We set $D = \{w \in \mathbb{C}, |w| < 1\}$ and $D^* = \{w \in \mathbb{C}, 0 < |w| < 1\}$. Let $\pi : D \to D^*$ be the universal covering, where $\pi(w) = e^{(w+1)/(w-1)}$.

Definition 5.12.1. A holomorphic mapping f from D^* into a complex analytic space N is said to be normal if $f \circ \pi : D \to N$ is a normal holomorphic mapping.

Theorem 5.12.2. *Let N be a paracompact complex manifold. If $f : D^* \to N$ is a normal holomorphic mapping, then f can be extended to a holomorphic mapping from D into N.*

Proof. First of all, we show that there exists a sequence $\{z_n\}_{n=1}^{\infty} \subset D^*$ converging to the origin 0 such that $\{f(z_n)\}_{n=1}^{\infty}$ converges to some point p_0 in N. Take a sequence $\{z_n\}_{n=1}^{\infty} \subset D^*$ converging to 0, points $\{w_n\}$ such that $\pi(w_n) = z_n$ and take $g_n \in \mathrm{Aut}(D)$ such that $g_n(0) = w_n$. Consider the family $\mathcal{F} = \{f \circ \pi \circ g_n\}_{n=1}^{\infty}$. By the hypothesis, $\mathcal{F}$ is normal. We can take a subsequence $\{f \circ \pi \circ g_{n_k}\}_{n=1}^{\infty}$ of $\{f \circ \pi \circ g_n\}_{n=1}^{\infty}$ such that $\{f \circ \pi \circ g_{n_k}\}_{k=1}^{\infty}$ converges to some $h \in \mathcal{O}(D, N)$. Particularly $f \circ \pi \circ g_{n_k}(0) \to h(0) = p_0$. We set $\pi \circ g_{n_k}(0) = z_k$. Then $f(z_k) \to p_0$.

Secondly, we take a Hermitian metric ds_N^2 of N. Since $f \circ \pi : D \to N$ is normal, according to Theorem 5.8.4

$$(f \circ \pi)^* ds_N^2 \leq C \cdot ds_N^2, \tag{5.1}$$

where C is a constant and $ds_N^2 = dzd\bar{z}/\left(1 - |z|^2\right)^2$. Now (5.1) implies

$$f^* ds_N^2 \leq C \cdot ds_{D^*}^2 \tag{5.2}$$

where

$$ds_{D^*}^2 = \frac{1}{4} \cdot \frac{dzd\bar{z}}{|z|^2 (\log 1/z)^2}.$$

We set $r_k = |z_k|$. It may be assumed that $\{r_k\}_{k=1}^{\infty}$ is a decreasing sequence. We denote by d_N the distance of N defined by ds_N^2. Then

$$d_N\left(f(z), p_0\right) \leq d_N\left(f(z), f\left(z_k\right)\right) + d_N\left(f\left(z_k\right), p_0\right). \tag{5.3}$$

When $|z| = r_k$, according to (5.2),

$$d_N\left(f(z), f\left(z_k\right)\right) = \frac{\sqrt{C}}{2} \int_{z_k}^{z} \frac{|dz|}{|z| \log 1/|z|} \leq \sqrt{C} \cdot \frac{\pi}{\log 1/r_k}. \tag{5.4}$$

The right-hand side converges to 0 as $k \to \infty$. According to (5.3) and (5.4), we see that, for every ε-neighborhood W of p_0, when k_0 is sufficiently large, $f\left(\{|z| = r_k\}\right)$ is contained in W for $k \geqq k_0$. By the method in the proof of Theorem 3.1 in S. Kobayashi [135, Chapter VI] we can see that, for every neighborhood U of p_0, if $\delta > 0$ is sufficiently small, $f(\{z \in D^* : |z| < \delta\})$ is contained in U, and so we can define a holomorphic extension of f to D by $f(0) = p_0$. □

Remark 5.12.3. The above theorem remains valid for a paracompact connected complex analytic space N, because N has a Hermitian metric h in the extended sense, which induces the original topology of N, and the discussion of Section 5.1 and the above proof are available for such (N, h) (see the introduction of P. Kiernan [129]).

Remark 5.12.4. Let N be a paracompact connected complex analytic space and M be hyperbolically embedded subspace of N. Then a holomorphic mapping $f : D^* \to N$ with $f\left(D^*\right) \subset M$ is normal. Hence, Theorem 5.12.2 is a generalization of a big Picard theorem given in S. Kobayashi [135, Chapter V, Theorem 6.1].

6

Some Classical Theorems and Families of Normal Maps in Several Complex Variables

Preamble: We show that normal maps as defined by authors in various settings are, as singleton sets, uniformly normal families and that, in these settings, the maps in uniformly normal families are normal. We give an example of a family of functions which is not uniformly normal in the framework of the original Noshiro-Lehto-Virtanen definition of normal maps, even though each of its members is a normal function. In this chapter we provide significant generalizations of theorems of Brody, Lehto and Virtanen, Hahn, and Zaidenberg. A number of results for uniformly normal families on hyperbolic manifolds are shown to carry over to uniformly normal families on arbitrary complex spaces.

A Zaidenberg generalization (to complex manifolds) of a single-variable version of Schottky's theorem due to Hayman is extended to uniformly normal families of functions on arbitrary complex spaces and provided with a simpler proof, a several variables form of a classical lemma of Bohr.

Proved theorems include **(1)** extension theorems of big Picard type for such families—defined on complex manifolds having divisors with normal crossings—which encompass results of Järvi, Kiernan, Kobayashi, and Kwack as special cases, and **(2)** generalizations to such families of an extension-convergence theorem due to Noguchi.

6.1 Preliminaries

The type of a Riemann surface can be characterized by the sign of the curvature of the natural metric it carries. Namely, elliptic with curvature $+1$, parabolic with curvature 0, and hyperbolic with curvature -1.

We are interested in complex spaces of hyperbolic type since they represent the general case. The good introduction to the theory of Complex Space are [98, Chapter 5] or [88, Chapters 1, 2].

DOI: 10.1201/9781032669861-6

Let Z be a complex manifold. By a closed complex subspace X we mean a closed subset which can be locally defined by a finite number of analytic equations.

By a complex subspace of X we mean a locally closed complex subspace, that is a closed complex subspace of an open subset of X.

We do not want to go through a systematic treatment of the foundations of complex spaces. For this we refer to Gunning and Rossi [98]. Gunning and Rossi defined an analytic space as a topological space, with an additional structure provided by a distinguished sheaf of germs of continuous functions. Thus the "geometry" of an analytic space is derived from the function theory. In this sense geometry cannot be separated from function theory, and in fact the distinction they make is one of terminology rather than substance. This will be observable in the many "geometric" facts which have function-theoretic significance, and vice versa.

There are several ways to extend the concept of hyperbolicity to higher dimensional complex spaces. Kobayashi hyperbolicity is based on the existence of a certain intrinsic distance, and this intrinsic distance was originally introduced to generalize Schwarz's lemma to higher dimensional complex spaces. Schwarz's lemma, reformulated by Pick, says that every holomorphic map from the unit disk D of $\mathbb{C}$ into itself is distance-decreasing with respect to the Poincaré distance ρ, and is at the heart of geometric function theory.

Our discussion of tangent spaces to analytic spaces is also not as thorough as it could be, and serves only as an introduction. The reader is urged to consult the paper of Whitney [285] which discusses more geometric candidates for the tangential structure and has many interesting examples.

Given a complex space X, Kobayashi [134, Chapter 3, Section 5, p. 86] has defined an intrinsic pseudometric, i.e., the infinitesimal form $\widehat{F}_X$ of the Kobayashi pseudodistance d_X, in the following way. All necessary algebraic results on norms are summarized in [134, Chapter 1, Section 2, p. 86].

Let T_xX and T_x^*X denote the Zariski tangent space and cotangent space of X at x. Since there are some unresolved technical problems when X has singularities, we often have to assume that X is a complex manifold. We first define a quasinorm F_X^* on the Zariski cotangent space T_x^*X by setting

$$F_X^*(\lambda) = \sup \|f^*\lambda\| \qquad \text{for} \qquad \lambda \in T_x^*X$$

where $\|f^*\lambda\|$ is the length of the cotangent vector $f^*\lambda \in T^*D$ measured by the Poincaré metric ds^2 of the unit disk D, and the supremum is taken over all $f \in \mathrm{Hol}(D, X)$ with $x \in f(D)$. Because of the homogeneity of D, it suffices to take the supremum over all $f \in \mathrm{Hol}(D, X)$ with $f(0) = x$.

The notation $\mathbf{T}(X, Y)$ ($\mathbf{C}(X, Y)$) ($\mathbf{H}(D^n, Y)$) denotes the space of maps (continuous maps) (holomorphic maps) from a complex space X to a (topological) (complex) space Y (with the compact-open topology). Also $Y^* = Y \cup \{\infty\}$ denotes the one-point compactification of a noncompact space Y and $Y^* = Y$ otherwise.

If $F \subset C(X,Y)$ we say that F is *evenly continuous* from $p \in X$ to $q \in Y$ if for each U open in Y about q, there exist V, W open in X, Y about p, q respectively such that

$$\{f \in F : f(p) \in W\} \subset \{f \in F : f(V) \subset U\}.$$

If F is evenly continuous from each $p \in X$ to each $q \in Y$, then we say that F is evenly continuous (from X to Y) [126].

We will have occasion to rely on the following topological version of the Ascoli-Arzelà theorem.

Proposition 6.1.1. *Let X be a regular locally compact space and let Y be a regular space. Then $F \subset C(X,Y)$ is relatively compact in $C(X,Y)$ if and only if*

(a) F is evenly continuous, and

(b) $F(x)$ is relatively compact in Y for each $x \in X$.

It is readily derived from a Kelley-Morse theorem (Theorem 7.21 in [126]) since $\overline{F(x)} = \overline{\overline{F}(x)}$ for each $x \in X$ and, under the hypothesis, F is evenly continuous if and only if $\overline{F}$ is evenly continuous, that $F(x) = \{f(x) : f \in F\}$, where $\overline{A}$ = topological closure of A.

If $F \subset C(Y,Z)$ and $G \subset C(X,Y)$, then we write $F \circ G$ for $\{f \circ g : g \in G, f \in F\}$.

A family of maps $F \subset \mathbf{H}(X,Y)$ is called a *normal family* if $F \circ H(D,X)$ is relatively compact in $\mathbf{H}(D,Y) \cup \{\infty\} \subset \mathbf{C}(D,Y^*)$.

We now recall the definition of uniformly normal families and normal maps [124].

Definition 6.1.2. Let X and Y be complex spaces. A family $F \subset H(X,Y)$ is *uniformly normal* if $F \circ H(M,X)$ is relatively compact in $C(M,Y^*)$ for each complex manifold M. We say that f is a *normal map* if $\{f\}$ is uniformly normal.

Here $Y^* = Y \cup \{\infty\}$ represents the Alexandroff one-point compactification of the topological space Y, $Y^* = Y$ if Y is compact.

It is clear from Definition 6.1.2 that each member of a uniformly normal family is a normal map in that setting.

Example 6.1.3 shows that a family of normal functions might fail to be uniformly normal.

Example 6.1.3. Define $F \subset H\left(D, \mathbf{P}^1(\mathbb{C})\right)$ by $F = \{f_n : n = 1, 2, \ldots\}$ where $f_n(z) = 1/n(nz+1)$. Then $|f_n(z)| \geq 1/n(n+1)$ on D and f_n is a normal map in the framework of Lehto-Virtanen by the Picard Theorem. Define $\varphi_n \in A(D)$ by $\varphi_n(z) = \left(n^3 z + \left(1 - n^2\right)\right) / \left(\left(1 - n^2\right) z + n^3\right)$. Then $f_n \circ \varphi_n(0) \nrightarrow 0$ while $f_n \circ \varphi_n\left(n^{-1} - n^{-3}\right) \to 0$. Hence F is not uniformly normal.

Definition 6.1.2 encompasses and unifies the concepts of normal maps defined by authors in various settings (see [47], [81], [119] [112], [164]). The observation that a family of holomorphic maps from $(D \setminus \{0\})^n$ into a complex space Y is uniformly normal if and only if there is a distance E on Y such that each member is distance decreasing with respect to E together with the hyperbolic distance of $(D^*)^n$ leads to extensions of the big Picard theorem. The Kiernan-Kobayashi-Kwack Theorem and Noguchi's convergence theorems for uniformly normal families [157] tell us that each element of a uniformly normal family of maps from $(D^*)^n$ to a relatively compact subspace of a complex space Y extends to a holomorphic map and the family of extensions is also uniformly normal [122].

Abate [1] showed that a complex space Y is hyperbolic iff $\mathbf{H}(D, Y)$ is a relatively compact subset of $C(D, Y^*)$. Hence $\mathbf{H}(D, Y)$ is a uniformly normal family if Y is hyperbolic. Where $Y^* = Y \cup \{\infty\}$ represents the Alexandroff one-point compactification of the topological space $Y, Y^* = Y$ if Y is compact.

The proofs of Propositions 6.1.4 and 6.1.5 are omitted since they are evident from Definition 6.1.2.

Proposition 6.1.4. *If M is a complex manifold and Y is a complex space and $F \subset H(M, Y)$ is uniformly normal, then F is relatively compact in $C(M, Y^*)$.*

Proposition 6.1.5. *If X and Y are complex spaces, then the following statements are equivalent for $F \subset H(X, Y)$:*

1. *F is uniformly normal.*
2. *If Z is a complex space and $G \subset H(Z, X)$, then $F \circ G$ is uniformly normal.*
3. *If Z is a complex subspace of X, then the family of restrictions of the elements of F to Z is uniformly normal.*

In Proposition 6.1.6 we establish that the requirement that all complex manifolds M satisfy the condition in Definition 6.1.2 may be replaced by a condition on D.

Proposition 6.1.6. *If X, Y are complex spaces, then the following are equivalent for $F \subset H(X, Y)$*

(1) F is uniformly normal.

(2) $F \circ H(D, X)$ is uniformly normal.

(3) $F \circ H(D, X)$ is relatively compact in $C(D, Y^*)$.

(4) The closure of F in $H(X, Y)$ is uniformly normal.

Proof. **(1)** $\Rightarrow$ **(2)** $\Rightarrow$ **(3)** and **(4)** $\Rightarrow$ **(1)**; from Propositions 6.1.5 and 6.1.4.

(3) $\Rightarrow$ **(1)**. If **(1)** does not hold we may assume $M = \{p \in C^m : \|p\| < 1\}$ and, by Proposition 6.1.1, that $F \circ M(M, X)$ is not evenly continuous from $0 \in M$ to $q \in Y^*$. There are sequences $\{p_n\}$ in $M \setminus \{0\}, \{f_n\}$ in $F, \{\varphi_n\}$ in $H(M, X)$ such that $\|p_n\| \to 0$, $f_n \circ \varphi_n(0) \to q$ and $f_n \circ \varphi_n(p_n) \nrightarrow q$. Define $\lambda_n \in H(D, X)$ by $\lambda_n(z) = \varphi_n(zp_n / \|p_n\|)$; $f_n \circ \lambda_n(0) \to q$ while $f_n \circ \lambda_n(\|p_n\|) \nrightarrow q$.

From Proposition 6.1.1, $F \circ H(D,X)$ is not relatively compact in $C(D,Y^*)$, so **(3)** does not hold and we complete the proof of **(3)** $\Rightarrow$ **(1)**.

(1) $\Rightarrow$ **(4)**. We show that, for any complex manifold M,

$$(\overline{F} \cap H(X,Y)) \circ H(M,X) \subset \overline{F \circ H(M,X)}.$$

Let $g \in \overline{F} \cap H(X,Y)$ and $\varphi \in H(M,X)$. There is a sequence $\{f_n\}$ in F such that $f_n \to g$. It follows that $f_n \circ \varphi \to g \circ \varphi$. □

From Proposition 6.1.6 it is not difficult to see that some important classes of complex spaces are defined by uniformly normal families. A complex subspace X of a complex space Y is *hyperbolically imbedded* in Y if, for $p, q \in \overline{X}$ and $p \neq q$, there are open sets V, W in Y about p, q respectively such that $k_X(V \cap X, W \cap X) > 0$, where k_X is Kobayashi's pseudo-distance on X [135].

In Proposition 6.1.8 we show that families of distance-decreasing maps between metric spaces are relatively compact as maps into the one-point compactifications of the ranges.

Example 6.1.7.

1. Kiernan [129] has shown that a relatively compact complex subspace of a complex space Y is hyperbolically imbedded in Y if and only if $H(D,X)$ is relatively compact in $H(D,Y)$, i.e. if and only if $H(D,X)$ is a uniformly normal subset of $H(D,Y)$.
2. Royden [254] showed that a complex manifold M is hyperbolic if and only if $H(D,M)$ is evenly continuous and Abate [1] showed that M is hyperbolic if and only if $H(D,M)$ is relatively compact in $C(D,M^*)$. Hence $H(D,M)$ is a uniformly normal family if and only if M is hyperbolic.
3. Joseph and Kwack showed in [123] that a complex subspace X of a complex space Y is hyperbolically imbedded in Y if and only if $H(D,X)$ is relatively compact in $C(D,Y^*)$, i.e. if and only if $H(D,X)$ is a uniformly normal subset of $H(D,Y)$.

In Proposition 6.1.8, we show that families of distance-decreasing maps between metric spaces are relatively compact as maps into the one-point compactifications of the range.

Proposition 6.1.8. *Let (Y,σ) be a locally compact metric space. Let X be a topological space and let ρ be a pseudometric on X which is continuous on $X \times X$. If each $f \in F \subset C(X,Y)$ is distance-decreasing with respect to ρ and σ (i.e. if $\sigma(f(x), f(y)) \leq \rho(x,y)$ for all $f \in F, x, y \in X$) then F is relatively compact in $C(X,Y^*)$.*

Proof. We show that F is evenly continuous from X to Y^*. If not, there exist $p \in X, q, s \in Y^*$ and nets $\{p_\alpha\}, \{f_\alpha\}$ in X, F respectively such that $p_\alpha \to p, s \neq q, f_\alpha(p_\alpha) \to s, f_\alpha(p) \to q$. If $q \in Y$, then for each α we have

$$\sigma(f_\alpha(p_\alpha), q) \leq \sigma(f_\alpha(p_\alpha), f_\alpha(p)) + \sigma(f_\alpha(p), q) \leq \rho(p_\alpha, p) + \sigma(f_\alpha(p), q).$$

So $\sigma(f_\alpha(p_\alpha), q) \to 0$ and $q = s$, a contradiction. If $s \in Y$ then for each α we have

$$\sigma(f_\alpha(p), s) \leq \rho(p, p_\alpha) + \sigma(f_\alpha(p_\alpha), s)$$

Hence $\sigma(f_\alpha(p), s) \to 0$ and $s = q$, a contradiction. □

6.2 Uniformly Normal Families on Hyperbolic Manifolds

In this section we establish some properties of uniformly normal families defined on hyperbolic manifolds. We apply some of these properties to generalize theorems of Brody, Hahn, and Zaidenberg; these properties are also used to demonstrate that normal maps as defined by various authors in a variety of settings are uniformly normal as singleton sets and that maps in any uniformly normal family (in particular our normal maps) are normal maps in the prevailing setting. Furthermore, with these properties we set the stage for those main results which come in the next section.

Let M be a hyperbolic manifold. Let F_M^K represent the infinitesimal form of the Kobayashi distance k_M on M, that is,

$$F_M^K(p, v) = \inf\{r > 0 : \varphi(0) = p, d\varphi(0, re) = v \text{ for some } \varphi \in H(D, M)\},$$

where $p \in M$, $v \in T_p(M)$—the tangent space of M at p—$d\varphi$ is the tangent map for φ, and e is the unit vector 1 at $0 \in D$ (see [157, pp. 88–94]). For $r > 0$, let $D_r = \{z \in \mathbb{C} : |z| < r\}$ and let $D_r^* = D_r \setminus \{0\}$. Brody has proved the following (see [22], p. 68).

Theorem 6.2.1. *Let X be a relatively compact complex subspace of a complex space Y. If X is not hyperbolically imbedded in Y then there exist sequences $\{r_n\}, \{g_n\}$ such that $r_n > 0$ and $g_n \in H(D_{r_n}, X)$ and a nonconstant $g \in H(\mathbb{C}, Y)$ such that $r_n \uparrow \infty$ and $g_n \to g$ on compact subsets of $\mathbb{C}$.*

Remark 6.2.2. In Theorem 6.2.1 we may assume that $r_n = n$; to see this we first assume $r_1 > 1$ and that $r_{n+1} - r_n > 1$ by taking a subsequence. If k is a positive integer and $k \leq r_1$, define $f_k = g_1$; if $r_n < k \leq r_{n+1}$, define $f_k = g_{n+1}$. Then $f_k \in H(D_k, X)$ and $f_k \to g$ on compact subsets of $\mathbb{C}$.

Definition 6.2.3. Let X be a complex space, let Y be a complex space, and let $F \subset H(X, Y)$.

1. A *Brody sequence* for F is a sequence $\{f_n \circ g_n\}$ where $f_n \in F$ and $g_n \in H(D_n, X)$.
2. A map $h \in C(\mathbb{C}, Y^*)$ is a *Brody limit* for F if there is a Brody sequence $\{h_n\}$ for F such that $h_n \to h$ on the compact subsets of $\mathbb{C}$.

Remark 6.2.4. If Y is a relatively compact complex subspace of a complex space Z, and X is a complex space, then Brody sequences for $F \subset H(X,Y)$ coincide with Zaidenberg's holomorphic curves and Brody limits for F coincide with Zaidenberg's F-limiting mappings [290].

Theorem 6.2.5 will be used extensively in the sequel.

Theorem 6.2.5. *Let M be a hyperbolic manifold, and let Y be a complex space. The following statements are equivalent for $F \subset H(M,Y)$:*

(1) F is uniformly normal.

(2) For each complex manifold $\Omega, F \circ H(\Omega, M)$ is an evenly continuous subset of $H(\Omega, Y)$.

(3) $F \circ H(D,M)$ is an evenly continuous subset of $H(D,Y)$.

(4) There is a length function E on Y such that $|df|_E \leq 1$ for each $f \in F$.

(5) There is a length function E on Y such that $E(h_n(0), dh_n(0,e)) \to 0$ for each Brody sequence $\{h_n\}$ for F.

(6) There is a length function E on Y such that $E(h_n(0), dh_n(0,e)) \to 0$ for each Brody sequence $\{h_n\}$ for F with a Brody limit.

Proof. It is obvious that **(1)** $\Rightarrow$ **(2)** $\Rightarrow$ **(3)** and that **(5)** $\Rightarrow$ **(6)**. To show that **(3)** $\Rightarrow$ **(4)** we will first show that if **(3)** holds then for each length function E on Y and compact $Q \subset Y$ there is a $c > 0$ such that $|df(p)| \leq c$ on $f^{-1}(Q)$ for each $f \in F$. The proof that **(3)** $\Rightarrow$ **(4)** may then be completed by arguments similar to those in ([157, p. 34]). To facilitate the demonstration that each compact $Q \subset Y$ satisfies the stated property for each length function E on Y, we prove a lemma which establishes a property of hyperbolic manifolds that is interesting in its own right. □

Lemma 6.2.6. *Let $f \in H(M,Y)$ where M is a hyperbolic manifold and Y is a complex space with a length function E. Then*

$$
\begin{aligned}
|df(p)| &= \sup\{|df \circ \varphi(0)| : \varphi \in H(D,M), \varphi(0) = p\} \quad \textit{for } p \in M, \\
\textit{and} \quad |df| &= \sup\{|df \circ \varphi(0)| : \varphi \in H(D,M)\} \\
&= \sup\{|df \circ \varphi| : \varphi \in H(D,M)\}.
\end{aligned}
$$

Proof. Let $p \in M$ and $v \in T_p(M)$ such that $F_M^K(p,v) = 1$ and let $\epsilon > 0$. There exist $\varphi \in H(D,M)$ and $r > 0$ such that $\varphi(0) = p, d(\varphi(0), re) = v$ and $r < 1 + \epsilon$. Now

$$
\begin{aligned}
E(f(p), df(p,\nu)) = E(f \circ \varphi(0), df \circ \varphi(0, re)) &\leq (1+\epsilon)|df \circ \varphi(0)| \\
&\leq (1+\epsilon)\sup\{|df \circ \varphi(0)| : \varphi \in H(D,M), \varphi(0) = p\} \\
&\leq (1+\epsilon)\sup\{|df \circ \varphi| : \varphi \in H(D,M)\} \\
&\leq (1+\epsilon)|df|.
\end{aligned}
$$

The promised equalities are now immediate. □

We now return to the proof of Theorem 6.2.5.

Proof. **(3)** $\Rightarrow$ **(4)**. If a compact $Q \subset Y$ fails the stated condition for the length function E, then there are sequences $\{p_n\}, \{f_n\}, \{v_n\}$, and $q \in Q$, with $p_n \in M, f_n \in F, v_n \in T_{p_n}(M)$, $f_n(p_n) \in Q, F_M^K(p_n, v_n) = 1, f_n(p_n) \to q$ and $E(f_n(p_n), df_n(p_n, v_n)) > n$. It follows that $|df_n(p_n)| \to \infty$ and, by Lemma 6.2.6, some sequence $\{\varphi_n\}$ in $H(D, M)$ satisfies $\varphi_n(0) = p_n$ and $|df_n \circ \varphi_n(0)| \to \infty$. Let V be a relatively compact neighborhood of q hyperbolically imbedded in Y. Since $F \circ H(D, M)$ is an evenly continuous subset of $H(D, Y)$, there is an $0 < r < 1$ such that $f_n \circ \varphi_n(D_r) \subset V$ eventually; the sequence of restrictions of $\{f_n \circ \varphi_n\}$ to D_r, which we call again $\{f_n \circ \varphi_n\}$, is uniformly normal and is consequently relatively compact in $H(D_r, Y)$. Hence some subsequence of $\{f_n \circ \varphi_n\}$ converges to $h \in H(D_r, Y)$, contradicting $|df_n \circ \varphi_n(0)| \to \infty$.

(4) $\Rightarrow$ **(5)**. Let E be a length function as promised in **(4)**. If $\{f_n \circ \varphi_n\}$ is a Brody sequence for F we have

$$E(f_n \circ \varphi_n(0), df_n \circ \varphi_n(0, e)) \le F_M^K(\varphi_n(0), d\varphi_n(0, e)) \le K_{D_n}(0, e) = \frac{1}{n}$$

and **(5)** holds.

(4) $\Rightarrow$ **(1)**. From **(4)** there is a distance function d_E on Y such that each $f \in F \circ H(D, M)$ decreases distances from k_D to d_E and **(1)** follows from Proposition 6.1.6 and 6.1.8.

(6) $\Rightarrow$ **(4)**. If **(4)** does not hold, then for any length function E on Y there are sequences $\{f_n\}$ in F, $\{\varphi_n\}$ in $H(D, M)$ such that $|df_n \circ \varphi_n(0)| \to \infty$. Utilizing Brody's arguments (see [22], pp. 68–71) we may obtain a Brody sequence $\{g_n\}$ and Brody limit g for F such that $g_n \to g$ on the compact subsets of $\mathbb{C}$ and such that $E(g_n(0), dg_n(0)) = 1$, contradicting **(6)**. □

We point out that equivalence **(4)** of Theorem 6.2.5 generalizes the following theorem of Lehto and Virtanen [164] since all length functions on compact complex spaces are equivalent [157]. Hahn [101] generalized the theorem to $f \in H(\Omega, \mathbf{P}^n(\mathbb{C}))$ where Ω is a bounded homogeneous domain in $\mathbb{C}^m$.

Theorem 6.2.7. *A meromorphic function $f : D \to \mathbf{P}^1(\mathbb{C})$ is normal if and only if $|df| < \infty$.*

Corollary 6.2.8. *Let M be a hyperbolic manifold and let $F \subset H(M, Y)$ be uniformly normal. Then*

(1) Each Brody sequence for F has a subsequence which converges to a Brody limit for F on the compact subsets of $\mathbb{C}$, and

(2) Each Brody limit for F is constant.

Proof. From **(4)** in Theorem 6.2.5 there is a length function E on Y such that F decreases distances from k_M to d_E. As for **(1)**, if m is a positive integer and $\{g_n\}$ is a Brody sequence for F, then each $g \in G = \{g_n : n \geq m\}$ decreases distances from k_{D_m} to d_E, and G is therefore by Proposition 6.1.8 relatively compact in $C(D_m, Y^*)$. For **(2)**, let $\{g_n\}$ be a Brody sequence for F and suppose g is a Brody limit for F such that $g_n \to g$ on the compact subsets of $\mathbb{C}$. If $p, q \in \mathbb{C}$ and $g(p), g(q) \in Y$, for n large enough we have $d_E(g_n(p), g_n(q)) \leq k_{D_n}(p, q)$. Since $k_{D_n}(p, q) \to 0$ we have $g(p) = g(q)$. If $g(p) = \infty$, then the continuity of g and connectedness of $\mathbb{C}$ force $g(q) = \infty$ since $g(\mathbb{C}) \cap Y$ is at most a singleton. □

Corollary 6.2.9. *Let M be a hyperbolic manifold, let Y be a complex space, and let $F \subset H(M, Y)$ satisfy $F(x)$ is relatively compact in Y for each $x \in M$. Then F is uniformly normal if and only if each Brody limit for F is constant.*

Proof. The necessity follows from **(2)** of Corollary 6.2.8. As for the sufficiency, if F is not uniformly normal, then the Brody limit g constructed in **(6)** $\Rightarrow$ **(4)** of Theorem 6.2.5 is nonconstant since Y is a complex space and $g(0) \in Y$. □

In the case of $\widehat{\mathbb{C}}$ we have the following result. Here of course we identify the Riemann sphere with projective space. The proof is omitted as it is readily obtained as a consequence of previous results and Hurwitz's lemma.

Corollary 6.2.10. *Let M be a hyperbolic manifold. The following are equivalent for $F \subset H(M, \mathbb{C})$*

(1) F is uniformly normal.

(2) F is uniformly normal as a subset of $H\left(M, \widehat{\mathbb{C}}\right)$.

(3) If g is a Brody limit for F and $g \in H(\mathbb{C}, \mathbb{C})$ then g is constant.

A complex space Y is said to be *Brody hyperbolic* if every $f \in H(C, Y)$ is constant. We observe that Y is Brody hyperbolic if and only if each Brody limit for the identity $i : Y \to Y$ with values in Y is constant, i.e. if $f \in H(C, Y)$ and $\{f_n\}$ is a sequence such that $f_n \in H(D_n, Y)$ and $f_n \to f$ on the compact subsets of $\mathbb{C}$, then f is constant.

In Corollaries 6.2.11–6.2.13 we give characterizations of hyperbolic and hyperbolically imbedded spaces in terms of Brody sequences.

Corollary 6.2.11. *A complex space Y is hyperbolic if and only if there is a length function E on Y such that $E(f_n(0), df_n(0, e)) \to 0$ for each sequence $\{f_n\}$ satisfying $f_n \in H(D_n, Y)$ and $f_n \to g \in C(\mathbb{C}, Y^*)$ on compact subsets of $\mathbb{C}$. We note that g is necessarily constant.*

Corollary 6.2.12. *A complex subspace Y of a complex space Z is hyperbolically imbedded in Z if and only if there is a length function E on Z such that $E(f_n(0), df_n(0, e)) \to 0$ for each sequence $\{f_n\}$ satisfying $f_n \in H(D_n, Y)$ and $f_n \to g \in C(\mathbb{C}, Z^+)$ on compact subsets of $\mathbb{C}$. Again we note that g is necessarily constant.*

In view of Corollaries 6.2.9 and 6.2.12 we have Corollary 6.2.13 (compare with Theorem 2.3, Chapter III in [157]).

Corollary 6.2.13. *Let Y be a relatively compact complex subspace of the complex space Z. Then Y is not hyperbolically imbedded in Z if and only if there is a nonconstant $g \in H(\mathbb{C}, Z)$ and a sequence $\{g_n\}$ such that $g_n \in H(D_n, Y)$ and $g_n \to g$ on compact subsets of $\mathbb{C}$.*

If M is a homogeneous hyperbolic manifold, then we represent its space of automorphisms by $A(M)$. Theorem 6.2.14 is a characterization theorem for uniformly normal families on such manifolds and Corollary 6.2.15 gives an indication of why we adopted the terminology "uniformly normal families".

Theorem 6.2.14. *The following statements are equivalent for $F \subset H(M,Y)$ when M is a homogeneous hyperbolic manifold and Y is a complex space:*

(1) F is uniformly normal.

(2) $F \circ A(M)$ is an evenly continuous subset of $H(M,Y)$.

(3) $F \circ A(M)$ is relatively compact in $C(M,Y^*)$.

(4) $F \circ H(M,M)$ is relatively compact in $C(M,Y^*)$.

Proof. **(1)** $\Rightarrow$ **(2)**. This is clear from Definition 6.1.2 and Proposition 6.1.1.

(2) $\Rightarrow$ **(3)**. We need show that $F \circ A(M)$ is evenly continuous from M to Y^*. Suppose $x \in M, p \in Y$ and $\{x_n\}, \{f_n\}$, and $\{\varphi_n\}$ are sequences in M, F and $A(M)$ respectively such that $x_n \to x$, $f_n \circ \varphi_n(x) \to \infty$, and $f_n \circ \varphi_n(x_n) \to p$. Let $\lambda_n \in A(M)$ satisfy $\lambda_n(x) = x_n$. It follows that $f_n \circ \varphi_n \circ \lambda_n$ is not evenly continuous from x to p so **(2)** does not hold.

(3) $\Rightarrow$ **(1)**. We show that $F \circ H(D,M)$ is an evenly continuous subset of $H(M,Y)$ and use equivalence **(3)** of Theorem 6.2.5. Suppose that $F \circ H(D,M)$ is not evenly continuous from $0 \in D$ to $p \in Y$. There are sequences $\{z_n\}, \{f_n \circ \varphi_n\}$ in $D, F \circ H(D,M)$ respectively and a neighborhood U of p in Y such that $z_n \to 0$, $f_n \circ \varphi_n(0) \to p$, and $f_n \circ \varphi_n(z_n) \notin U$ is satisfied. Let $a \in M$ and let $\{\lambda_n\}$ be a sequence in $A(M)$ such that $\lambda_n(a) = \varphi_n(0)$. Then $f_n \circ \lambda_n(a) \to p$ while $f_n \circ \lambda_n\left(\lambda_n^{-1}(\varphi_n(z_n))\right) \notin U$. Since each $f \in A(M)$ preserves hyperbolic distances and each $f \in H(D,M)$ is distance decreasing with respect to F_D^K and F_M^K we see that $\lambda_n^{-1}(\varphi_n(z_n)) \to a$ and this contradicts our hypothesis **(3)**.

The proof of the theorem may be completed by observing that the equivalence of **(4)** to the other statements follows from the equivalence of **(1)** and **(3)**, Propositions 6.1.4 and 6.1.5, and the inclusion $F \circ A(M) \subset F \circ H(M,M)$. □

Corollary 6.2.15. *Let M be a homogeneous hyperbolic manifold, let Y be a complex space, and let $F \subset H(M,Y)$ satisfy $F = F \circ A(M)$. Then*

(1) The family F is uniformly normal if and only if F is relatively compact in $C(M,Y^*)$;

(2) If $M = D$ and $Y = \mathbb{C}$, then F is uniformly normal if and only if F is normal in the sense of Wu [287].

Proof. Assertion **(1)** is an easy consequence of Proposition 6.1.4 and **(3)** of Theorem 6.2.14. Assertion **(2)** can be seen to be a product of **(1)** and Hurwitz's lemma. □

Remark 6.2.16. Hayman [111] called $F \subset H\left(D, \mathbf{P}^1(\mathbb{C})\right)$ *invariant* if $F = F \circ A(D)$ and called an invariant family *uniformly normal* if it is normal in the sense of Montel. In view of this and Corollary 6.2.15, we used the terminology "uniformly normal family".

6.3 Uniformly Normal Families Defined on Arbitrary Complex Spaces. Higher Dimensional Versions of Classical Results of Schottky, Lappan, and Bohr for Uniformly Normal Families

In this section we apply results in Chapter 6 and Section 6.2 to establish properties of uniformly normal families on arbitrary complex spaces and we generalize theorems of Schottky, Hayman, and Lappan by replacing the bounded domains in $\mathbb{C}$ by arbitrary complex spaces and the maps under consideration by uniformly normal families. A Lindelöf principle shown by Cima and Krantz to be satisfied by normal maps from a bounded domain in $\mathbb{C}^n$ to $\mathbf{P}^1(\mathbb{C})$ is generalized by replacing $\mathbf{P}^1(\mathbb{C})$ by an arbitrary complex space and normal maps by uniformly normal families.

The following proposition on Brody sequences and limits will be useful.

Proposition 6.3.1. *Let X, Y be complex spaces and let $F \subset H(X, Y)$. Then*

(1) The sequence $\{g_n\}$ is a Brody sequence for F if and only if $\{g_n\}$ is a Brody sequence for $F \circ H(D, X)$, and

(2) The mapping g is a Brody limit for F if and only if g is a Brody limit for $F \circ H(D, X)$.

Proof. Conclusion **(2)** is an easy consequence of conclusion **(1)**. As for **(1)**, if $g_n = f_n \circ \varphi_n$, where $f_n \in F$ and $\varphi_n \in H(D_n, X)$, then $g_n = f_n \circ \varphi_n \circ m_n \circ m_n^{-1}$ where $m_n \in H(D, D_n)$ is multiplication by n. Then $f_n \circ \varphi_n \circ m_n \in F \circ H(D, X)$ and $m_n^{-1} \in H(D_n, D)$, so $\{g_n\}$ is a Brody sequence for $F \circ H(D, X)$ if $\{g_n\}$ is a Brody sequence for F. On the other hand, if $g_n = h_n \circ \varphi_n$ where $h_n \in F \circ H(D, X)$ and $\varphi_n \in H(D_n, D)$, then $g_n = f_n \circ \alpha_n \circ \varphi_n$ where $f_n \in F$ and $\alpha_n \circ \varphi_n \in H(D_n, X)$. So each Brody sequence for $F \circ H(D, X)$ is a Brody sequence for F. □

Theorem 6.3.2 is a consequence of Definition 6.1.2, Propositions 6.1.6, 6.3.1, and Theorem 6.2.5.

Theorem 6.3.2. *If X, Y are complex spaces, the following statements are equivalent for $F \subset H(X,Y)$*

(1) The set F is uniformly normal.

(2) The set $F \circ H(M,X)$ is an evenly continuous subset of $H(M,Y)$ for each complex manifold M.

(3) $F \circ H(D,X)$ is an evenly continuous subset of $H(D,Y)$.

(4) There is a length function E on Y such that $|dg|_E \leq 1$ for each $g \in F \circ H(D,X)$.

(5) There is a length function E on Y such that $E(h_n(0), dh_n(0,e)) \to 0$ for each Brody sequence $\{h_n\}$ for F.

(6) There is a length function E on Y such that $E(h_n(0), dh_n(0,e)) \to 0$ for each Brody sequence $\{h_n\}$ for F with a Brody limit.

Propositions 6.1.6, 6.3.1 and Corollary 6.2.8 lead to the following result.

Theorem 6.3.3. *Let X be a complex space and let $F \subset H(X,Y)$ be uniformly normal. Then*

(1) Each Brody sequence for F has a subsequence which converges to a Brody limit for F on compact subsets of $\mathbb{C}$.

(2) Each Brody limit for F is constant.

Theorem 6.3.4 follows from Propositions 6.1.6, 6.3.1 and Corollary 6.2.9.

Theorem 6.3.4. *Let X, Y be complex spaces and let $F \subset H(X,Y)$ have the property that $F(x)$ is relatively compact in Y for each $x \in X$. Then F is uniformly normal if and only if each Brody limit for F is constant.*

For uniformly normal families of functions, we have the following theorem.

Theorem 6.3.5. *The following statements are equivalent for a complex space X and $F \subset H(X,\mathbb{C})$*

(1) F is a uniformly normal family.

(2) F is a uniformly normal family as a subset of $H\left(X, \mathbf{P}^1(\mathbb{C})\right)$.

(3) If $g \in H(C,C)$ is a Brody limit for F then g is constant.

Proof. See Corollary 6.2.10. □

Hayman ([111, p. 165]) proved the following strong version of Schottky's theorem.

Theorem 6.3.6. *Let $F \subset H(D, \mathbb{C})$ be a normal invariant family. Then there is a $c > 0$ depending only on F such that*

$$\sup_{|z|=r} |f(z)| \leq \mu^{(1+r)/(1-r)} \exp\left(\frac{2cr}{1-r}\right)$$

for each $f \in F, 0 \leq r < 1$, where $\mu = \max\{1, |f(0)|\}$.

If X is a complex space, $f \in H(X, \mathbb{C})$, and $x \in X$, we denote $\max\{1, |f(x)|\}$ by $\mu(f, x)$. It is obvious that the conclusion of Theorem 6.3.6 can be replaced by the existence of a $c > 0$ depending only on F such that

$$\mu(f, z) \leq [\mu(f, 0)]^{(1+|z|)/(1-|z|)} \exp\left(\frac{2c|z|}{1-|z|}\right)$$

for each $f \in F$ and $z \in D$.

Zaidenberg [290] extended Hayman's result to uniformly normal families of functions defined on complex manifolds. This generalization was effected by use of the Kobayashi-Royden-Green metric which was introduced in [90] and studied further in [289]. We utilize parts of Hayman's proof of Theorem 6.3.6 to extend Zaidenberg's result, using more elementary proofs, to uniformly normal families of functions defined on complex spaces.

Theorem 6.3.7. *Let X be a complex space. The following statements are equivalent for $F \subset H(X, \mathbb{C})$:*

(1) F is uniformly normal.

(2) There is a $c > 0$ such that all $f \in F, x, y \in X$ satisfy the inequality

$$\mu(f, x) \leq [\mu(f, y)]^{\exp[2k_X(x,y)]} \exp\left[c\left[\exp\left[2k_X(x, y)\right] - 1\right]\right]$$

(3) There is a $c > 1$ such that all $f \in F, x, y \in X$ satisfy the inequality

$$\log(c\mu(f, x)) \leq \left[\exp\left[2k_X(x, y)\right]\right] \log(c\mu(f, y)).$$

(4) There is a $c > 1$ such that all $f \in F$, $x \in X$ and $\varnothing \neq Q \subset X$ satisfy the inequality

$$\log(c\mu(f, x)) \leq \left[\sup_{y \in Q} \log(c\mu(f, y))\right] \exp\left[2k_X(x, Q)\right]$$

(5) There is a $c > 1$ such that all $f \in F$, $x, y \in X$ satisfy the inequality

$$c\mu(f,x) \leq [c\mu(f,y)]^{\exp[2k_X(x,y)]}.$$

(6) There is a $c > 1$ such that all $f \in F$, $\varphi \in H(D,X)$, $x, y \in D$ satisfy the inequality

$$c\mu(f \circ \varphi, x) \leq [c\mu(f \circ \varphi, y)]^{\exp[2k_D(x,y)]}.$$

Proof. **(1)** $\Rightarrow$ **(2)**. Clearly $F \circ H(D,X)$ is uniformly normal and invariant, and is consequently normal in the sense of Montel by Corollary 6.2.15. Hayman ([111, p. 165]) has shown that there is a $c > 0$ such that for all $g \in F \circ H(D,X)$,

$$|g'(0)| \leq 2\mu_0 (\log \mu_0 + c)$$

where $\mu_0 = \max\{1, |g(0)|\}$. For $z \in D$ define $\psi_z \in A(D)$ by $\psi_z(w) = (w + z)/(1 + \overline{z}w)$. Then, for each $g \in F \circ H(D,X)$, we have

$$\left(1 - |z|^2\right) |g'(z)| = \left|(g \circ \psi_z)'(0)\right| \leq 2\mu_z (\log \mu_z + c),$$

where $\mu_z = \max\{1, |g(z)|\}$. Let $x, y \in X$, let $f \in F$ and let $\epsilon > 0$. There exist an integer $j > 1, \varphi_1, \varphi_2, \ldots, \varphi_j \in H(D,X)$, and $a_1, \ldots, a_j \in (0,1)$ satisfying $\varphi_1(0) = y, \varphi_i(a_i) = \varphi_{i+1}(0)$ for $i = 1, \ldots, j-1$ and $\varphi_j(a_j) = x$. We may assume that $|f(x)| > 1$, that $g_i([0,a_i]) \subset \overline{D}$ or $g_i([0,a_i]) \subset \mathbb{C} - D$, where $g_i = f \circ \varphi_i$, and that $\sum_i k_D(0,a_i) < k_X(x,y) + \epsilon/2$. Let $I = \{i : g_i([0,a_i]) \subset \mathbb{C} \setminus D\}$. For each $i \in I$,

$$\frac{|g_i'(z)|}{|g_i(z)| (\log |g_i(z)| + c)} \leq \frac{2}{1 - |z|^2} \quad \text{for} \quad z \in [0, a_i]$$

hence for each $i \in I$ we have

$$\log \left[\frac{\log |g_i(a_i)| + c}{\log |g_i(0)| + c}\right] \leq 2k_D(0, a_i).$$

If $1 \in I$, then

$$\log \left[\frac{\log |f(x)| + c}{\log |f(y)| + c}\right] \leq 2k_X(x,y) + \epsilon.$$

If $1 \notin I$ let α be the smallest element of I. Then $|g_\alpha(0)| = 1$ and

$$\log \left[\frac{\log |f(x)| + c}{\log |g_\alpha(0)| + c}\right] \leq 2k_X(x,y) + \epsilon.$$

The proof is easy to complete.

(2) $\Rightarrow$ **(3)**. Replace c with $\log c$ in the conclusion of **(2)**.

(3) $\Rightarrow$ **(4)**. If $\varnothing \neq Q \subset X$ and $x \in X$, let $\{y_n\}$ be a sequence in Q such that $d_X(x, y_n) \to d_X(x, Q)$. The conclusion of **(4)** is then evident.

(4) $\Rightarrow$ **(5)**. Obvious.

(5) ⇒ (6). If $f \in F$, $\varphi \in H(D,X)$, and $x, y \in D$ we have, from **(5)**,

$$\begin{aligned} c\mu(f \circ \varphi, x) = c\mu(f, \varphi(x)) &\leq [c\mu(f, \varphi(y))]^{\exp[2k_X(\varphi(x),\varphi(y))]} \\ &\leq [c\mu(f, \varphi(y))]^{\exp[2k_D(x,y)]} \end{aligned}$$

(6) ⇒ (1). In view of Proposition 6.1.6 and the homogeneity of D, it suffices to show that if $\{f_n\}$ and $\{\varphi_n\}$ are sequences in F and $H(D,X)$, respectively, then some subsequence of $\{f_n \circ \varphi_n\}$ converges to $g \in C(V, Y^*)$ on some neighborhood V of 0. If, for each compact $K \subset Y$ and neighborhood V of 0, $f_n \circ \varphi_n(V) \cap K = \varnothing$ eventually, we are finished. Otherwise we may assume a sequence $\{z_n\}$ in $D_{1/2}$ such that $\{f_n \circ \varphi_n(z_n)\}$ is bounded. There is a $c > 1$ such that, for each $z \in D_{1/2}$,

$$c\mu(f_n \circ \varphi_n, z) \leq [c\mu(f_n \circ \varphi_n, z_n)]^{\exp[2d_D(z,z_n)]}$$

and $\{f_n \circ \varphi_n\}$ is uniformly bounded on $D_{1/2}$. □

In Corollary 6.3.9 we apply equivalence **(2)** in Theorem 6.3.7 to extend the following lemma of Bohr to holomorphic functions defined on complex spaces (see [111, p. 50]).

Lemma 6.3.8. *Suppose that* $w = f(z)$ *is regular in* $|z| \leq 1$, *satisfies* $f(0) = 0$, *and*

$$\max_{|z|=1/2} |f(z)| \geq 1\,.$$

Then $f(z)$ *assumes in* $|z| < 1$ *all values on some circle* $|w| = r$, *where* $r > A$ *and* A *is a positive absolute constant.*

Corollary 6.3.9. *Let* X *be a complex space, and let* $B \subset X$, $f \in H(X, \mathbb{C})$ *satisfy*

(1) B *is bounded with respect to the pseudometric* k_X,

(2) $\sup_{x \in B} |f(x)| > 1$ *and*

(3) $0 \in f(B)$.

Then there is an $r > 0$ *independent of* f *such that*

$$\{w \in \mathbb{C} : r < |w| < 2r\} \subset f(X) \quad \text{or} \quad \{w \in \mathbb{C} : 4r < |w| < 5r\} \subset f(X).$$

Proof. We note that $H(X, \mathbb{C} \setminus \{0,1\})$ is a uniformly normal family in $H(X, \mathbb{C})$. By Theorem 6.3.7, there is a $c > 0$ such that

$$|\alpha(p)| \leq \exp\left[c\left[\exp\left(2k_X(p,q)\right) - 1\right]\right]$$

for all $p, q \in B$, $\alpha \in H(X, \mathbb{C} \setminus \{0,1\})$ such that $|\alpha(q)| \leq 1$; let

$$A = \exp\left[c\left[\exp\left(2\sup_{x,y \in B} k_X(x,y)\right) - 1\right]\right] \quad \text{and} \quad r = (7A+2)^{-1}.$$

Suppose

$$w_1 \in \{w \in \mathbb{C} : r < |w| < 2r\} \setminus f(X), \quad w_2 \in \{w \in \mathbb{C} : 4r < |w| < 5r\} \setminus f(X),$$

and define $\varphi \in H(X, \mathbb{C} \setminus \{0,1\})$ by $\varphi(p) = (f(p) - w_1) / (w_2 - w_1)$. There is a $q \in B$ such that $f(q) = 0$; hence $|\varphi(q)| \leq 1$ and, for each $p \in B$, $|\varphi(p)| \leq A$ and

$$|f(p)| = |\varphi(p)\,(w_2 - w_1) + w_1| \leq A\,(|w_2| + |w_1|) + |w_1| \leq (7A + 2)r = 1\,.$$

This is a contradiction. □

Remark 6.3.10. Equivalences **(1)** ⇒ **(3)** ⇒ **(4)** ⇒ **(1)** of Theorem 6.3.7 are due to Zaidenberg [290] for uniformly normal families of functions defined on complex manifolds. In Theorem 6.3.14, we provide a generalization of the following five-point theorem of Lappan [160].

Theorem 6.3.11. *Let A be a subset of $\mathbf{P}^1(\mathbb{C})$ with at least 5 elements. Then*

$$f \in H\left(D, \mathbf{P}^1(\mathbb{C})\right)$$

is normal if and only if

$$\sup\left\{|f'(z)|\left(1 - |z|^2\right) / \left(1 + |f(z)|^2\right) : z \in f^{-1}(A)\right\} < \infty.$$

We extend this theorem of Lappan to uniformly normal families from arbitrary complex spaces to $\mathbf{P}^n(\mathbb{C})$ by first extending the theorem to such families defined on hyperbolic manifolds and then using equivalence **(2)** of Propositions 6.1.6 and 6.3.1 (also see [101], [100], [103]).

We say that $g \in H\left(\mathbb{C}^m, \mathbf{P}^n(\mathbb{C})\right)$ is *degenerate* if $g\left(\mathbb{C}^m\right) \subset \pi$ for some hyperplane π in $\mathbf{P}^n(\mathbb{C})$.

Theorem 6.3.12. *Let M be a hyperbolic manifold, let σ be a collection of at least $2n+3$ hyperplanes in general position in $\mathbf{P}^n(\mathbb{C})$ and let $A = \bigcup_\sigma \pi$.*

Then $F \subset H\left(M, \mathbf{P}^n(\mathbb{C})\right)$ is uniformly normal if and only if the following two conditions hold:

(1) We have that $\sup\left\{|df(p)| : p \in \bigcup_F f^{-1}(A)\right\} < \infty$, *and*

(2) Each degenerate Brody limit for F is constant.

Proof. The necessity of the two conditions is obvious from equivalences **(4)** and **(6)** of Theorem 6.2.5 since all length functions on $\mathrm{P}^n(\mathbb{C})$ are equivalent. To establish the sufficiency we will show that if condition **(2)** holds and F is not uniformly normal then condition **(1)** fails. To effect this demonstration we use some additional terminology and the following lemma of Hahn [101] which generalizes a well-known result in Nevanlinna theory for meromorphic functions (see [116, Corollary 3, p. 231]). If π is a hyperplane in $\mathbf{P}^n(\mathbb{C})$ let T_π be a nonzero linear functional on $\mathbb{C}^{n+1}$ for which π is the kernel. If σ

is a collection of hyperplanes in general position in $\mathbf{P}^n(\mathbb{C})$, we say that a nondegenerate $g \in H(\mathbb{C}^m, \mathbf{P}^n(\mathbb{C}))$ is *totally ramified* over σ if for each $\pi \in \sigma$, $(T_\pi \circ g)'(\zeta) = 0$ whenever $T_\pi \circ g(\zeta) = 0$. We need this lemma to complete the proof:

Lemma 6.3.13. *Let $\sigma = \{\pi_1, \ldots, \pi_q\}$ be any collection of hyperplanes in general position in $\mathbf{P}^n(\mathbb{C})$. If $g \in H(\mathbb{C}^m, \mathbf{P}^n(\mathbb{C}))$ is totally ramified over σ, then $q \leq 2n+2$.*

Now we complete the proof. Since $\mathbf{P}^n(\mathbb{C})$ is compact and F is not uniformly normal, there is a nonconstant Brody limit $g \in H(\mathbb{C}, \mathbf{P}^n(\mathbb{C}))$ for F by equivalence **(6)** of Theorem 6.2.5. Let $\{f_k\}$, $\{\psi_k\}$ be sequences such that $f_k \in F, \psi_k \in H(D_k, M)$, and $f_k \circ \psi_k \to g$. From **(2)**, g is nondegenerate and from Lemma 6.3.13 g is not totally ramified over $\{\pi\}$ for some $\pi \in \sigma$. Choose homogeneous coordinates $[{}^\circ w, \ldots, {}^n w]$ of $\mathbf{P}^n(\mathbb{C})$ so that π is defined by ${}^\circ w = 0$. If $g_k = f_k \circ \psi_k$, we represent g and g_k by the homogeneous coordinates $[{}^\circ g, \ldots, {}^n g]$ and $[{}^\circ g_k, \ldots, {}^n g_k]$, respectively, where ${}^s g$ and ${}^s g_k$ are holomorphic and ${}^\circ g_k \to {}^\circ g$. The equation ${}^\circ g(z) = 0$ has a solution $z_0 \in \mathbb{C}$ for which ${}^\circ g'(z_0) \neq 0$, and hence if E is a length function on $\mathbf{P}^n(\mathbb{C})$, we have $E(g(z_0), dg(z_0, e)) = \alpha \neq 0$. By Hurwitz's lemma there is a sequence $\{z_k\}$ in $\mathbb{C}$ such that $z_k \to z_0$, ${}^\circ g_k(z_k) = 0$, and $E(g_k(z_k), dg_k(z_k, e)) \to \alpha$. Let $p_k = \psi_k(z_k)$. Then

$$|df_k(p_k)| \geq E\left(f_k \circ \psi_k\left(z_k, k\left(1 - |\tfrac{z_k}{k}|^2\right) e\right)\right) \geq \\ k\left(1 - |\tfrac{z_k}{k}|^2\right) E(g_k(z_k), dg_k(z_k, e)),$$

so $|df_k(p_k)| \to \infty$. Since $p_k \in f_k^{-1}(\pi) \subset f_k^{-1}(A)$, condition **(1)** does not hold. □

We are now in a position to prove the following generalization of the theorem of Lappan.

Theorem 6.3.14. *Let X be a complex space, let σ be a collection of at least $2n+3$ hyperplanes in general position in $\mathbf{P}^n(\mathbb{C})$ and let $A = \bigcup_\sigma \pi$. Then $F \subset H(X, \mathbf{P}^n(\mathbb{C}))$ is uniformly normal if and only if the following two conditions hold:*

(1) We have $\sup\left\{|df \circ \varphi(0)| : \varphi \in H(D, X), \varphi(0) \in \bigcup_F f^{-1}(A)\right\} < \infty$, *and*

(2) Each degenerate Brody limit for F is constant.

Proof. From Theorem 6.3.12 and Proposition 6.1.6, we have that F is uniformly normal if and only if the following two conditions hold:

(**f1**°) We have
$\sup\left\{|dg(p)| : g \in F \circ H(D, X), p \in \bigcup g^{-1}(A)\right\} < \infty$, and

(**f2**°) Each degenerate Brody limit function for $F \circ H(D, X)$ is constant.

Conditions (**f1°**) and (**f2°**) are equivalent to conditions **(1)** and **(2)** respectively in view of Proposition 6.3.1 and the following set equality:

$$\left\{ | \operatorname{dg}(p)| : g \in F \circ H(D,X), p \in \bigcup g^{-1}(A) \right\}$$
$$= \left\{ |d(f \circ \varphi)(0)| : \varphi \in H(D,X), \varphi(0) \in \bigcup_F f^{-1}(A) \right\}.$$

□

The next two corollaries are now readily available.

Corollary 6.3.15. *Let X be a complex space. Then $F \subset H\left(X, \mathbf{P}^1(\mathbb{C})\right)$ is uniformly normal if and only if*

$$\sup \left\{ |d(f \circ \varphi)(0)| : \varphi \in H(D,X), \varphi(0) \in \bigcup_F f^{-1}(A) \right\} < \infty,$$

where $A \subset \mathbf{P}^1(\mathbb{C})$ has more than four elements (two finite elements if $F \subset H(X,\mathbb{C})$).

Corollary 6.3.16. *Let M be a hyperbolic manifold. Then $F \subset H\left(M, \mathbf{P}^1(\mathbb{C})\right)$ is uniformly normal if and only if*

$$\sup \left\{ |df(p)| : p \in \bigcup_F f^{-1}(A) \right\} < \infty,$$

where $A \subset \mathbf{P}^1(\mathbb{C})$ has more than four elements (two finite elements if $F \subset H(M,\mathbb{C})$).

6.4 Extension and Convergence Theorems

A *pseudo-length function* [157] on a complex space X is an upper-semicontinuous non-negative function H on the tangent cone, $T(X)$, such that $H(av) = |a|H(v)$ for $a \in \mathbb{C}$ and $v, av \in T(X)$.

A *length function* is a pseudo-length function which is continuous and $H(v) > 0$ for all nonzero $v \in T(X)$.

Every pseudo-length function H on a complex space X gives rise to an inner pseudo-distance d_H by

$$d_H(x,y) = \inf_{\gamma} \int_a^b H\left(\gamma'(t)\right) \mathrm{d}t,$$

where $\gamma : [a, b] \to X$ is a piecewise C^1 curve joining x to y and $\gamma'(t)$ is the velocity vector of γ at $\gamma(t)$. If H is a length function, then d_H is a distance function.

A Hermitian metric on a complex manifold X defines a length function. However, a length function is much more general than a Hermitian metric or even a Finsler metric.

The distance function d_H is known to generate the topology on X [157]. If X is a complex space, k_X (K_X) will denote the Kobayashi pseudodistance (Kobayashi-Royden semi-length function [157]) on X. The notation f^*E represents the pull-back of the semi-length function E by f (i.e. $f^*E(v) = E(df(v))$ for $v \in T(X)$).

If Y is a complex space and E is a length function on Y we denote by d_E the distance function generated on Y by E [157].

Let X be a complex subspace of a complex space Y. Set

$$\mathcal{F}_{X,Y} = \left\{ f \in \mathcal{H}(D, Y) : f^{-1}(Y \setminus X) \text{ is at most a singleton } \right\} .$$

Then, using the family $\mathcal{F}_{X,Y}$ instead of $\mathcal{H}(D, X)$, Kobayashi [137] defined another intrinsic pseudodistance $k_{X,Y}$ on $\overline{X}$ and its infinitesimal form $K_{X,Y}$ when X is a complex submanifold of a complex manifold Y.

If no $f \in \mathcal{F}_{X,Y}$ satisfies $f(0) = p$ and $(df)_0(re) = v$, then $K_{X,Y}(p, v)$ is defined to be ∞. It was noted by Kobayashi in [137] that $k_{D^*,D} = k_D$, and, indeed, he proves in his book that $k_{(D^*)^m,D^m} = k_{D^m}$ for each positive integer m.

The norm $|df|_E$ of the tangent map for $f \in H(M, Y)$ with respect to E is defined by

$$\begin{aligned} |df|_E &= \sup \{ |df(p)|_E : p \in M \} \qquad \text{where} \\ |df(p)|_E &= \sup \left\{ E(f(p), df(p, v)) : F_M^K(p, v) = 1 \right\} . \end{aligned}$$

(We will use simply $|df|$ and $|df(p)|$ when no confusion may arise).

Recall that a complex subspace X of a complex space Y is *hyperbolically imbedded* in Y if, for $p, q \in \overline{X}, p \neq q$, there are sets V, W open in Y about p, q respectively such that $k_X(V \cap X, W \cap X) > 0$ [129].

Example 6.4.1. Royden showed (Theorem 2 in [254]) that a complex manifold M is hyperbolic if and only if $\mathcal{H}(D, M)$ is evenly continuous, and Abate showed (Theorem 1.3 in [1]) that M is hyperbolic if and only if $\mathcal{H}(D, M)$ is relatively compact in $\mathbb{C}(D, M^*)$. Hence $\mathcal{H}(D, M)$ is a uniformly normal family if and only if M is hyperbolic.

Example 6.4.2. It is shown in Theorem 2 in [129] that a complex subspace X of a complex space Y is hyperbolically imbedded in Y if and only if $\mathcal{H}(D, X)$ is relatively compact in $\mathbb{C}(D, Y^*)$, i.e. if and only if $\mathcal{H}(D, X)$ is a uniformly normal subfamily of $\mathcal{H}(D, Y)$. This is a generalization of Kiernan's theorem [129] which deals with the case when X is relatively compact in Y.

For $r > 0$, let $D_r = \{z \in \mathbb{C} : |z| < r\}$ and $D_r^* = D_r \setminus \{0\}$.

Theorem 6.4.3. *Let M be a hyperbolic manifold and let Y be a complex space. Then $\mathcal{F} \subset \mathcal{H}(M,Y)$ is uniformly normal if and only if there is a length function E on Y such that $|df|_E \leq 1$ for each $f \in \mathcal{F}$.*

Proof. **Necessity.** Clearly $\mathcal{F} \circ \mathcal{H}(D,M)$ is an evenly continuous subset of $\mathcal{H}(D,Y)$. We will show first that, for each length function E on Y and compact $Q \subset Y$, there exists $c > 0$ such that $|df| \leq c$ on $f^{-1}(Q)$ for each $f \in \mathcal{F}$. If $Q \subset Y$ is compact and fails the stated condition for the length function E, we choose sequences $\{p_n\}$, $\{f_n\}$, $\{v_n\}$ and $q \in Q$, such that $p_n \in M, f_n \in \mathcal{F}, v_n \in T_{p_n}(M), f_n(p_n) \in Q, K_M(p_n, v_n) = 1, f_n(p_n) \to q$ and $E\left((df_n)_{p_n}(v_n)\right) > n$. It follows that $\left|(df_n)_{p_n}\right| \to \infty$ and we choose a sequence $\{\varphi_n\}$ in $\mathcal{H}(D,M)$ satisfying $\varphi_n(0) = p_n$ and $|(df_n \circ \varphi_n)_0| \to \infty$. Let V be a relatively compact neighborhood of q hyperbolically imbedded in Y. Since $\mathcal{F} \circ \mathcal{H}(D,M)$ is an evenly continuous subset of $\mathcal{H}(D,Y)$, we choose $0 < r < 1$ such that $f_n \circ \varphi_n(D_r) \subset V$; the sequence of restrictions of $\{f_n \circ \varphi_n\}$ to D_r, which we call again $\{f_n \circ \varphi_n\}$, is uniformly normal and is consequently relatively compact in $\mathcal{H}(D_r, Y)$. Some subsequence of $\{f_n \circ \varphi_n\}$ converges to $h \in \mathcal{H}(D_r, Y)$, contradicting $|(df_n \circ \varphi_n)_0| \to \infty$. Now, to complete the proof of the necessity, choose sequences $\{V_n\}, \{c_n\}$ such that V_n is open and relatively compact in $Y, \overline{V_n} \subset V_{n+1}, \bigcup_1^\infty V_n = Y, c_n > 0$ and $|df|_E \leq c_n$ on $f^{-1}\left(\overline{V_n}\right)$ for each $f \in \mathcal{F}$. Choose a positive continuous function μ on Y such that $\mu(q)c_n \leq 1$ on $\overline{V_n}$. The length function H on Y defined by $H(v) = \mu(q)E(v)$ for $v \in T_q(Y)$ satisfies $|df|_H \leq 1$ for each $f \in \mathcal{F}$ ([157, p. 34]).

Sufficiency. Each $f \in \mathcal{F} \circ \mathcal{H}(D,M)$ is distance decreasing with respect to k_D and d_E and the desired conclusion follows from Propositions 6.1.6 and Theorem 3.1.5. □

Example 6.4.4. If $f \in \mathcal{H}\left(D, \mathbf{P}^1(\mathbf{C})\right)$, and $\overline{\Delta} \subset D$ is a closed disk and $\partial\overline{\Delta}$ denotes the boundary of $\overline{\Delta}$, let $J(f(\overline{D}))$ and $L(f(\overline{\Delta}))$ be respectively the spherical area of $f(\overline{\Delta})$ and spherical length of $f(\partial\overline{\Delta})$. Let $h > 0$ and

$$\mathcal{F}(h) =$$

$$\left\{f \in \mathcal{H}\left(D, \mathbf{P}^1(\mathbf{C})\right) : J(f(\overline{\Delta})) \leq hL(f(\overline{\Delta})) \text{ for each closed disk } \overline{\Delta} \subset D\right\}.$$

Hayman ([111, p. 164]) showed that $\mathcal{F}(h)$ is invariant relative to $\mathcal{A}(D)$, the group of automorphisms of D, and normal in the sense of Montel. The family $\mathcal{F}(h)$ is uniformly normal.

Example 6.4.5. Let M be a complex manifold, $r > 0$, and $\mathcal{F} \subset \mathcal{H}\left(M, \mathbf{P}^1(\mathbf{C})\right)$ be a family of maps such that for each $f \in \mathcal{F}$ three points $a_f, b_f, c_f \in \mathbf{P}^1(\mathbf{C}) - f(M)$ satisfy $\chi(a_f, b_f)\chi(c_f, b_f)\chi(c_f, a_f) \geq r$ where χ represents the spherical metric. Carathéodory ([36, p. 202]) showed in this situation that $\mathcal{F} \circ \mathcal{H}(D,M)$ is normal in the sense of Montel. So $\mathcal{F}$ is uniformly normal.

The normal maps studied in [81], [105], and [144] are all normal maps in our sense.

Theorem 6.4.6. *Let N be a hyperbolically imbedded complex submanifold of a complex manifold M and let Y be a complex space. The following are equivalent for $\mathcal{F} \subset \mathcal{H}(N, Y)$:*

(1) The family $\mathcal{F}$ is uniformly normal.

(2) If $p \in Y$ and $\{g_n\}, \{z_n\}$ are sequences in $\mathcal{F} \circ \mathcal{H}(D^*, N), D^*$, respectively, such that $z_n \to 0$ and $g_n(z_n) \to p$, then for each neighborhood U of p there is an r, $0 < r < 1$, satisfying $g_n(D_r^*) \subset U$.

(3) There is a length function E on Y such that $f^*E \leq K_{N,M}$ for each $f \in \mathcal{F}$.

Proof. **(1)** $\Longrightarrow$ **(2)**. From Theorem 6.4.3 and the fact that N is hyperbolic, there exists a length function E on Y such that each $f \in \mathcal{F} \circ \mathcal{H}(D^*, N)$ is distance decreasing with respect to k_{D^*} and d_E. The proof that **(2)** holds may be completed by arguments similar to those in [128] and **(1)** $\Longrightarrow$ **(2)** of Theorem 1 in [123].

(2) $\Longrightarrow$ **(3)**. We show that for any compact $Q \subset Y$ and length function E on Y there exists $c > 0$ such that $cE((df)_p(v)) \leq 1$ when $f \in \mathcal{F}, f(p) \in Q, v \in T_p(N)$ and $K_{N,M}(p, v) = 1$. The proof may then be completed as in the proof of the necessity of Theorem 6.4.3. Suppose $Q \subset Y$ is compact and fails the stated condition for the length function E. We choose $q \in Q$ and sequences $\{f_n\}, \{p_n\}, \{v_n\}$ such that $f_n \in \mathcal{F}, f_n(p_n) \in Q, v_n \in T_{p_n}(N), E\left((df_n)_{p_n}(v_n)\right) > n, K_{N,M}(p_n, v_n) = 1$, and such that $f_n(p_n) \to q$. We choose sequences $\{\varphi_n\}$ in $\mathcal{F}_{N,M}, \{r_n\}$ in **(1)**, **(2)** satisfying $\varphi_n(0) = p_n, (d\varphi_n)_0(r_n e) = v_n$ and $E((df_n \circ \varphi_n)_0(r_n e)) > n$. Suppose there exists $r, 0 < r < 1$, such that a subsequence of the sequence of restrictions of $\{f_n \circ \varphi_n\}$ to D_r, called again $\{f_n \circ \varphi_n\}$, satisfies $f_n \circ \varphi_n \in \mathcal{F} \circ \mathcal{H}(D_r, N)$; for such r it follows from **(2)** that $\mathcal{F} \circ \mathcal{H}(D_r, N)$ is evenly continuous and since $f_n \circ \varphi_n(0) \to q$ we again obtain a contradiction as in the proof of the necessity of Theorem 6.4.3. Alternatively we can choose a sequence $\{z_n\}$ in D^* such that $z_n \to 0$ and $\varphi_n(z_n) \in M \setminus N$, and a sequence $\{\alpha_n\}$ in $\mathcal{A}(D)$ such that $\alpha_n(0) = z_n$; let $h_n = \varphi_n \circ \alpha_n$ on D^*. Then $h_n \in \mathcal{H}(D^*, N), f_n \circ h_n\left(\alpha_n^{-1}(0)\right) \to q$ and $\alpha_n^{-1}(0) \to 0$. Let $g_n = f_n \circ h_n$ and let V be a neighborhood of q which is relatively compact and hyperbolically imbedded in Y. There exists r, $0 < r < 1$, such that $g_n(D_r^*) \subset V$; so g_n extends to $\widetilde{g}_n \in \mathcal{H}(D, Y)$. From Theorem 2 in [123] there exists a subsequence of $\{\widetilde{g}_n\}$, called again $\{\widetilde{g}_n\}$, satisfying $\widetilde{g}_n \to g \in \mathcal{H}(D, Y)$, a contradiction since $\left|(d\widetilde{g}_n)_{\alpha_n^{-1}(0)}\right| = E((df_n \circ \varphi_n)_0(e))$.

(3) $\Longrightarrow$ **(1)**. This follows easily from Theorem 6.4.3 since $K_{N,M} \leq K_N$ on N. □

Proof. Recall that a divisor A on a complex manifold M has *normal crossings* ([164, p. 58]) if at each point of A there exists a system of complex coordinates $z_1, \ldots, z_m$ for M such that, locally, $M \setminus A = (D^*)^r \times D^s$ with $r + s = m$. □

Lemma 6.4.7. *Let $\mathcal{F} \subset \mathcal{H}((D^*)^m, Y)$ be uniformly normal. If $\{w_n\}, \{f_n\}$ are sequences in $(D^*)^m, \mathcal{F}$ respectively such that $w_n \to w_0 \in D^m$ and $f_n(w_n) \to p \in Y$, then for each neighborhood U of p there is a neighborhood W of w_0 in D^m such that $f_n(W \cap (D^*)^m) \subset U$.*

Proof. The proof is by induction on m. Equivalence **(2)** of Theorem 6.4.6 establishes the result for $m = 1$. Suppose the statement is true for the integer k but not for the integer $k+1$. Let $\mathcal{F} \subset \mathcal{H}\left((D^*)^{k+1}, Y\right)$ be uniformly normal, let $\{w_n\}, \{w'_n\}$ be sequences in $(D^*)^{k+1}$ such that $w_n \to w_0 \in D^{k+1}, w'_n \to w_0$, and let $\{f_n\}$ be a sequence in $\mathcal{F}$ such that $f_n(w_n) \to p$ while $f_n(w'_n) \nrightarrow p$. Let U, V be open relatively compact neighborhoods of p such that $\overline{V} \subset U$ and assume that $f_n(w'_n) \in Y \setminus U$. Let $w_n = (s_n, t_n), w'_n = (s'_n, t'_n)$, and $w_0 = (s_0, t_0)$ where $s_n, s'_n, s_0 \in (D^*)^k$ and $t_n, t'_n, t_0 \in D^*$.

Let

$$\mathcal{F}_1 = \left\{\varphi_t \in \mathcal{H}\left((D^*)^k, (D^*)^{k+1}\right) : t \in D^*, \varphi_t(s) = (s,t)\right\} \quad \text{and}$$

$$\mathcal{F}_2 = \left\{\psi_s \in \mathcal{H}\left(D^*, (D^*)^{k+1}\right) : s \in (D^*)^k, \psi_s(t) = (s,t)\right\}.$$

Then $\mathcal{F} \circ \mathcal{F}_1 \subset \mathcal{H}\left((D^*)^k, Y\right)$ and $\mathcal{F} \circ \mathcal{F}_2 \subset \mathcal{H}(D^*, Y)$ are both uniformly normal families; $\{f_n \circ \varphi_{t_n}\}$ is a sequence in $\mathcal{F} \circ \mathcal{F}_1$, $s_n \to s_0$, and $f_n \circ \varphi_{t_n}(s_n) \to p$. By the induction hypothesis we choose a neighborhood N_1 of s_0 such that $f_n \circ \varphi_{t_n}\left(N_1 \cap (D^*)^k\right) \subset V$ and $f_n \circ \varphi_{t_n}(s'_n) \in V$. There exists a subsequence of $\{f_n \circ \varphi_{t_n}(s'_n)\}$, called again $\{f_n \circ \varphi_{t_n}(s'_n)\}$, such that $f_n \circ \varphi_{t_n}(s'_n) \to q \in \overline{V}$; $f_n \circ \varphi_{t_n}(s'_n) = f_n \circ \psi_{s'_n}(t_n); t'_n \to t_0$ and we choose a neighborhood N_2 of t_0 in D such that $f_n \circ \psi_{s'_n}(N_2 \cap D^*) \subset U$. We see that $f_n \circ \psi_{s'_n}(t'_n) \in U$, a contradiction. □

If $\{A_n\}$ is a sequence of subsets of a topological space, then we define the *limit superior* of the sequence A_n to be the collection of elements x of the space with the property that each neighborhood of x intersects A_n for infinitely many n. We use the notation $\limsup A_n$ for this set.

Theorem 6.4.8. *Let M be a complex manifold, let A be a divisor on M with normal crossings, let $\mathcal{F} \subset \mathcal{H}(M \setminus A, Y)$ be uniformly normal and let $\overline{\mathcal{F}}$ be the closure of $\mathcal{F}$ in $\mathbb{C}(M \setminus A, Y^*)$.*

Then

(1) Each $f \in \overline{\mathcal{F}}$ extends to $\widetilde{f} \in \mathbb{C}(M, Y^*)$.

(2) The space $\mathbb{C}\left[M, Y^*; \overline{\mathcal{F}}\right]$ is compact in $\mathbb{C}(M, Y^*)$.

(3) If $\{f_n\}$ is a sequence in $\overline{\mathcal{F}}$ and $f_n \to f$, then $\widetilde{f}_n \to \widetilde{f}$.

(4) For each sequence $\{f_n\}$ in $\overline{\mathcal{F}}$ there is a subsequence $\{f_{n_k}\}$ of $\{f_n\}$ such that $\limsup f_{n_k}^{-1}(P) \cap \limsup f_{n_k}^{-1}(Q) = \emptyset$ in the topology of M for each pair P, Q of disjoint subsets of Y with P compact in Y and Q closed in Y.

(5) If M is hyperbolic and $K_{M\setminus A,M} = K_M$, then $\mathcal{H}\left[M, Y^*; \overline{\mathcal{F}}\right]$ is uniformly normal.

Proof. As for the proofs of **(1)** and **(2)**, we show first that each $f \in \mathcal{F}$ extends to $\widetilde{f} \in \mathbb{C}(M, Y^*)$ and that $\mathbb{C}[M, Y^*; \mathcal{F}]$ is relatively compact in $\mathbb{C}(M, Y^*)$. Since the considerations involved are local in nature, we may assume that $M = D^m$, $\mathcal{F} \subset \mathcal{H}((D^*)^m, Y)$, and we show that each $f \in \mathcal{F}$ extends to $\tilde{f} \in \mathbb{C}(D^m, Y^*)$ and that $\mathbb{C}[(D)^m, Y^*; \mathcal{F}]$ is evenly continuous from D^m to Y^*; Lemma 6.4.7 leads to these conclusions as in **(1)** $\Longrightarrow$ **(2)** of Theorem 8 and Corollary 7 in [123]. To finish the proof of **(1)**, if $f \in \overline{\mathcal{F}}$, then there exists a sequence $\{f_n\}$ in $\mathcal{F}$ such that $f_n \to f$. There exists a subsequence $\{f_{n_k}\}$ of $\{f_n\}$ such that $\widetilde{f}_{n_k} \to g \in \mathbb{C}(M, Y^*); now g = \widetilde{f}$ so **(1)** holds. To establish **(2)**, we show that $\mathbb{C}\left[M, Y^*; \overline{\mathcal{F}}\right] = \overline{\mathbb{C}}[M, Y^*; \mathcal{F}]$. If $g \in \overline{\mathcal{F}}$, choose a sequence $\{f_n\}$ in $\mathcal{F}$ such that $f_n \to g$. It follows that $\widetilde{f}_{n_k} \to \widetilde{g}$ for some subsequence $\{f_{n_k}\}$ of $\{f_n\}$ and one inclusion is established. For the other inclusion, if $\{f_n\}$ is a sequence in $\mathcal{F}$ and $\widetilde{f}_n \to g$, then $f_n \to g$ on $M \setminus A$. To see that **(3)** holds note that, from **(2)**, each subsequence of $\left\{\widetilde{f}_n\right\}$ has a convergent subsequence; and that if $\left\{\widetilde{f}_{n_k}\right\}$ is a convergent subsequence, then $\widetilde{f}_{n_k} \to \widetilde{f}$. For the proof of **(4)** we see from **(2)** that there is a subsequence $\{f_{n_k}\}$ of $\{f_n\}$ such that $\widetilde{f}_{n_k} \to g \in \mathbb{C}(M, Y^*)$. If P, Q are respectively compact and closed subsets of Y and $x \in \limsup f_{n_k}^{-1}(P) \cap \limsup f_{n_k}^{-1}(Q)$, then $\widetilde{f}_{n_k}(x) \to g(x)$ and for each V open about x, $f_{n_k}(V \setminus A) \cap P \neq \emptyset$ and $f_{n_k}(V \setminus A) \cap Q \neq \emptyset$ occur frequently. Hence $g(x) \in \overline{P} \cap \overline{Q}$ in Y^*. Since P is compact in Y and Q is closed in $Y, g(x) \in P \cap Q$.

Finally we prove **(5)**. From Theorem 6.4.6 part **(3)**, let E be a length function on Y such that $f^*E \le K_{M-A,M}$ for each $f \in \mathcal{F}$. Let $\widetilde{f} \in \mathcal{H}\left[M, Y^*; \overline{\mathcal{F}}\right]$. There is a sequence $\{f_n\}$ in $\mathcal{F}$ such that $f_n \to f$. It follows that $\widetilde{f}^*E \le K_{M\setminus A,M} = K_M$. □

Remark 6.4.9. Theorem 6.4.8 parts **(1)**, **(2)**, and **(3)** extend work of the authors [123] to uniformly normal families from complex manifolds having divisors with normal crossings in complex spaces. The results in [123] generalize work of Järvi [119], Kobayashi [135], Kwack [152], Kiernan [128], and Noguchi [222, 223]. Theorem 2.3 part **(5)** generalizes a second result in [119]. In particular we have the following corollary.

Corollary 6.4.10. *Let Y be a complex space, and let M and $M \setminus A$ be hyperbolic manifolds as defined in any one of the following four cases:*

(1) We have $M \setminus A = (D^*)^{n-k} \times D^k$ and $M = D^n$.

(2) We have $M \setminus A = (D^*)^n$ and $M = D^n$.

(3) We have that A is a closed analytic subset of M of codimension at least 2.

(4) We have that M is n-dimensional and A is a closed subset of M with $(2n-2)$-dimensional Hausdorff measure equal to zero.

Let $\mathcal{F} \subset \mathcal{H}(M \setminus A, Y)$ be uniformly normal, and let $\overline{\mathcal{F}}$ be the closure in $\mathbb{C}(M \setminus A, Y^*)$. Then $\mathcal{H}\left[M, Y^*; \overline{\mathcal{F}}\right]$ is uniformly normal.

Proof. The proof follows from Theorem 6.4.8 part **(5)** since in each case $K_M = K_{M\setminus A, M}$ (see [34] for **(4)**). □

Theorem 6.4.11. *The following statements are equivalent for complex spaces X, Y and $\mathcal{F} \subset \mathcal{H}(X,Y)$, where $\overline{\mathcal{F} \circ \mathcal{H}(D^*, X)}$ is the closure of $\mathcal{F}$ in $\mathbb{C}(D^*, Y^*)$:*

(1) The family $\mathcal{F}$ is uniformly normal.

(2) The family $\mathcal{F} \circ \mathcal{H}(D^*, X)$ is uniformly normal.

(3) The space $\mathbb{C}[D, Y^*; \mathcal{F} \circ \mathcal{H}(D^*, X)]$ is relatively compact in $\mathbb{C}(D, Y^*)$.

(4) For each sequence $\{f_n\}$ in $\overline{\mathcal{F} \circ \mathcal{H}(D^*, X)}$ there is a subsequence $\{f_{n_k}\}$ of $\{f_n\}$ such that $\limsup f_{n_k}^{-1}(P) \cap \limsup f_{n_k}^{-1}(Q) = \emptyset$ in the topology of D for each pair of disjoint subsets P, Q of Y with P compact in Y and Q closed in Y.

(5) The family $\mathcal{F}$ satisfies the following three conditions:

(a) The family $\mathcal{F} \circ \mathcal{H}(D^*, X)$ is relatively compact in $\mathbb{C}(D^*, Y^*)$,

(b) Each $f \in \overline{\mathcal{F}} \circ \mathcal{H}(D^*, X)$ extends to $\widetilde{f} \in \mathbb{C}(D, Y^*)$, and

(c) If $\{f_n\}$ is a sequence in $\mathcal{F} \circ \mathcal{H}(D^*, X)$ such that $f_n \to f$, then $\widetilde{f}_n \to \widetilde{f}$.

Proof. **(1)** $\Longrightarrow$ **(2)**. Follows from Proposition 6.1.6 part **(2)**.

(2) $\Longrightarrow$ **(3)**. From **(2)** of Theorem 6.4.8 and the set inclusion

$$\mathcal{H}\left[D, Y^*; \overline{\mathcal{F} \circ \mathcal{H}(D^*, X)}\right] \cup \mathbb{C}[D, Y^*; \mathcal{F} \circ \mathcal{H}(D^*, X)]$$

$$\subset \mathbb{C}\left[D, Y^*; \overline{\mathcal{F} \circ \mathcal{H}(D^*, X)}\right].$$

(3) $\Longrightarrow$ **(1)**. From the fact that $\mathcal{F} \circ \mathcal{H}(D, X)$ is a subset of the collection of extensions in **(3)**.

(2) $\Longrightarrow$ **(4)**. From **(2)** and **(4)** of Theorem 6.4.8.

(4) $\Longrightarrow$ **(3)**. We show that $\mathbb{C}[D, Y^*; \mathcal{F} \circ \mathcal{H}(D^*, X)]$ is evenly continuous. If this is not the case choose sequences $\{f_n\}$ in $\mathcal{F} \circ \mathcal{H}(D^*, X)$, $\{v_n\}$, $\{x_n\}$ in D^*, $x \in D$, $y \in Y$, and open sets W_1, W_2 in Y about y such that $\overline{W}_1 \subset W_2$, $\overline{W}_1$ compact, $v_n \to x$, $x_n \to x$, $f_n(v_n) \to y$, and $f_n(x_n) \in Y \setminus W_2$. For any subsequence $\{f_{n_k}\}$ of $\{f_n\}$ we have $x \in \limsup f_{n_k}^{-1}\left(\overline{W}_1\right) \cap \limsup f_{n_k}^{-1}(Y \setminus W_2)$ even though $\overline{W}_1 \cap (Y \setminus W_2) = \emptyset$, so **(4)** does not hold.

(2) $\Longrightarrow$ **(5)**. Condition **(a)** follows from definition, condition **(b)** from **(1)** of Theorem 6.4.8 and condition **(c)** from **(3)** of Theorem 6.4.8.

(5) $\Longrightarrow$ **(3)**. Let $\{f_n\}$ be a sequence in $\mathcal{F} \circ \mathcal{H}(D^*, X)$. By condition **(a)** there exists a subsequence$\{f_{n_k}\}$ of $\{f_n\}$ such that $f_{n_k} \to f \in \mathbb{C}(D^*, Y^*)$; $\widetilde{f}_{n_k}, \widetilde{f}$ exist for each k by condition **(b)** and $\widetilde{f}_{n_k} \to \widetilde{f}$ by condition **(c)**. □

Kiernan [129] gave further insight into the concept of hyperbolic imbeddedness by showing that a relatively compact complex subspace X of a complex space Y is hyperbolically imbedded in Y if and only if there is a length function E on Y such that $f^*E \leq K_D$ for each $f \in \mathcal{H}(D, X)$. Our Theorem 6.4.12 provides additional insight into the role of hyperbolic imbeddedness in Kobayashi's generalization of the big Picard theorem ([135, Theorem 6.1]). Zaidenberg has provided a number of other criteria for hyperbolic imbeddedness and hyperbolicity in [289].

Theorem 6.4.12. *Let X be a complex subspace of a complex space Y. The following statements are equivalent:*

(1) The space X is hyperbolically imbedded in Y.

(2) The family $\mathcal{H}(D^*, X)$ is a uniformly normal subfamily of $\mathcal{H}(D^*, Y)$

(3) There exists a length function E on Y such that each $f \in \mathcal{H}(D^*, X)$ satisfies $f^*E \leq K_{D^*}$.

(4) There exists a distance function d on Y such that each $f \in \mathcal{H}(D^*, X)$ is distance decreasing with respect to k_{D^*} and d.

Proof. **(1)** $\Longrightarrow$ **(2)**. It is obvious that $\mathcal{H}(D^*, X) \circ \mathcal{H}(D, D^*) \subset \mathcal{H}(D, X)$ and hence $\mathcal{H}(D^*, X)$ is a uniformly normal subfamily of $\mathcal{H}(D^*, Y)$ (see Example 1.5).

(2) $\Longrightarrow$ **(3)**. Since D^* is hyperbolically imbedded in D, by Theorem 6.4.6 part **(3)** there exists a length function E on Y satisfying $f^*E \leq K_{D^*,D} = K_D \leq K_{D^*}$ for each $f \in \mathcal{H}(D^*, X)$.

(3) $\Longrightarrow$ **(4)**. The distance function induced on Y by the length function E promised in **(3)** meets the requirements of **(4)**.

(4) $\Longrightarrow$ **(1)**. It follows from Proposition 3.1.5 that $\mathcal{H}(D^*, X)$ is relatively compact in $\mathbb{C}(D^*, Y^*)$. Let $\{f_n\}$ be a sequence in $\mathcal{H}(D, X)$ such that $f_n \to f \in \mathbb{C}(D^*, Y^*)$ on D^*. We show that f extends to $\widetilde{f} \in \mathbb{C}(D, Y^*)$ and that $f_n \to \widetilde{f}$ on D. This will complete the proof of **(4)** $\Longrightarrow$ **(1)** in view of the result referred to in Example 6.4.2. If for each compact $Q \subset Y$ there exists a neighborhood $V$of 0 in D satisfying $f_n(V) \cap Q = \emptyset, then f$ may be extended to $\widetilde{f} \in \mathbb{C}(D, Y^*)$ by defining $\widetilde{f}(0) = \infty$, and $f_n \to \widetilde{f}$ on D. Otherwise, choose a subsequence of $\{f_n\}$, called again $\{f_n\}$, a sequence $\{z_n\}$inD^*, and $p \in Y$ such that $|z_n| \downarrow 0, \quad f_n(z_n) \to p$. If $r_n \downarrow 0$, the hyperbolic length of $\sigma_{r_n} = \{z \in D : |z| = r_n\}$ in D^* converges to 0; it follows from **(4)** and winding number arguments modeled after those of Grauert and Reckziegel ([87, p. 120]) and found in [123] and [128], that $f_n(0) \to p, \quad f_n(z'_n) \to p$ for any sequence $\{z'_n\}$ for which $z'_n \to 0$. Hence f may be extended to $\widetilde{f} \in \mathbb{C}(D, Y^*)$ by defining $\widetilde{f}(0) = p$, and $f_n \to \widetilde{f}$ on D. □

Remark 6.4.13. The equivalence **(1)** $\Longleftrightarrow$ **(3)** in Theorem 6.4.12 shows surprisingly that the statement obtained by dropping the relative compactness of X and replacing D by D^* in Kiernan's result is valid.

Remark 6.4.14. Let X be a complex subspace of a complex space Y. Kwack's generalization of the big Picard theorem establishes that $f \in \mathcal{H}(D^*, X)$ is extendable to $\widetilde{f} \in \mathcal{H}(D, Y)$ if **(1)** There exists a distance function d on Y such that f is distance decreasing with respect to k_{D^*} and d, and **(2)** There exists a sequence $\{z_n\}$ in D^* and a $p \in Y$ such that $z_n \to 0$ and $f(z_n) \to p$ (Theorem 3 in [152]). We observe from **(1)** $\Longleftrightarrow$ **(3)** in Theorem 6.4.12 that, under the hypothesis of Kobayashi's generalization [135] of Kwack's theorem, all $f \in \mathcal{H}(D^*, X)$ satisfy conditions **(1)** and **(2)**.

Remark 6.4.15. The equivalence **(1)** $\Longleftrightarrow$ **(2)** of Theorem 6.4.12 establishes that a complex space X is hyperbolic if and only if $\mathcal{H}(D^*, X)$ is a uniformly normal subfamily of $\mathcal{H}(D^*, X)$ (compare with Abate's characterization cited in Example 6.4.1).

6.5 Separately Normal Maps

Barth [19] proved that separately holomorphic maps into hyperbolic space are holomorphic and also a separately normal family of maps into hyperbolic space is a normal family. Extending these results, we prove that normal maps and uniformly normal families satisfy similar results.

Let X, Y be complex spaces. A family of maps $F \subset \mathbf{T}(D^n, Y)$ is said to satisfy *separately a property* P if at each point $p = (p_{(1)}, p_{(2)}, \ldots, p_{(n)}) \in D^n$, the family of restrictions of members of F to each i-th coordinate disk $D_i(p) = \{p_{(1)}\} \times \{p_{(2)}\} \times \cdots \times \{p_{(i-1)}\} \times D \times \{p_{(i+1)}\} \times \cdots \times \{p_{(n-1)}\} \times \{p_{(n)}\}$ satisfies the property P. A family of maps $F \subset \mathbf{T}(X, Y)$ is said to satisfy *separately a property* P if F_{D^n}, the family of restrictions of maps of F to each coordinates disk, D^n, satisfies separately a property P. If $v \in T_p(D^n) = \prod_i T_{p_{(i)}}(D)$ where $p \in D^n$, then its coordinates will be denoted by $v = (v_{(1)}, v_{(2)}, \ldots, v_{(n)})$. If $v \in T_p(D_i(p))$, then $v_{(j)} = 0$ for $j \neq i$. It is to be noted that $K_{D^n}(v) = \max\{K_D(v_{(i)})\}$ [134].

Let $f : D^2 \to \widehat{\mathbb{C}}$, the Riemann sphere, be given by

$$f(z, w) = \begin{cases} \frac{(z+w)^2}{z-w} & \text{if } z \neq w \\ \infty & \text{if } z = w \neq 0 \\ 0 & \text{if } z = w = 0 \end{cases}$$

Theorem 6.5.1 extends the result of Barth [19] that every separately holomorphic map into a hyperbolic space is holomorphic and that every separately normal holomorphic family of maps into a hyperbolic space is normal. A coordinate disk in D^n of the form $D_i(p) = \{p_{(1)}\} \times \{p_{(2)}\} \times \cdots \times \{p_{(i-1)}\} \times D \times \{p_{(i+1)}\} \times \cdots \times \{p_{(n-1)}\} \times \{p_{(n)}\}, p \in D^n$, will be identified with the unit disk D under the projection onto the i-th coordinate.

Theorem 6.5.1. *Let Y be a complex space and let F be a collection of maps from D^n to Y. Suppose that each $f \in F$ is separately holomorphic and that there is a length function E on Y such that for $i = 1, \ldots, n, p \in D^n, q \in D_i(p)$,*

$$d_E(f(p), f(q)) \leq k_D\left(p_{(i)}, q_{(i)}\right), \quad \textit{for all } f \in F.$$

Then F is uniformly normal.

Proof. From the triangle inequality and the property of hyperbolic distances that for $p, q \in D^n$

$$k_{D^n}(p, q) = \max\left\{k_D\left(p_{(i)}, q_{(i)}\right)\right\}$$

it follows that

$$d_E(f(p), f(q)) \leq \sum k_D\left(p_{(i)}, q_{(i)}\right) \leq n k_{D^n}(p, q), \quad \text{for all } f \in F.$$

Then since $f \in F$ is distance-decreasing with respect to the distances d_E and nk_{D^n}, $f \in F$ is continuous. Since f is also separately holomorphic, it is holomorphic. Now it follows from Theorem 6.4.12 that F is uniformly normal. This function f is separately normal but not continuous at $(0, 0)$. □

This chapter is based on the articles of M. H. Kwack and J. E. Josef [122, 150, 151, 153].

7

Normal Families of Holomorphic Functions and Mappings on a Banach Space

Preamble: This chapter treats the theory of normal families in the context of infinite-dimensional spaces. The material presented here sets our book apart from other treatments of normal families. It is quite modern, and it is fascinating.

Holomorphic function theory in infinitely many variables is complicated because standard devices used in the finite-dimensional setting are no longer valid. For instance, it is not the case in a Banach space that a domain may be exhausted by compact sets. It is also no longer the case that the compact-open topology is the most useful topology on holomorphic functions.

So we will have to give a number of new definitions in this chapter. We will provide copious examples to illustrate how things have changed in this new context.

But we will also give a number of positive results. The theory is especially rich on Hilbert space.

The authors lay the foundations for the study of normal families of holomorphic functions and mappings on an infinite-dimensional normed linear space. Characterizations of normal families, in terms of value distribution, spherical derivatives, and other geometric properties are derived. Montel-type theorems are established. A number of different topologies on spaces of holomorphic mappings are considered. Theorems about normal families are formulated and proved in the language of these various topologies.

Normal functions are also introduced. Characterizations in terms of automorphisms and also in terms of invariant derivatives are presented.

7.1 Introduction

The theory of holomorphic functions of infinitely many complex variables is at least seventy years old. Pioneers of the subject were Nachbin [16–18, 203–217],

DOI: 10.1201/9781032669861-7

Gruman and Kiselman [96], and Mujica [202]. After a quiet period of nearly two decades, the discipline is now enjoying a rebirth thanks to the ideas of Lempert [168, 169, 171–173]. Lempert has taught us that it is worthwhile to restrict attention to particular Banach spaces, and he has directed our efforts to especially fruitful questions.

The work of the present chapter is inspired by the results of [130]. That paper studied domains in a Hilbert space with an automorphism group orbit accumulating at a boundary point. As was the case in even one complex variable, normal families played a decisive role in that study. With a view to extending those explorations, it now seems appropriate to lay the foundations for normal families in infinitely many complex variables.

One of the thrusts of the present book is to demonstrate that normal families may be understood from several different points of view. These include:

2. Hyperbolic geometry
3. Functional analysis
4. Distribution theory
5. Currents
6. Comparison of different topologies and norms on the space of holomorphic functions

It is our intention to explain these different approaches to the subject, in the infinite-dimensional context, and to establish relationships among them.

A second thrust is to relate the normality of a family on the entire space X (or on a domain in X) to the normality of the restriction of the family to slices (suitably formulated). This point of view was initiated in [47], and it has proved useful and intuitively natural.

Throughout this paper, X is a separable Banach space over the scalar field $\mathbb{C}, \Omega$ is a domain (a connected open set) in $X, \mathcal{F} = \{f_\alpha\}_{\alpha \in \mathcal{A}}$ is a family of holomorphic functions on Ω, and $D \subseteq \mathbb{C}$ is the unit disk. If Ω' is another domain in some other separable Banach space Y, then we will also consider families $\{F_\alpha\}$ of holomorphic mappings from Ω to Ω'. Although separability of X is not essential to all of our results, it is a convenient tool in many arguments. Certainly, in the past, the theory of infinite dimensional holomorphy has been hampered by a tendency to shy away from such useful extra hypotheses.

Part of the beauty and utility of studying infinite-dimensional holomorphy is that the work enhances our study of finite-dimensional holomorphy. Indeed, it is safe to say that the present study has caused us to re-invent what a normal family of holomorphic functions ought to be.

One of the interesting features of the present work, making it different from more classical treatments in finite dimensions, is that compact sets now play a different role. If W is a given open set in our space X (say the unit ball in a separable Hilbert space), then W cannot be exhausted by an increasing union of compact sets in any obvious way. Another feature is that, in finite dimensions, all reasonable topologies on the space of holomorphic functions on a given domain are equivalent. In infinitely many variables this is no longer the case, and we hope to elucidate the matter both with examples and results relating the different topologies. We will define holomorphic functions on a Banach space below.

A final note is that there are interesting underlying questions, throughout our study, about the geometry of Banach spaces. We sidestep most of these by concentrating our efforts on separable Banach spaces; most of our deepest results are in separable Hilbert spaces. We intend to study the deeper questions of the geometry of Banach space, and their impact on normal families, in a future work.

It is a pleasure to thank John McCarthy for helpful conversations about various topics in this paper. Eric Bedford also pointed us in some interesting directions.

7.2 Basic Definitions

We will now define holomorphic functions and mappings and normal families. We refer the reader to the paper [130] and the book [203] for background on complex analysis in infinite dimensions.

Definition 7.2.1. A domain $\Omega \subseteq X$ is a connected open set.

Definition 7.2.2. Let X be a separable Banach space. Let $\Omega \subseteq X$ be an open set. Let $u : \Omega \to Y$ be a mapping, where Y is some other separable Banach space. For $q \in \Omega$ and $\mathbf{v}_1, \ldots, \mathbf{v}_k \in X$, we define the derivatives

$$du\,(\mathbf{q}; \mathbf{v}_j) = \lim_{\mathbb{R} \ni \epsilon \to 0} \frac{u\,(\mathbf{q} + \epsilon \mathbf{v}_j) - u(\mathbf{q})}{\epsilon}$$

and

$$\overline{D}u(\mathbf{q}; \mathbf{v}) = \frac{du(\mathbf{q}; \mathbf{v}) + i du(\mathbf{q}; i\mathbf{v})}{2}.$$

In what follows, we use the word "function" to refer to a (complex) scalar-valued object and "mapping" to refer to a Banach-space-valued object as in Definition 7.2.2.

A function (mapping) f on Ω is said to be continuously differentiable, or C^1, if $df(\mathbf{q}; \mathbf{v})$ exists for every point $\mathbf{q} \in \Omega$ and every vector $\mathbf{v}$, and if the resulting function $(\mathbf{q}, \mathbf{v}) \mapsto df(\mathbf{q}; \mathbf{v})$ is continuous.

Definition 7.2.3. Let $\Omega \subseteq X$ be an open set and f a C^1-smooth function or mapping defined on Ω. We say that f is *holomorphic* on Ω if $\overline{D}f \equiv 0$ on Ω.

The definition just given of "holomorphic function" or "holomorphic mapping" is equivalent, in the C^1 category, to requiring that the restriction of the function or mapping to every complex line be holomorphic in the classical sense of the function theory of one complex variable. We shall have no occasion, in the present paper, to consider functions that are less than C^1 smooth, but holomorphicity can, in principle, be defined for rougher functions.

Definition 7.2.4. Let $\mathcal{F} = \{f_\alpha\}_{\alpha \in \mathcal{A}}$ be a family of holomorphic functions on a domain $\Omega \subseteq X$. We say that $\mathcal{F}$ is a *normal family* if every subsequence $\{f_j\} \subseteq \mathcal{F}$ either

7.2.4.1 (**normal convergence**) has a subsequence that converges uniformly on compact subsets of Ω;

or

7.2.4.2 (**compact divergence**) has a subsequence $\{f_{j_k}\}$ such that, for each compact $K \subseteq \Omega$ and each compact $L \subseteq \mathbb{C}$, there is a number N so large that $f_{j_k}(K) \cap L = \emptyset$ whenever $k \geq N$.

It is convenient at this juncture to define a type of topology that will be of particular interest for us. If $\Omega \subseteq X$ is a domain and $\mathcal{O}(\Omega)$ the space of holomorphic functions on Ω, then we let $\mathcal{B}$ denote the topology on $\mathcal{O}(\Omega)$ of uniform convergence on compact sets. Of course a sub-basis for the topology $\mathcal{B}$ is given by the sets (with $\epsilon > 0, g \in \mathcal{O}(\Omega)$ arbitrary, and $K \subseteq \Omega$ a compact set)

$$B_{g,K,\epsilon} = \left\{ f \in \mathcal{O}(\Omega) : \sup_{z \in K} |f(z) - g(z)| < \epsilon \right\}$$

It is elementary to verify (or see [203]) that the limit of a sequence of holomorphic functions on Ω, taken in the topology $\mathcal{B}$, will be another holomorphic function on Ω. Indeed, it is clear that the limit of such a sequence is holomorphic on any finite-dimensional slice (this property is commonly called "G-holomorphic"). The limit function is clearly locally bounded. Now it follows (see [203, p. 74]) that the limit is holomorphic.

Definition 7.2.5. Let $\Omega \subseteq X$ be a domain. Let $\mathcal{U} \equiv \{U_\alpha\}_{\alpha \in \mathcal{A}}$ be a semi-norm topology on the space $\mathcal{O}(\Omega)$ of holomorphic functions on Ω.[1] A real-valued function $p : X \to \mathbb{R}$ is called a *seminorm* if it satisfies the following two conditions:

1 Triangle inequality: $p(x + y) \leq p(x) + p(y)$ for all $x, y \in X$.

2 Absolute homogenity: $p(sx) = |s|\, p(x)$ for all $x \in X$ and all scalars s.

These two conditions imply that $p(0) = 0$ and that every seminorm p also has the following property:

3 Nonnegativity $p(x) \geq 0$ for all $x \in X$.

Some authors include non-negativity as part of the definition of "seminorm" (and also sometimes of "norm"), although this is not necessary since it follows from the other two properties.

---

[1]Let X be a vector space over either real numbers $\mathbb{R}$ or the complex numbers $\mathbb{C}$.

By definition, a norm on X is a seminorm that also separates points, meaning that it has the following additional property:

4 Positive definite: whenever $x \in X$ satisfies $p(x) = 0$ then $x = 0$.

[2] We say that $\mathcal{U}$ is a *Montel topology* on $\mathcal{O}(\Omega)$ if the mapping

$$\begin{aligned} \mathrm{id} : [\mathcal{O}(\Omega), \mathcal{U}] &\longrightarrow [\mathcal{O}(\Omega), \mathcal{B}] \\ f &\longmapsto f \end{aligned}$$

is a compact operator.

Example 7.2.6. (1) Let $\Omega \subseteq \mathbb{C}^n$ be a domain in finite-dimensional complex space. The topology $\mathcal{B}$ is a Montel topology. This is the content of the classical Montel theorem on normal families (see [203]).

(2) Let $\Omega \subseteq \mathbb{C}$ be a domain in one-dimensional complex space. Let k be a positive integer, and let superscript (k) denote the kth derivative. The topology $\mathcal{C}_k$ with sub-basis given by the union of the sets (with $\epsilon > 0$, $g \in \mathcal{O}(\Omega)$ arbitrary, and $K \subseteq \Omega$ a compact set)

$$N^{(j)}_{g,K,\epsilon} = \left\{ f \in \mathcal{O}(\Omega) : \sup_{z \in K} \left| f^{(j)}(z) - g^{(j)}(z) \right| < \epsilon \right\}$$

for $j = 0, 1, \ldots, k$, is a Montel topology. Of course, by integration (and using the Cauchy estimates), the topology $\mathcal{C}_k$ is equivalent to the topology $\mathcal{B}$. [A similar topology can be defined for holomorphic functions on a domain in the finite-dimensional space $\mathbb{C}^n$.]

(3) Let $\Omega \subseteq \mathbb{C}$ be a domain in one-dimensional complex space. The topology $\mathcal{D}$ with sub-basis given by the sets (with $\epsilon > 0, g \in \mathcal{O}(\Omega)$ arbitrary, and $\widetilde{\gamma} \subseteq \Omega$ the compact image of a closed curve $\gamma : [0, 1] \to \Omega$)

$$M_{g,\widetilde{\gamma},\epsilon} = \left\{ f \in \mathcal{O}(\Omega) : \sup_{z \in \widetilde{\gamma}} |f(z) - g(z)| < \epsilon \right\}$$

is a Montel topology, as the reader may verify by using the Cauchy estimates. Of course, once again, the maximum principle may be used to check that the topology $\mathcal{D}$ is equivalent to the topology $\mathcal{B}$.

(4) In the reference [208], Leopoldo Nachbin defined the concept of a seminorm that is "ported" by a compact set. We review the notion here.

Let X and Y be separable Banach spaces as usual. Let $\Omega \subseteq X$ be a domain, and let $K \subseteq \Omega$ be a fixed compact subset. We consider the family $\mathcal{H}(\Omega, Y)$ of holomorphic mappings from Ω to Y. A seminorm ρ on $\mathcal{H}(\Omega, Y)$ is said to be

---

[2] A seminormed space is a pair (X, p) consisting of a vector space X and a seminorm p on X. If the seminorm p is also a norm then the seminormed space (X, p) is called a normed space.

ported by the set K if, given any open set V with $K \subseteq V \subseteq \Omega$, we can find a real number $c(V) > 0$ such that the inequality

$$\rho(f) \leq c(V) \cdot \sup_{x \in V} \|f(x)\| \tag{$*$}$$

holds for every $f \in \mathcal{H}(\Omega, Y)$.

We note that the holomorphic mapping f here need not be bounded on V. What is true, however (and we have noted this fact elsewhere in the present book), is that once Ω and K are fixed then there will exists some open set V as above on which f is bounded. So that, for this choice of V, the inequality $(*)$ will be non-trivial.

Now we use the notion of "seminorm ported by K" to define a topology on $\mathcal{H}(\Omega, Y)$ as follows: we consider the topology induced by all seminorms that are ported by compact subsets of Ω. It is to be noted that, in finite dimensions, this new topology is no different from the standard compact-open topology. But in infinite dimensions it is quite different. As an example, let $X = Y = \ell_2$ (the space of square-summable sequences), which is of course a separable Hilbert space. Let a typical element of ℓ_2 be denoted by $\{a_j\}_{j=1}^{\infty}$, and let the jth coordinate be z_j. Let $\Omega \subseteq X$ be a domain and let $K \subseteq \Omega$ be a compact set. Consider holomorphic functions $f : \Omega \to \mathbb{C}$. Define a semi-norm by

$$\rho^*(f) \equiv \sum_{j=1}^{\infty} \sup_K \left| \frac{\partial f}{\partial z_j} \right| .$$

Then it is clear, by the Cauchy estimates, that ρ^* is ported.

But it is also clear that a typical open set defined by ρ^* will not contain any non-trivial open set from the compact-open topology. Thus this topology is not Montel. Of course it is now a simple matter to generate many other interesting examples of ported seminorms.

Now we have

Theorem 7.2.7. *Let $\mathcal{F} = \{f_\alpha\}_{\alpha \in \mathcal{A}}$ be a family of holomorphic functions on a domain $\Omega \subseteq X$. Assume that there is a finite constant M such that $|f_\alpha(z)| \leq M$ for all $f_\alpha \in \mathcal{F}$ and all $z \in \Omega$. Let K be a compact subset of Ω. Then every sequence in $\mathcal{F}$ has itself a subsequence that converges uniformly on K.*

Proof. Of course the hypothesis of uniform boundedness precludes compact divergence. So we will verify 7.2.4.1. Fix a compact subset $K \subseteq \Omega$. Then there is a number $\eta > 0$ such that if $k \in K$ then $B(k, 3\eta) \subseteq \Omega$. Select $f_\alpha \in \mathcal{F}$. Now if $k \in K$ and ℓ is any point such that $\|k - \ell\| < \eta$ then we may apply the Cauchy estimates (on $B(k, 2\eta)$) to the restriction of f_α to the complex line through k and ℓ. We find that the f_α have bounded directional derivatives. Therefore they are (uniformly) Lipschitz and form an equicontinuous family of functions.

As a result of these considerations, the Arzelà-Ascoli theorem applies to the family $\mathcal{F}$ restricted to K. Thus any sequence in $\mathcal{F}$ has a subsequence convergent on K. □

In practice, it is useful to have a version of Theorem 7.2.7 that hypothesizes only uniform boundedness on compact sets. This is a tricky point in the infinite-dimensional setting for the following reason: Classically (in finite dimensions), one derives this new result from (the analog of) Theorem 7.2.7 by taking a compact set $K \subseteq \Omega$ and fattening it up to a slightly larger compact $L \subseteq \Omega$. Since the family $\mathcal{F}$ is uniformly bounded on L, an analysis similar to the proof of Theorem 7.2.7 may now be performed. In the infinite-dimensional setting this attack cannot work, since there is no notion of fattening up a compact set to a larger compact set.

Nonetheless, we have several different ways to prove a more general, and more useful, version of Montel's theorem. The statement is as follows.

Theorem 7.2.8 (Montel). *Let $\mathcal{F} = \{f_\alpha\}_{\alpha \in \mathcal{A}}$ be a family of holomorphic functions on a domain Ω in a separable Banach space X. Assume that $\mathcal{F}$ is uniformly bounded on compact sets, in the sense that for each compact $L \subseteq \Omega$ there is a constant $M_L > 0$ such that $|f_\alpha(z)| \leq M_L$ for every $z \in L$ and every $f_\alpha \in \mathcal{F}$. Then every sequence in $\mathcal{F}$ has itself a subsequence that converges uniformly on each compact set $K \subseteq \Omega$. [Note that we are saying that there is a single sequence that works for every set K.] Thus $\mathcal{F}$ is a normal family.*

Remark 7.2.9. We may rephrase Montel's theorem by saying that the topology $\mathcal{B}$ is a Montel topology.

Proof of the Theorem:. Fix a compact set $K \subseteq \Omega$. Of course the family $\mathcal{F}$ is bounded on K by hypothesis. We claim that $\mathcal{F}$ is bounded on some neighborhood U of K. To this end, and seeking a contradiction, we suppose instead that for each integer $N > 0$ there is a point $x_N \in \Omega$ such that $\operatorname{dist}(x_N, K) < 1/N$ and $|f_\alpha(x_N)| > N$. Then the set

$$L = K \cup \{x_N\}_{N=1}^{\infty}$$

is compact. So the family $\mathcal{F}$ is bounded on L. But that contradicts the choice of the x_N.

We conclude that, for some N, x_N does not exist. That means that there is a number $N_0 > 0$ such that the family $\mathcal{F}$ is uniformly bounded on $U \equiv \{x \in \Omega : \operatorname{dist}(x, K) < 1/N_0\}$. As a result, we may imitate the proof of Theorem 7.2.7, merely substituting U for Ω. □

Remark 7.2.10. We thank Laszlo Lempert for the idea of the proof of Theorem 7.2.8 just presented.

We now indulge in a slight digression, partly for interest's sake and partly because the argument will prove useful below. In fact we will provide a proof of Theorem 7.2.8 that depends on the Banach-Alaoglu theorem. This is philosophically appropriate, for it validates in yet another way that a normal families theorem is nothing other than a compactness theorem. After that we will sketch a proof that depends on the theory of currents.

Alternative (Banach-Alaoglu) Proof of Theorem 7.2.8:

For clarity and simplicity, we begin by presenting this proof in the complex plane $\mathbb{C}$. The reader who has come this far will have no trouble adapting the argument to finitely many complex variable space $\mathbb{C}^n$. We provide a separate argument below for the infinite dimensional case.

Now fix a domain $\Omega \subseteq \mathbb{C}$. Let $\mathcal{F} = \{f_\alpha\}_{\alpha \in \mathcal{A}}$ be a family of holomorphic functions on Ω which is bounded on compact sets. Fix a piecewise C^1 closed curve $\gamma : [0,1] \to \Omega$. Let $\tilde{\gamma}$ denote the image of γ, which is of course a compact set in Ω. Consider the functions

$$g_\alpha \equiv f_\alpha|_{\tilde{\gamma}}$$

Then each g_α is smooth on $\tilde{\gamma}$ and the family $\mathcal{G} \equiv \{g_\alpha\}_{\alpha \in \mathcal{A}}$ is bounded by some constant M. So we may think of $\mathcal{G} \subseteq L^\infty(\widetilde{\gamma})$ as a bounded set. Since $L^\infty(\widetilde{\gamma})$ is the dual of $L^1(\widetilde{\gamma})$, we may apply the Banach-Alaoglu theorem to extract a subsequence (which we denote by $\{g_j\}$ for convenience) that converges in the weak-* topology. Call the weak-* limit function g.

Now fix a point z that lies in the interior, bounded component of the complement of $\tilde{\gamma}$. Of course the function

$$t \longmapsto \frac{\gamma'(t)}{\gamma(t) - z}$$

lies in $L^1(\widetilde{\gamma})$. So, by weak-* convergence and the Cauchy integral formula, we know that

$$g_j(z) \equiv \frac{1}{2\pi i} \oint_\gamma \frac{g_j(\zeta)}{\zeta - z} d\zeta \to \frac{1}{2\pi i} \oint_\gamma \frac{g(\zeta)}{\zeta - z} d\zeta \equiv G(z)$$

Here the last equality defines the function G.

So we see that the functions g_j, which of course must agree with f_j at points inside the curve γ, tend pointwise to the function G; and the function G is perforce holomorphic inside the image curve $\tilde{\gamma}$. We will show that in fact the convergence is uniform on compact sets inside of $\tilde{\gamma}$.

So fix a compact set K that lies in the bounded open set interior to $\widetilde{\gamma}$. Fix a piecewise C^1, simple, closed curve γ^* whose image is disjoint from, and lies inside of, $\tilde{\gamma}$, and which surrounds K. Let $\eta > 0$ be the distance of K to $\widetilde{\gamma^*}$, the image of γ^*. Now fix a small $\epsilon > 0$ (here ϵ should be smaller than the length of γ^*). Choose a set $E \subseteq \widetilde{\gamma^*}$ such that E has linear measure less than ϵ and so that (by Lusin's theorem)

$$|g_j(\zeta) - g(\zeta)| < \epsilon$$

when j is sufficiently large ($j > N$, let us say) and $\zeta \in \widetilde{\gamma^*}\setminus E$. Then, for $j, k > N$ and $z \in K$, we have

$$\begin{aligned}|g_j(z) - g_k(z)| &\leq \frac{1}{2\pi}\int_{cE}\left|\frac{g_j(\zeta) - g_k(\zeta)}{\zeta - z}\right| d|\zeta| + \frac{1}{2\pi}\int_{E}\left|\frac{g_j(\zeta) - g_k(\zeta)}{\zeta - z}\right| d|\zeta| \\ &\leq \frac{1}{2\pi}\,\text{length}\,(\widetilde{\gamma^*})\frac{\epsilon}{\eta} + \frac{1}{2\pi}\cdot\epsilon\cdot\frac{2M}{\eta}.\end{aligned}$$

Since $\epsilon > 0$ may be chosen to be arbitrarily small, we conclude that $g_j \to g$ uniformly on the compact set K. That is what we wished to prove for the single compact set K.

We note that this proof may be performed when γ is a positively oriented curve describing any square inside Ω with sides parallel to the axes, rational center, and rational side length. Of course it is always possible to produce the curve γ^* as the union of finitely many such curves. As a result, the usual diagonalization procedure may be formed over these countably many curves, producing a single subsequence that converges uniformly on any compact set in Ω to a limit function G.

Our next proof depends on the theory of currents. For background in this important technique of geometric analysis, we refer the reader to [29, 72, 73, 131, 166, 167, 188].

Alternative (Currents) Proof of Theorem 7.2.8 for Separable Banach Spaces:

We refer to the very interesting paper [9] of Almgren. That paper gives a characterization of the dual of the space of all k-dimensional, real, rectifiable currents in $\mathbb{R}^N$. Remarkably, Almgren's proof uses both the Continuum Hypothesis and the Axiom of Choice. An examination of Almgren's proof reveals that the arguments are also valid when $\mathbb{R}^N$ is replaced by any separable Banach space. We take that result for granted and leave it to the reader to check the details in [9].

Accepting that assertion, we see that the hypothesis of uniform boundedness of a family $\mathcal{F}$ of holomorphic functions on compact subsets of a domain Ω in a separable Banach space X can be interpreted as a boundedness statement about one-dimensional holomorphic currents. Specifically, let $\mathcal{F}$ be a family of holomorphic functions on a domain $\Omega \subseteq X$, and assume that $\mathcal{F}$ is bounded on compact subsets of Ω. As we have seen (proof of Theorem 7.2.8), it follows that if $K \subseteq \Omega$ is any compact set, then there is a small neighborhood U of K, with $K \subset\subset U \subseteq \Omega$, such that $\mathcal{F}$ is bounded on U. As a result, we may apply Cauchy estimates to see that if $\mathcal{F} = \{f_\alpha\}_{\alpha\in\mathcal{A}}$ then $\mathcal{F}' = \{\partial f_\alpha\}_{\alpha\in\mathcal{A}}$ is bounded on K. But then, by the generalization of Almgren's theorem to infinite dimensions, we may think of $\mathcal{F}'$ as a bounded family in the dual of the space of 1-dimensional (complex) currents on Ω. By the Banach-Alaoglu theorem, we may therefore extract from any sequence in $\mathcal{F}'$ a weak-* convergent subsequence. Call it, for convenience, $\{f_j\}$.

But now it is possible to imitate the first alternative proof of Theorem 7.2.8 as follows. Fix a closed, piecewise C^1 curve $\gamma : [0,1] \to \Omega$ that bounds an analytic disk d in Ω[3].

Think of the elements ∂f_j restricted to the image $\tilde{\gamma}$ of this curve. They form a bounded family in $L^\infty(\tilde{\gamma})$. Thus the first alternative proof may be imitated, step by step, to produce a limit holomorphic function on the analytic disk d. In fact we may even take the argument a step further.

For $z, w \in \mathbb{C}^n$ the Bochner-Martinelli kernel $\omega(z, w)$ is a differential form in ζ of bidegree $(n, n-1)$ defined by

$$\omega(\zeta, z) = \frac{(n-1)!}{(2\pi i)^n} \frac{1}{|z-\zeta|^{2n}} \sum_{1 \le j \le n} (\overline{\zeta}_j - \overline{z}_j)\, d\overline{\zeta}_1 \wedge d\zeta_1 \wedge \cdots \wedge d\zeta_j \wedge \cdots \wedge d\overline{\zeta}_n \wedge d\zeta_n$$

(where the term $d\overline{\zeta}_j$ is omitted).

Suppose that f is a continuously differentiable function on the closure of a domain Ω in $\mathbb{C}^n$ with piecewise smooth boundary $\partial\Omega$. Then the Bochner-Martinelli formula states that if z is in the domain Ω, then

$$f(z) = \int_{\partial\Omega} f(\zeta)\omega(\zeta, z) - \int_{\Omega} \overline{\partial} f(\zeta) \wedge \omega(\zeta, z).$$

In particular, if f is holomorphic, the second term vanishes, so

$$f(z) = \int_{\partial\Omega} f(\zeta)\omega(\zeta, z).$$

(For more on this topic, see, for example, [143, Section 1.1, p. 19].)

Further, we may look at any k-dimensional slice of Ω and use the Bochner-Martinelli kernel instead of the 1-dimensional Cauchy kernel to find that there is a uniform limit on any compact subset of any k-dimensional slice of Ω.

This produces the required limit function G for the subsequence f_j. [Note that, because we are assuming the space to be separable, we can go further and even extract a subsequence that converges on every compact subset. More will be said about this point in the next remark and in what follows.]

Remark 7.2.11. This last is still not the optimal version of what we usually call Montel's theorem. In the classical, finite-dimensional formulation of Montel's result we usually derive a single subsequence that converges uniformly on every compact set. The question of whether such a result is true in infinite dimensions is complicated by the observation that it is no longer possible, in general, to produce a sequence of sets $K_1 \subset\subset K_2 \subset\subset \cdots \subset\subset X$ for our Banach space X with the property that each compact subset of X lies in some K_j.

[3]Recall that a holomorphic embedding of D into a Banach space E is a holomorphic map $f : D \to E$ such that $f(D)$ is a closed complex submanifold and $f|_D \to f(D)$ is biholomorphic. D can be embedded into $\mathbb{C}^2$, but there is no bounded holomorphic embedding $f : D \to \mathbb{C}^n$. Aurich [13] shows that the existence or nonexistence of bounded holomorphie embeddings $f : D \to E$ depends on E.

In fact the full-bore version of Montel's theorem, as just described, is false. The next example of Y. Choi illustrates what can go wrong, at least in a non-separable Banach space. [We note in passing that most of our examples are in separable spaces-which is the proper venue for the present study. But in some instances we only have examples in the non-separable case.]

Example 7.2.12. Consider the Banach space $X = \ell^\infty$. Let

$$e_j = (0, \ldots, 0, 1, 0, \ldots)$$

in which all components except the jth are zero. Let $e_j^* : X \to \mathbb{C}$ be defined by

$$e_j^*\left(\sum_{k=1}^{\infty} a_k e_k\right) = a_j.$$

This function is obviously holomorphic. However, the sequence $\{e_j^*\}$ does not have a subsequence that converges uniformly on compact subsets. To see this, let us assume to the contrary that $\{e_{j_m}^*\}_{m=1}^{\infty}$ is a subsequence that converges uniformly on compact subsets. Then in particular it should converge on the singleton set consisting of the point p that is given by

$$\mathrm{p} = \sum_{m=1}^{\infty} (-1)^m e_{j_m}$$

But, $e_{j_m}^*(\mathbf{p}) = (-1)^m$, and this sequence of scalars does not converge.

It should be noted that this example can be avoided if we demand in advance that the Banach space X be separable. One simply produces a countable, dense family of open balls, extracts a convergent sequence for each such ball, and then diagonalizes as usual. Mujica and Nachibin [203], in their treatment of normal families, achieves the full result by adding a hypothesis of pointwise convergence.

Proposition 7.2.13 ([203, p. 74]). *Let U be a connected open subset of an arbitrary Banach space X and let $\{f_n : U \to \mathbb{C}\}_{n=1,2,\ldots}$ be a bounded sequence of holomorphic functions in the compact-open topology. Suppose also that there exists a nonempty open subset V of U such that the sequence $\{f_n(x)\}_n$ converges in $\mathbb{C}$ for every $x \in V$. Then, the sequence $\{f_n\}_n$ converges to a holomorphic function of U uniformly on every compact subset of U.*

Now we turn our attention to characterizations of normal families that depend on invariant metrics. In what follows, we shall make use of the Kobayashi metric on a domain $\Omega \subseteq X$. It is defined as follows: If $p \in \Omega$ and $\xi \in X$ is a direction vector, then we set

$$F_K^{\Omega}(p;\xi) = \inf\left\{\frac{\|\xi\|}{\|\varphi'(0)\|} \mid \varphi : D \to \Omega, \varphi \text{ holomorphic}, \varphi(0) = p,\right.$$
$$\left.\varphi'(0) = \lambda\xi \text{ for some } \lambda \in \mathbb{R}\right\}.$$

Here $\|\eta\|$ is the norm of the vector $\eta \in X$.

One of the most useful characterizations of normal families, and one that stems naturally from invariant geometry, is Marty's criterion. We now establish such a result in the infinite dimensional setting.

Proposition 7.2.14. *Let X be a Banach space. Let $\Omega \subseteq X$ be a domain and let $\mathcal{F} = \{f_\alpha\}_{\alpha \in \mathcal{A}}$ be a family of holomorphic functions. The family $\mathcal{F}$ is normal if and only if there is a constant C such that, for each (unit) direction ξ,*

$$\frac{|Df_\alpha(z;\xi)|}{1 + |f_\alpha(z)|^2} \leq C \cdot F_K^\Omega(z;\xi).$$

Here $Df_\alpha(z;\xi)$ denotes the directional derivative of the function f_α at the point z in the direction ξ.

Proof. The proof follows standard lines. See the proof of Proposition 1.3 in [47, p. 306]. □

We next present a rather natural characterization of normal families that relates the situation on the ambient space to that on one-dimensional slices (more aptly, one-dimensional analytic disks):

Proposition 7.2.15. *Let X be a Banach space. Let $\Omega \subseteq X$ be a domain and let $\mathcal{F} = \{f_\alpha\}_{\alpha \in \mathcal{A}}$ be a family of holomorphic functions. The family $\mathcal{F}$ is normal if and only if the following condition holds:*

$(*)$ For each sequence $\varphi_j : D \to \Omega$ of holomorphic mappings and each sequence of indices $\alpha_j \in \mathcal{A}, j = 1, 2, \ldots$, the family $f_{\alpha_j} \circ \varphi_j$ is normal on the unit disk D.

Proof. The implication "$\mathcal{F}$ normal $\Rightarrow (*)$" is immediate from Marty's characterization of normal families. For the converse, notice that if condition $(*)$ holds then, for each sequence φ_j of mappings and each collection f_{α_j} the compositions $f_{\alpha_j} \circ \varphi_j$ satisfy the conclusion of Marty's theorem:

$$\frac{\left[f_{\alpha_j} \circ \varphi_j\right]'(\zeta)}{1 + \left|f_{\alpha_j} \circ \varphi_j\right|^2} \leq C \cdot \frac{1}{1 - |\zeta|^2}.$$

Here the constant C depends in principle on the choice of φ_j and also on the choice of f_{α_j}. But in fact a moment's thought reveals that the choice of C can be taken to be independent of the choice of these mappings; otherwise; there would be a sequence for which condition $(*)$ fails (this is just an exercise in logic).

But then, using the chain rule, we may conclude that Marty's criterion for holomorphic families on a Banach space holds for the family $\mathcal{F}$ (see also the proof of Proposition 1.4 in [47, p. 307]). As a result, $\mathcal{F}$ is normal. □

7.3 Other Characterizations of Normality

It is an old principle of Bloch, enunciated more formally by Abraham Robinson and actually recorded in mathematical notation by L. Zalcman (see [291]), that any "property" that would tend to make an entire function constant would also tend to make a family of holomorphic functions normal. Zalcman's formulation, while incisive, is rather narrowly bound to the linear structure of Euclidean space. The paper [6] finds a method for formulating these ideas that will even work on a manifold. Unfortunately, we must note that the paper [6] has an error, which was kindly pointed out to us by the authors of [275]. We shall include their correct formulation of the theorem, and also provide an indication of their proof.

Proposition 7.3.1. *Let X be a separable Banach space and let $\Omega \subseteq X$ be a hyperbolic domain (i.e., a domain on which the Kobayashi metric is non-degenerate). Let Y be another separable Banach space. Let $\mathcal{F} = \{f_\alpha\}_{\alpha \in \mathcal{A}} \subseteq$ $\mathrm{Hol}(\Omega, Y)$. The family $\mathcal{F}$ is not normal if and only if there exists a sequence $\{p_j\} \subseteq \Omega$ with $p_j \to p_0 \in \Omega$, a sequence $f_j \in \mathcal{F}$, and $\{\rho_j\} \subseteq \mathbb{R}$ with $\rho_j > 0$ and $\rho_j \to 0$ such that*

$$g_j(\xi) = f_j\left(p_j + \rho_j \xi\right), \quad \xi \in X$$

satisfies one of the following assertions:

(i) The sequence $\{g_j\}_{j \geq 1}$ is compactly divergent on Ω;

(ii) The sequence $\{g_j\}_{j \geq 1}$ converges uniformly on compact subsets of Ω to a nonconstant holomorphic mapping $g : \Omega \to Y$.

Remark 7.3.2. The error in [6] is that the authors did not take into account the compactly divergent case in the theorem. Consider the example (also from [275]) of the family $\mathcal{F}$ of mappings $f_j : D \to \mathbb{C}^2$ given by

$$f_j(\zeta) = (\alpha_j, \zeta)$$

where $1 > \alpha_j > 0$ and $\alpha_j \to 0$. Then the family $\mathcal{F}$ is not normal, but $\mathcal{F}$ also does not satisfy the conclusions of part (ii) of Proposition 7.3.1 above, which is the sole conclusion of the Aladro/Krantz theorem.

Proof. Sketch of the Proof of Proposition 7.3.1:

We first need a definition. We say that a non-negative, continuous function E defined on the tangent bundle TY is a length function if it satisfies

(a) $E(v) = 0$ if and only if $v = 0$.

(b)

$$E(\alpha v) = |\alpha| E(v) \text{ for all } \alpha \in \mathbb{C} \text{ and all } v \in TX. \tag{7.1}$$

Now we have: Let $\mathcal{F} \subseteq \mathrm{Hol}(\Omega, Y)$. Then

(1) If $\mathcal{F}$ is normal then, for each length function E on Y, and for each compact subset K of Ω, there is a constant $c_K > 0$ such that

$E(df(z)\xi) \leq c_K \cdot \|f\|$ for all $z \in K, \xi \in X \backslash \{0\}, f \in \mathcal{F}$.

(2) If Y is complete and the family $\mathcal{F}$ is not compactly divergent and satisfies (7.1), then $\mathcal{F}$ is normal.

This result is standard and can be found in [287] or [275]. Now we treat the result by cases:

Necessity

Case 1. The family $\mathcal{F}$ is compactly divergent. We treat this case in some detail since it is new and does not appear in [6]. There is a sequence $\{f_j\} \subseteq \mathcal{F}$ that is compactly divergent. Take $p_0 \in \Omega$ and $r_0 > 0$ such that $B(p_0, r_0) \subset\subset \Omega$. Take $p_j = p_0$ for all $j \geq 1$ and $p_j > 0$ for all $j \geq 1$ such that $\rho_j \to 0^+$and define

$$g_j(\xi) = f_j(p_j + \rho_j \xi), \qquad \text{all } j \geq 1 .$$

Observe that each g_j is defined on

$$S_j = \left\{\xi \in X : \|\xi\| \leq R_j = \frac{1}{\rho_j} \operatorname{dist}(p_0, \partial\Omega)\right\}$$

If $K \subseteq X$ is compact and L is a compact subset of Y, then there is an index $j_0 \geq 1$ such that $p_0 + \rho_j K \subseteq B(p_0, r_0)$ for all $j \geq j_0$. This implies that $g_j(K) \subseteq f_j\left(\overline{B}(p_0, r_0)\right)$ for each $j \geq j_0$. Since the sequence $\{f_j\}$ is compactly divergent, there is an index $j_1 > j_0$ such that $f_j\left(\overline{B}(p_0, r_0)\right) \cap L = \emptyset$ for all $j \geq j_1$. Thus $g_j(K) \cap L = \emptyset$ for all $j \geq j_1$. This means that the family $\{g_j\}$ is compactly divergent.

Case 2. The family $\mathcal{F}$ is not compactly divergent. This follows standard lines, as indicated in [6].

Sufficiency

Case 1. The sequence $g_j \to g$ with g not a constant function. By direct estimation, one shows that

$$\lim_{j \to \infty} E(g_j(\xi), dg_j(\xi)(t)) = E(g(\xi), dg(\xi)(t)) = 0$$

for $\xi, t \in X$. Hence $g' \equiv 0$ and so g is constant, a clear contradiction.

So the family $\mathcal{F}$ cannot be normal.

Case 2. The sequence $\{g_j\}$ is compactly divergent. We may assume that $\{f_j\} \subseteq \mathcal{F}$ and $f_j \to f$. For $\xi \in X$ we then have

$$g_j(\xi) = f_j(p_j + \rho_j \xi) \to f(p_0) \in Y$$

since $\rho_j \to 0$. This implies that the family $\{g_j\}$ is not compactly divergent, a clear contradiction.

That completes our outline of the proof of Proposition 7.3.1. □

Constantin Carathéodory produced a geometric characterization of normal families that is quite appealing (see [260, p. 68]). It has never been adapted even to several complex variables. We take the opportunity now to offer an infinite dimensional version (which certainly specializes down to any finite number of dimensions).

We begin with a little terminology. Let Ω_j be domains in a separable Banach space X. If some Euclidean ball $B(0,r), r > 0$, is contained in all the domains Ω_j, then $\ker\{\Omega_j\}$ is the largest domain containing 0 and so that every compact subset of $\ker\{\Omega_j\}$ lies in all but finitely many of the Ω_j. We say that $\{\Omega_j\}$ converges to $\Omega_0 \equiv \ker\{\Omega_j\}$, written $\Omega_j \to \Omega_0$, if every subsequence $\{\Omega_{j_k}\}$ of these domains has the property that $\ker\{\Omega_{k_j}\} = \Omega_0$.

Theorem 7.3.3. *2.3 Fix a separable Banach space X. Let $\{f_n\}$ be a sequence of univalent, holomorphic mappings from the unit ball $\mathbf{B} \subset X$ to another separable Banach space Y with the properties that*

1. $f_n(0) = 0$;
2. $\langle df_n(0)1, 1\rangle > 0$.

[Here 1 is the unit vector $(1,0,0,\ldots)$.] Set $\Omega_n \equiv f_n(B), n = 1,2,\ldots$. Then the f_n converge normally in B to a univalent function f if and only if

1. $\Omega_0 = \ker\{\Omega_n\}$ *is hyperbolic and is not* $\{0\}$.
2. $\Omega_n \to \Omega_0$
3. $\Omega_0 = f(B)$.

Proof. Sketch of Proof: We first establish that it is impossible for $\ker\{\Omega_n\} = \{0\}$. Consider the Kobayashi metric ball $\mathbf{B} = B_{\mathcal{B}}(0,1)$. Then

$$f_n : B \to B_{\mathcal{B}}(0,1)$$

since of course each f_n will be a Kobayashi isometry onto its image. Assume that $f_n \to f$ normally (i.e., uniformly on compact sets) in $\mathcal{B}$. Clearly, under the hypothesis that $\ker\{\Omega_n\} = \{0\}$, there is no $\epsilon > 0$ such that $b(0,\epsilon) \subseteq \Omega_n$ for n large. Here b denotes a Euclidean ball. Thus B_{Ω_n} must shrink to a set with no interior. It follows that the sequence $\{f_n\}$ collapses any compact subset of B to a set without interior. Thus $df \equiv 0$ on B hence f is identically constant. Since $f(0) = 0, f \equiv 0$ (a clear contradiction).

Now we begin proving the theorem proper. Suppose that $f_n \to f$ normally on $\mathcal{B}$ with f univalent. Since, by the preceding paragraph, f is not identically 0, we may conclude that $\Omega_0 = \ker(\Omega_n) \neq \{0\}$.

$CLAIM : f(\mathcal{B}) = \Omega_0$

It would follow from this claim that $\Omega_0 \neq X$, for if $\Omega_0 = X$ then $f^{-1} : X \to B$ univalently, violating Liouville's theorem [203, p. 39]. [It would also contradict the distance-decreasing property of the Kobayashi metric.] Since every subsequence of $\{f_n\}$ converges to f, it follows that every subsequence of $\{\Omega_n\}$ has kernel Ω_0. We write $\Omega_n \to \Omega_0$.

$SUBCLAIMI : f(\mathcal{B}) \subseteq \Omega_0$

For consider any closed metric ball $\overline{B}(0, R) \subseteq f(\mathcal{B})$. We may restrict attention to any finite-dimensional slice L of this ball, which will of course be compact. Then $f_n|_L \to f_L$. Thus $f_n|_{L\cap\partial\mathcal{B}} \to f|_{L\cap aB}$. As a result, for n large, we apply the argument principle to any curve in $L \cap \partial\mathcal{B}$ to see that each value in $f(\overline{B}(0, R))$ is attained just once by f_n for n large. But this just says that $f(\mathcal{B}) \subseteq \Omega_0$.

$SUBCLAIMI : \Omega_0 \subseteq f(\mathcal{B})$

For consider $\Omega_0 \neq \{0\}$ and assume Ω_0 is hyperbolic. Let $\Omega_n \to \Omega_0$. If $b(0, \epsilon) \subseteq \Omega_n$ for all n large, then

$$\bar{b}(0, \epsilon) \subseteq B_{\Omega_n}(0, R) \subseteq \Omega_n \text{ for } n \text{ large.}$$

So

$$f_n : B_{\mathcal{B}}(0, R) \to B_{\Omega_n}(0, R) \supset b(0, \epsilon)$$

Hence we have a bound from below on the eigenvalues of df_n.

Obversely, we also claim that the eigenvalues of df_n are bounded above. If not, then there exist (Euclidean) unit vectors ξ_n such that

$$df_n\,(\xi_n) \to \infty$$

After a rotation and passing to a subsequence, we can assume that the ξ_n all point in the direction 1. The result would then be that Ω_0 cannot be hyperbolic, a contradiction.

Thus the $\{f_n\}$ are locally bounded and $\{f_n\}$ forms a normal family, as required. Thus some subsequence converges to a univalent f such that $f(0) = 0$. This last follows from the argument principle (see any standard complex analysis text). □

7.4 A Budget of Counterexamples

We interrupt our story to provide some examples that exhibit the limitations of the theory of normal families in infinitely many variables.

Example 7.4.1. There is no Montel theorem for holomorphic mappings of infinitely many variables. Indeed, let $\mathbf{B}$ be the open unit ball in the Hilbert space ℓ_2. Define

$$\varphi_j(\{a_m\}) = \left(\frac{\sqrt{3/4}a_1}{1-a_j/2}, \frac{\sqrt{3/4}a_2}{1-a_j/2}, \ldots, \right.$$
$$\left. \frac{\sqrt{3/4}a_{j-1}}{1-a_j/2}, \frac{a_j - 1/2}{1-a_j/2}, \frac{\sqrt{3/4}a_{j+1}}{1-a_j/2}, \ldots \right).$$

Then each φ_j is an automorphism of B.

Now fix an index j. Let $K = K_j$ be the compact set

$$\{(0, 0, \ldots, 0, \zeta, 0, \ldots, 0) : |\zeta| \leq c\},$$

where the non-zero entry is in the j-th position and $1/2 < c < 1$ is a constant. Define the point $p \in K$ to be $p = (0, 0, \ldots, 0, c, 0, \ldots)$, where the non-zero entry is in the j-th position. Then

$$\sup_K \|\varphi_j - \varphi_k\| \geq \|\varphi_j(p) - \varphi_k(p)\| \geq \left|\frac{c - 1/2}{1 - c/2} - 0\right| = \left|\frac{2c-1}{2-c}\right| > 0.$$

As a result, we see that the sequence $\{\varphi_j\}$ can have no convergent subsequence. It also cannot have a compactly divergent subsequence.

Example 7.4.2. 3.2 *There are no taut domains in infinite dimensional space.* First we recall H.H. Wu's notion of "taut". Let N be a complex manifold. We say that N is taut if, for every complex manifold M, the family of holomorphic mappings from M to N is normal. We now demonstrate that there are no such manifolds in infinite dimensions.

We begin by studying the ball $\mathbf{B}$ in the Hilbert space ℓ_2. We let $N = B$ and $M = D$, the disk in $\mathbb{C}$ (in fact it is easy to see that, when testing tautness, it always suffices to take M to be the unit disk). Consider the mappings

$$\varphi_j(\zeta) = \left(0, 0, \ldots, 0, \frac{1}{3} + \frac{\zeta}{4}, 0, \ldots, 0\right)$$

Here the non-zero entry is in the jth position. Then

$$|\text{image}(\varphi_j) - \text{image}(\varphi_k)| \geq \sqrt{\left(\frac{1}{12}\right)^2 + \left(\frac{1}{12}\right)^2} = \sqrt{\frac{1}{72}} > 0$$

Also

$$\text{dist}(\text{image}(\varphi_j), \partial D) = \frac{5}{12} > 0$$

As a result, the sequence $\{\varphi_j\}$ has no convergent subsequence and no compactly divergent subsequence.

Of course the same argument shows that there is no taut domain in Hilbert space, nor is there any taut Hilbert manifold.

It should be noted that the Arzelà-Ascoli theorem will fail for families of functions (mappings) taking values in an infinite dimensional space. For example, if X is the separale Hilbert space ℓ_2 and $f_j : X \to X$ is given by $f_j(\{x_j\}) = x_j$, then the f_j are equicontinuous and equibounded on bounded sets, yet no compact set supports a uniformly convergent subsequence. Thus the preceding examples do not come as a great surprise.

It is worth noting that there are results for weak or weak-* normal families that can serve as a good substitute when the regular (or strong) Montel theorem fails. We explore some of these in Section 7.7.

7.5 Normal Functions

Normal functions were created by Lehto and Virtanen in [164] as a natural context in which to formulate the Lindelöf principle.

A Stolz angle at $\xi \in \partial D$ is of the form

$$S(\xi) = \left\{z \in D : \left|\arg(1 - \bar{\xi}z)\right| < \alpha, |z - \xi| < \rho\right\} \quad (0 < \alpha < \frac{\pi}{2}, \rho < 2\cos\alpha).$$

The exact shape is not relevant. The important fact is that all points of $S(\xi)$ have bounded non-Euclidean distance from the radius $[0, \xi]$.

For ξ on the unit circle D and $\alpha > 1$, we define a nontangential approach region at ξ by

$$\Gamma(\xi, \alpha) = \{z \in D : |z - \xi| < \alpha(1 - |z|)\}.$$

Of course, the term "non-tangential" refers to the fact that the boundary curves of $\Gamma(\xi, \alpha)$ have a corner at ξ, with angle less than π.

Recall that the Lindelöf principle says this

Theorem 7.5.1. *4.1 (Lindelöf) Let f be a bounded holomorphic function on the disk D. If f has radial limit ℓ at a point $\xi \in \partial D$ then f has non-tangential limit ℓ at ξ, i.e., $\lim_{\Gamma(\xi,\alpha)\ni z\to\xi} f(z) = \ell$.*

Lehto and Virtanen realized that boundedness was too strong a condition, and not the natural one, to guarantee that Lindelöf's phenomenon would hold. They therefore defined the class of normal functions as follows:

Definition 7.5.2. Let f be a holomorphic (meromorphic) function on the disk $D \subseteq \mathbb{C}$. Suppose that, for any family $\{\varphi_j\}$ of conformal self-maps of the disk it holds that $\{f \circ \varphi_m\}$ is a normal family. Then we say that f is a normal function.

Clearly a bounded holomorphic function, a meromorphic function that omits three values, or a univalent holomorphic function (all in one complex dimension) will be normal according to this definition.

Unfortunately, the original definition given by Lehto and Virtanen is rather limited. One-connected domains in $\mathbb{C}^1$ have compact automorphism groups; finitely connected domains in $\mathbb{C}^1$, of connectivity at least two, have finite automorphism group. Generic domains in $\mathbb{C}^n, n \geq 2$, even those that are topologically trivial, have automorphism group consisting only of the identity (such domains are called rigid). Thus, for most domains in most dimensions, there are not enough automorphisms to make a working definition of "normal function" possible. In [47], Cima and Krantz addressed this issue and developed a new definition of normal function. We now adapt that definition to the infinite dimensional case.

Definition 7.5.3. Let X be a Banach space and let Ω be a domain in X. A holomorphic function f on Ω is said to be normal if

$$\frac{|Df(z;\xi)|}{1+|f(z)|^2} \leq C \cdot F_K^{\Omega}(z;\xi) \quad \text{for all } z \in \Omega, \xi \in X$$

Proposition 7.5.4. *Let f be a holomorphic function on a domain Ω in a Banach space X. The function f is normal if and only if $f \circ \varphi$ is normal for each holomorphic $\varphi : D \to \Omega$.*

Proof. The proof is just the same as that in Section 1 of [47]. □

Remark 7.5.5. It is a straightforward exercise, using for example Proposition 7.5.4 (or Marty's characterization of normality), to see that a holomorphic or meromorphic function on the unit ball $\mathbf{B}$ in a separable Hilbert space H is normal if and only if, for every family $\{\varphi_\alpha\}_{\alpha \in \mathcal{A}}$ of biholomorphic self maps of $\mathcal{B}$, it holds that $\{f \circ \varphi_\alpha\}$ is a normal family.

Now let $\mathbf{B} \subseteq X$ be the unit ball in a separable Banach space X. We define a holomorphic function f on $\mathbf{B}$ to be *Bloch* if

$$\|Df(z;\xi)\| \leq C \cdot F_K^{\Omega}(p;\xi)$$

for every $z \in \mathcal{B}$ and every vector ξ. Then it is routine, following classical arguments, to verify.

Proposition 7.5.6. *If f on $\mathcal{B}$ is a Bloch function, then f is normal.*

7.6 Different Topologies of Holomorphic Functions

One way to view a "normal families" theorem is that it is a compactness theorem. But another productive point of view is to think of these types of results as relating different topologies on spaces of holomorphic functions. We begin our discussion of this idea by recalling some of the standard topologies, as well as a few that are more unusual.

The Compact-Open Topology. In the language of analysis, this is the topology of uniform convergence on compact sets. Certainly in finite-dimensional complex analysis this is, for many purposes, the most standard topology on general spaces of holomorphic functions. In infinite dimensions this topology is often too coarse (just because compact sets are no longer very "fat").

The Topology of Pointwise Convergence. Here we say that a sequence f_j of functions or mappings converges if $f_j(x)$ converges for each x in the common domain X of the f_j.

The Weak Topology for Distributions. Here we think of a space of holomorphic functions as a subspace of the space $\mathcal{E}$ of testing functions for the compactly supported distributions. We say that a sequence f_j of holomorphic functions converges if $\psi(f_j)$ converges for each such distribution ψ. Of course a similar definition can (and should) be formulated for nets.

The Topology of Currents. Let f_j be holomorphic functions and consider the 1-forms ∂f_j. Then we may think of these forms as currents lying in the dual of the space of rectifiable 1-chains; we topologize the ∂f_j accordingly.

A holomorphic p-chain on a complex manifold M is a formal, locally finite

$$T = \sum k_j A_j,$$

where the A_j are pairwise distinct pure p-dimensional irreducible analytic subsets in M and $k_j > 0$ are integers. Here, locally finite means that for every compact set $K \subset M$ there are only finitely many indices j for which $K \cap A_j$ is nonempty.

Each holomorphic 1-chain

$$T = \sum k_j A_j,$$

on a complex manifold M determines a current

$$\mathbb{T} = \sum k_j [A_j],$$

on it; usually one will identify this current with the chain T.

This current is of measure type, is 1-closed and has bidegree $(1, 1)$. Thus, holomorphic 1-chains can be thought of as belonging to the space $\mathcal{D}'_2(\Omega)$ of currents of dimension 2, dual to $\mathcal{D}_2(\Omega)$ and endowed with the topology of weak-* convergence.

It was observed by P. Lelong [166] that a pure 1-dimensional analytic subvariety V of an open set Ω contained in $\mathbb{C}^n$ determines a current $\mathbb{T}$ on Ω via integration over the regular points of V. The current $\mathbb{T}$ has several special properties: It is of type $(n-1, n-1)$, positive, d-closed, and locally rectifiable.

For more on this topic see, for example, [44].

7.7 A Functional Analysis Approach to Normal Families

In the classical setting of the unit disk $D \subseteq \mathbb{C}$, it is straightforward to prove that

$$H^\infty(D) = \left(L^1(D)/\overline{H^1(D)}\right)^* \qquad (*)$$

Thus H^∞ is a dual space in a natural way. Properly viewed, the classical Montel theorem is simply weak-* compactness (i.e., the Banach-Alaoglu theorem) for this dual space. Using the Cauchy integral formula as usual, one can see that convergence in the dual norm certainly dominates uniform convergence on compact subsets of the disk.

Alternatively, one can think of the elements of $H^\infty(D)$, with D the disk, as the collection of all operators (by multiplication) on H^2 that commute with multiplication by z. This was Beurling's point of view. The operator topology turns out to be equivalent (although this is nontrivial to see) to the weak-* topology as discussed in the last paragraph, and this in turn is equivalent to the classical sup-norm topology on H^∞.

The classical arguments go through to show that there is still a Beurling theorem on the unit ball in Hilbert space. It is a purely formal exercise to verify that $(\star)$ still holds on the unit ball in ℓ_2, our usual separable Hilbert space. As a result, one can think of the Montel theorem even in infinite dimensions either in the operator topology or as an application of the Banach-Alaoglu theorem to H^∞, thought of as a dual space.

Now we would like to present an effective weak-normal family theorem in this context.

Let Z be a Banach space and let $Y = Z^*$ be its dual Banach space. Let Ω be an open subset of a Banach space X and let $\mathcal{O}(\Omega, Y)$ be the set of all holomorphic mappings from Ω into Y. Then we consider the topology on $\mathcal{O}(\Omega, Y)$ generated by the sub-basic open sets given by

$$G(K, U) \equiv \{f \in \mathcal{O}(\Omega, Y) \mid f(K) \subset U\}$$

where K is a compact subset of Ω and U is a weak-* open subset of Y. Let us call this topology the *compact-weak-*-open topology.*

Theorem 7.7.1. *6.1 Let Ω be a domain in a separable Banach space X. Let Z be a separable Banach space with a countable Schauder basis*[4]*, and let*

[4]Let V denote a topological vector space over the field F. A Schauder basis is a sequence $\{v_n\}$ of elements of V such that for every element $v \in V$ there exists a unique sequence $\{\alpha_j\}$ of scalars in F so that

$$v = \sum_{j=1}^{\infty} \alpha_j v_j .$$

The convergence of the infinite sum is implicitly that of the ambient topology, i.e.,

$Y = Z^*$. Further, let W be a bounded domain in Y. Then the compact-weak--open topology is a Montel topology. In particular, the family $\mathcal{O}(\Omega, Y)$ is normal with respect to the compact-weak-*-open topology.*

Proof. Let $\{f_j : j = 1, 2, \ldots\} \subset \mathcal{O}(\Omega, W)$ be given. We would like to show that there exists a subsequence that converges in the compact-weak-*-open topology.

Let $\{\mathbf{e}_j \mid j = 1, 2, \ldots\}$ be a Schauder basis for Z. For $z \in Y$, we define the linear functional $\psi_k : Y \to \mathbb{C}$ by $\psi_k(z) = z(\mathbf{e}_k)$. Now we define

$$\Psi_{k,j} \equiv \psi_k \circ f_j$$

Then we see for each k that the sequence $\{\Psi_{k,j}\}_j$ is normal by Theorem 7.2.8. Therefore, we may select subsequences inductively so that

(1) $\{\Psi_{1,\sigma_1(j)}\}_{j=1}^{\infty}$ is a subsequence of $\Psi_{1,j}$ which converges in the compact-weak-*-open topology, and

(2) $\{f_{\sigma_{k+1}(j)}\}_{j=1}^{\infty}$ is a subsequence of $\{f_{\sigma_k(j)}\}_{j=1}^{\infty}$ for every $k = 1, 2, \ldots$

Notice that the diagonal sequence $\Psi_{k,\sigma_k(k)} = \psi_k \circ f_{\sigma_k(k)}(k = 1, 2, \ldots)$ converges in the compact-weak-*-open topology. Since the weak-* topology separate points, we may denote the weak-* limit of the sequence $f_{\sigma_k(k)}(z)$ by $f(z)$ for each $z \in \Omega$. Then the map $f : \Omega \to Y$ is Gateaux holomorphic.[5]

Since the range of f is bounded, it follows that f is in fact holomorphic. Thus the proof is complete. □

Notice that this theorem works for mappings from the spaces ℓ^p or c_0 into the space ℓ^∞, for each p with $1 \le p < \infty$. Therefore this result may be useful for a characterization of the infinite-dimensional polydisk by its automorphism group in the space c_0 of sequences of complex numbers converging to zero, for instance. On the other hand, not only is this theorem a generalization of the weak-normal family theorem in the works of Kim/Krantz and Byun/Gaussier/Kim; it also provides an easier and shorter proof even in the case of separable Hilbert spaces. See [130] and [108].

$$v = \sum_{j=1}^{\infty} \alpha_j v_j,$$

but can be reduced to only weak convergence in a normed vector space (such as a Banach space).

[5]Let X and Y be complex Banach spaces and let $\mathcal{D}$ be an open subset of X. We use λ to denote a complex variable. A function $h : \mathcal{D} \to Y$ is called *Gâteaux holomorphic* if it is locally bounded and if for each $x \in \mathcal{D}$, $y \in X$, and linear functional $l \in Y^*$, the function

$$f(\lambda) = l(h(x + \lambda y))$$

is holomorphic at $\lambda = 0$. See, for example, [109].

We conclude this section with some examples, due to Jisoo Byun [32], that suggest some of the limitations of normal families in the infinite dimensional setting. These examples all relate to the failure of convexity.

Let Ω_1 and Ω_2 be bounded domains in a Banach space X. We point out that the holomorphic weak-* limit mapping $\widehat{f} : \Omega \to X$ of a sequence of holomorphic mappings $f_j : \Omega_1 \to \Omega_2$ may in general show a surprising behavior in contrast with the finite dimensional cases. In the finite dimensional case, $\hat{f}\,(\Omega_1)$ should be contained in the closure of Ω_2. Here we demonstrate that weak-* closure is about the best one can do, even with the nicest candidates such as sequences of biholomorphic mappings from the ball.

Example 7.7.2. Let B be the unit open ball in ℓ^2. Let $\{\mathbf{e}_j \mid j = 1, 2, \ldots\}$ be the standard orthonormal basis for ℓ^2. Let $f_k : B \to \ell^2$ be defined by

$$f_k(\mathbf{z}) = \sum_{j=1}^{k-1} z_j \mathbf{e}_j + \left(z_k + z_1^2\right) \mathbf{e}_k + \sum_{j=k+1}^{\infty} z_j \mathbf{e}_j \,,$$

where $\mathbf{z} = z_1 \mathbf{e}_1 + \ldots$. Notice that none of $f_k(B)$ is convex. In fact, the ball centered at $\frac{77}{80} e_1$ with radius $1/100$ never meets $f_k(B)$, while it is obvious that the origin and the point e_1 are clearly in the norm closure of the union of $f_k(B)$. Moreover, the weak limit $\widehat{f}$ of the sequence f_k is the identity map. Hence $\widehat{f}(B) = B$, which is convex. This shows that the weak limit can gain in its image more than the norm closure of the union of the images of $f_k(B)$.

Example 7.7.3. In general the weak-* limit does not automatically make the range convex. Consider $g_k : B \to \ell^2$ defined by

$$g_k(\mathbf{z}) = \left(z_1 + z_2^2\right) \mathbf{e}_1 + \sum_{j=2}^{k-1} z_j \mathbf{e}_j + \left(z_k + \frac{1}{2} z_2^2\right) \mathbf{e}_k + \sum_{j=k+1}^{\infty} z_j \mathbf{e}_j$$

for $k = 3, 4, \ldots$. Then each g_k and the weak limit

$$\hat{g}(\mathbf{z}) = \left(z_1 + z_2^2\right) \mathbf{e}_1 + \sum_{j=2}^{\infty} z_j \mathbf{e}_j$$

are biholomorphisms of the ball B onto its image. Notice that $\hat{g}(B)$ is not convex.

7.8 Many Approaches to Normal Families

It is natural to try to relate the infinite-dimensional case to the well-known case of finite dimensions. In particular, let $\mathcal{F}$ be a family of holomorphic functions on a domain Ω in a Banach space X. Is it correct to say that $\mathcal{F}$ is normal if and only if the restriction of $\mathcal{F}$ to any finite-dimensional subspace is normal? Obversely, if the post-composition of the elements of $\mathcal{F}$ with each finite-dimensional subspace projection operator is normal then can we conclude that $\mathcal{F}$ is normal? We would like to treat some of these questions here.

Example 7.8.1. Suppose that if $\mathcal{F}$ is a family of maps of a domain Ω in a separable Hilbert space H to itself, and assume that

$$\{\pi_j \circ f : f \in \mathcal{F}\}$$

is normal for each $\pi_j : H \to H_j$ the projection of H to the one-dimensional subspace H_j spanned by the unit vector in the j-th direction. Then it does not necessarily follow that $\mathcal{F}$ is a normal family.

To see this, let $H = \ell_2$, and let $f_j(\{x_\ell\}) = x_j$. Consider each f_j as a map from the unit ball $\mathbf{B} \subseteq H$ to itself. Then, for each fixed k, $\left\{(\pi_k \circ f)_j\right\}_{j=1}^{\infty}$ is a normal family, yet the family $\mathcal{F} = \{f_j\}_{j=1}^{\infty}$ is definitely not normal.

Example 7.8.2. Suppose that if $\mathcal{F}$ is a family of maps of a domain Ω in a separable Hilbert space H to itself. Suppose that, for each k, the collection

$$\{f \circ \mu_k : f \in \mathcal{F}\}$$

is normal for each $\mu_j : \mathbb{C} \to H$ the injection of $\mathbb{C}$ to H in the j-th variable. Then it does not necessarily follow that $\mathcal{F}$ is a normal family.

For this result, again consider $H = \ell_2$, and let $f_j(\{x_\ell\}) = x_j$. Consider each f_j as a map from the unit ball $\mathbf{B} \subseteq H$ to itself. Then, for each fixed k, the family $\{f \circ \mu_k\}_{f \in \mathcal{F}}$ is clearly normal. Yet the entire family $\mathcal{F}$ is plainly not normal—as we discussed in Example 7.2.1. The reader should compare this example to Proposition 7.2.14, which gives a positive result along these lines.

One of the main lessons of the classic paper [287] by H.H. Wu is that the normality or non-normality of a family of mappings depends essentially on the target space (this is the provenance of the notion of taut manifold). With this point in mind, we now formulate a counterpoint to Example 7.4.1:

Proposition 7.8.3. *Let* $\mathcal{C} = \left\{\{x_j\}_{j=1}^{\infty} : |x_j| \leq 1/j\right\}$ *be the Hilbert cube. Let* $H = \ell_2$ *be the canonical separable Hilbert space. Then any family* $\mathcal{F}$ *from a domain* $\Omega \subseteq H$ *to* $\mathcal{C}$ *will be normal.*

Proof. It suffices to prove that the correct formulation of the Arzelà-Ascoli theorem holds. In particular, we establish this result:

If $\mathcal{G} = \{g_\alpha\}_{\alpha \in \mathcal{A}}$ is a family of functions from a domain $\Omega \subseteq H$ into $\mathcal{C}$ which is (i) equibounded and (ii) equicontinuous, then $\mathcal{G}$ has a uniformly convergent subsequence.

In fact the usual proof of Arzelà-Ascoli, that can be found in any text (see, for instance, [146, p. 284]), goes through once we establish this basic fact: If $g_\alpha : \Omega \to \mathcal{C}$ and $x_0 \in \Omega$ is fixed, then $\{g_\alpha(x_0)\}$ has a convergent subsequence. Of course this simple assertion is the consequence of a standard diagonalization argument. □

8

A Montel Theorem for Holomorphic Functions on Infinite Dimensional Spaces That Omit the Values 0 and 1

Preamble: In a plane domain, locally uniform convergence of a sequence of functions is equivalent to uniform convergence on compact sets. That notion of convergence defines the compact-open topology τ_0, which is natural for the standard convergence theorems in classical function theory.

The situation in infinite dimensional holomorphy is more complicated. As the domains of the functions are not locally compact, the topology τ_0 is too weak for many purposes. There is an extensive literature about stronger topologies on spaces of holomorphic functions with infinite dimensional domains. See, for example, [43], [55], and their bibliographies.

In spite of its deficiencies, the topology τ_0 is convenient for extending some aspects of the theory of normal families to the infinite dimensional setting. For example, [43, Chapter 17, Theorem 17.1] and [55, Chapter 3, Lemma 3.25] use that topology to extend Montel's theorem about locally bounded families of holomorphic functions.

In this chapter we present the proof by Clifford J. Earle [69] that the set of holomorphic functions on a connected complex Banach manifold X that omit the values 0 and 1 is a normal family, and we shall closely follow the presentation there.

The material of this chapter based on articles [70] and [69].

8.1 Schwarz-Pick Systems

A pseudometric space is a generalization of a metric space in which the distance between two distinct points can be zero. Pseudometric spaces were introduced by Đurađ Kurepa [149] in 1934. Their particular importance for numerical analysis has been pointed out by J. Schröder [263]. In the same

DOI: 10.1201/9781032669861-8

way as every normed space is a metric space, every seminormed space is a pseudometric space.

General systems of pseudometrics with the Schwarz-Pick property were first studied systematically in [110] by Harris, who coined the term Schwarz-Pick system for them. These systems can be studied on various classes of complex spaces. We choose to use the class of complex manifolds modeled on complex Banach spaces of positive, possibly infinite, dimension. All our manifolds are assumed to be connected Hausdorff spaces.

If X and Y are complex Banach manifolds, we shall denote the set of all holomorphic maps of X into Y by $\mathcal{O}(X, Y)$.

Definition 8.1.1 (see Harris [110]). A *Schwarz-Pick system* is a functor, denoted by $X \to d_X$, that assigns to each complex Banach manifold X a pseudometric d_X so that the following conditions hold:

(a) The pseudometric assigned to D is the Poincarè metric

$$d_D(z_1, z_2) = \tanh^{-1}\left|\frac{z_1 - z_2}{1 - z_1\overline{z_2}}\right| \quad \text{if } z_1 \in D \text{ and } z_2 \in D\,. \tag{8.1}$$

(b) If X and Y are complex Banach manifolds then

$$d_Y(f(x_1), f(x_2)) \le d_X(x_1, x_2) \text{ if } x_1 \in X, x_2 \in X \text{ and } f \in \mathcal{O}(X, Y)\,. \tag{8.2}$$

Remark 8.1.2. Because of conditions (a) and (b), the sets $\mathcal{O}(D, X)$ and $\mathcal{O}(X, D)$ provide upper and lower bounds for d_X. The upper bound leads to the definition of the Kobayashi pseudometric.

8.2 The Classical Definition of Kobayashy Pseudometric

In what follows, d_D will always be the Poincaré metric (8.1) on the unit disk D.

Definition 8.2.1. A Schwarz-Pick pseudometric on the complex Banach manifold X is a pseudometric d such that

$$d(f(z), f(w)) \le d_D(z, w) \quad \text{for all } z \text{ and } w \text{ in } D \text{ and } f \text{ in } \mathcal{O}(D, X)\,. \tag{8.1}$$

If $X \mapsto d_X$ is a Schwarz-Pick system, then d_X is obviously a Schwarz-Pick pseudometric on X for every complex Banach manifold X.

Definition 8.2.2. The Kobayashi pseudometric K_X is the largest Schwarz-Pick pseudometric on the complex Banach manifold X.

As Kobayashi observed (see for example [136] or [134]), K_X is easily described in terms of the function $\delta_X : X \times X \to [0, \infty]$ defined by

$$\delta_X(x, x') = \inf \{d_D(0, z) : x = f(0) \text{ and } x' = f(z) \text{ for some } f \in \mathcal{O}(D, X)\}$$

for all x and x' in X. (As usual the infimum of the empty set is ∞.)

In fact Definition 8.2.1, the definition of δ_X, and the triangle inequality imply that any Schwarz-Pick pseudometric d on X satisfies

$$d(x, x') \le \inf \left\{ \sum_{j=1}^{n} \delta_X(x_{j-1}, x_j) \right\} \quad \text{for all } x \text{ and } x' \text{ in } X \tag{8.2}$$

where the infimum is taken over all positive integers n and all $(n+1)$-tuples of points $x_0, \ldots, x_n$ in X such that $x_0 = x$ and $x_n = x'$. The infimum on the right side of the inequality (8.2) defines a function on $X \times X$ that is obviously a Schwarz-Pick pseudometric on X, so (8.2) implies that

$$K_X(x, x') = \inf \left\{ \sum_{j=1}^{n} \delta_X(x_{j-1}, x_j) \right\} \quad \text{for all } x \text{ and } x' \text{ in } X. \tag{8.3}$$

(The infimum is of course taken over the same set as in (8.2) above.)

Equation (8.3) is Kobayashi's definition of the pseudometric K_X. It follows readily from (8.3) and the Schwarz-Pick lemma that the functor assigning the Kobayashi pseudometric K_X to each complex Banach manifold X is a Schwarz-Pick system. A slightly stronger property of this functor will follow from the arc length description of K_X that we shall explain in the remainder of next section.

8.3 The Infinitesimal Kobayashi Pseudometric and its Integrated Form

Every complex Banach manifold X has an infinitesimal Kobayashi pseudometric k_X, first introduced (in the finite dimensional case) by Kobayashi in [136]. Since k_X is a function on the tangent bundle $T(X)$ of X, we shall briefly review some properties of tangent bundles.

For the moment let X be a C^1 manifold modeled on a real Banach space V. For each x in X the tangent space to X at x will be denoted by $T_x(X)$. The tangent bundle $T(X)$ of X consists of the ordered pairs (x, v) such that $x \in X$ and $v \in T_x(X)$ (see [158]).

If X is an open set in V with the C^1 structure induced by the inclusion map, then each $T_x(X), x$ in X, is naturally identified with V, and $T(X) = X \times V$.

If X and Y are C^1 manifolds and x is a point of X, every C^1 map $f : X \to Y$ induces a linear map $f_*(x)$ from $T_x(X)$ to $T_{f(x)}(Y)$ (see [158]). If X and Y are subregions of Banach spaces V and W and the tangent spaces $T_x(X)$ and $T_{f(x)}(Y)$ are identified with V and W in the natural way, then $f_*(x)$ is the usual Fréchet derivative of f at x.

The tangent bundle $T(X)$ has a natural C^0 manifold structure modeled on $V \times V$. A convenient atlas for $T(X)$ consists of the charts $T(\varphi)$ defined by the formula

$$T(\varphi)(x, v) = (\varphi(x), \varphi_*(x)v), \quad (x, v) \in T(U),$$

where U is an open set in $X, T(U)$ is the open subset $\{(x, v) \in T(X) : x \in U\}$ of $T(X)$, and φ is a chart on X with domain U. The image of $T(U)$ under $T(\varphi)$ is the open set $\varphi(U) \times V$ in $V \times V$.

If X is a complex Banach manifold modeled on a complex Banach space V, then each tangent space $T_x(X)$ has a unique complex Banach space structure such that the map $\varphi_*(x)$ from $T_x(X)$ to V is a $\mathbb{C}$-linear isomorphism whenever φ is a (holomorphic) chart defined in some neighborhood of x. Furthermore, $T(X)$ has a unique complex Banach manifold structure such that the map $T(\varphi)$ from $T(U)$ to $\varphi(U) \times V$ is biholomorphic for every (holomorphic) chart φ on X with domain U (see [57]).

Now we are ready for Kobayashi's definition of k_X.

Definition 8.3.1. The infinitesimal Kobayashi pseudometric on the complex Banach manifold X is the function k_X on $T(X)$ defined by the formula

$$k_X(x, v) = \inf\{|z| : x = f(0) \text{ and } v = f_*(0)z \text{ for some } f \in \mathcal{O}(D, X)\} \quad (8.1)$$

Obviously $k_X(x, v) \geq 0$ and $k_X(x, cv) = |c| k_X(x, v)$ for all complex numbers c. The following Schwarz-Pick property is also an immediate consequence of the definition (see [253] or Theorem 1.2.6 in [224]).

Proposition 8.3.2. *If X and Y are complex Banach manifolds and $f \in \mathcal{O}(X, Y)$, then*

$$k_Y(f(x), f_*(x)v) \leq k_X(x, v) \quad \textit{for all } (x, v) \in T(X)$$

In particular, if f is biholomorphic, then $k_Y(f(x), f_*(x)v) = k_X(x, v)$.

Corollary 8.3.3. *We have that $k_D(w, z) = \frac{|z|}{1-|w|^2}$ for all (w, z) in $D \times \mathbb{C}$.*

Proof. Definition 8.3.1 and Schwarz's lemma imply that $k_D(0, z) = |z|$ for all complex numbers z. To prove the formula for $k_D(w, z)$, apply Proposition 3.4 with $X = Y = D$ and $f(\zeta) = (\zeta - w)/(1 - \zeta\bar{w}), \zeta$ in D. □

We shall use the function k_X to measure the lengths of piecewise C^1 curves in X. As usual, if the curve $\gamma : [a, b] \to X$ is differentiable at t in $[a, b]$, then the symbol $\gamma'(t)$ denotes the tangent vector $\gamma_*(t)1$ to X at $\gamma(t)$. If γ is piecewise C^1, it is natural to define the Kobayashi length of γ by integrating the function $k_X(\gamma(t), \gamma'(t))$ over the parameter interval of γ. That function is upper semicontinuous when X is either a domain in a complex Banach space (see [110] or [56]) or a finite dimensional complex manifold (see [254] and [255] or [224]), but the case of infinite dimensional complex manifolds is harder to deal with. In [253] Royden evades that difficulty by using the upper Riemann integral. Venturini [281] gets more refined results by using upper and lower Lebesgue integrals. We shall follow Royden's example, as it allows the very elementary arguments that we shall now present.

The required upper Riemann integrals exist because the function k_X is locally bounded on $T(X)$. To prove this we use special charts on X. By definition, a standard chart at x in X is a biholomorphic map φ of an open neighborhood of x onto the open unit ball of V with $\varphi(x) = 0$.

Lemma 8.3.4. *If φ is a standard chart at the point x in X, then*

$$k_X(y, v) \leq 2 \|\varphi_*(y)v\|$$

for all (y, v) in $T(X)$ such that y is in the domain of φ and $\|\varphi(y)\| \leq 1/2$.

Proof. If $v = 0$, the inequality is trivial. If $v \neq 0$, we derive it from (8.1) by setting

$$f(z) = \varphi^{-1}\left(\varphi(y) + z\frac{\varphi_*(y)v}{2\|\varphi_*(y)v\|}\right), \quad z \in D,$$

so that $f \in \mathcal{O}(D, X)$, $f(0) = y$, and $f_*(0)c = \frac{c}{2\|\varphi_*(y)v\|}v$ for all c in $\mathbb{C}$. □

Corollary 8.3.5. *The function k_X is locally bounded in $T(X)$.*

Proof. Let U be the domain of the standard chart φ in Lemma 8.3.4. Since holomorphic maps are C^1, the function $(y, v) \mapsto 2\|\varphi_*(y)v\|$ is locally bounded in the open set $T(U) = \{(y, v) \in T(X) : y \in U\}$. □

Following Royden [253], we can now define the arc length $L_X(\gamma)$ of a piecewise C^1 curve $\gamma : [a, b] \to X$ in X to be the upper Riemann integral

$$L_X(\gamma) = \int_a^b k_X(\gamma(t), \gamma'(t))\, dt \tag{8.2}$$

and the distance $\rho_X(x, y)$ to be the infimum of the lengths of all piecewise C^1 curves joining x to y in X. The resulting pseudometric ρ_X on X is the integrated form of the Kobayashi infinitesimal pseudometric k_X. By Proposition 8.3.2,

$$\rho_Y(f(x_1), f(x_2)) \leq \rho_X(x_1, x_2) \quad \text{for all } x_1 \text{ and } x_2 \text{ in } X \tag{8.3}$$

whenever $f \in \mathcal{O}(X, Y)$. In fact even more is true. If $f \in \mathcal{O}(X, Y)$ and γ is a piecewise C^1 curve in X, then $L_Y(f \circ \gamma) \leq L_X(\gamma)$.

Remark 8.3.6. The upper Riemann integral of an upper semicontinuous function equals its Lebesgue integral, so we can use a Lebesgue integral in (8.2) if X is finite dimensional or a region in a complex Banach space, but that is an unnecessary luxury.

Remark 8.3.7. By Corollary 8.3.3, ρ_D is the Poincaré metric d_D on D. Therefore, by (8.3), the functor that assigns ρ_X to each complex Banach manifold X is a Schwarz-Pick system. In particular ρ_X is a Schwarz-Pick pseudometric on X, so $\rho_X(x, y) \le K_X(x, y)$ for all x and y in X. In the next subsection we shall use methods of Harris [110] to prove that the pseudometrics ρ_X and K_X are in fact equal.

8.4 A Montel Theorem

In a plane domain, locally uniform convergence of a sequence of functions is equivalent to uniform convergence on compact sets. That notion of convergence defines the compact-open topology τ_0, which is natural for the standard convergence theorems in classical function theory.

The situation in infinite dimensional holomorphy is more complicated. As the domains of the functions are not locally compact, the topology τ_0 is too weak for many purposes. There is an extensive literature about stronger topologies on spaces of holomorphic functions with infinite dimensional domains. See, for example, [43, 55], and their bibliographies.

In spite of its deficiencies, the topology τ_0 is convenient for extending some aspects of the theory of normal families to the infinite dimensional setting. For example, [43, 2, Chapter 17, Theorem 17.1] and [55, 4, Chapter 3, Lemma 3.25] use that topology to extend Montel's theorem about locally bounded families of holomorphic functions.

The following proposition is a special case of the cited lemma in [55]. We shall use it in the proof of Theorem 8.4.2 below.

Proposition 8.4.1. *If U is an open subset of a complex Banach space, the set $\mathcal{B}$ of holomorphic functions f on U such that $|f(x)| \le 1$ for all x in U is compact with respect to the compact-open topology τ_0 and the topology τ_p of pointwise convergence, and these topologies coincide on $\mathcal{B}$.*

Here, τ_p is the topology induced by identifying $\mathcal{B}$ with a subset of the product space $\widehat{\mathbb{C}}^U$, where $\widehat{\mathbb{C}}$ is the Riemann sphere, and τ_0 can be thought of as the topology of uniform convergence on compact sets with respect to the chordal metric on $\widehat{\mathbb{C}}$.

Proposition 8.4.1 is our model for the following theorem, which extends Montel's theorem about holomorphic functions that omit the values 0 and 1. We shall denote the set of such functions on a complex manifold X by $\mathcal{O}(X, \widehat{\mathbb{C}} \backslash \{0, 1\})$.

Theorem 8.4.2. *Let X be a connected complex Banach manifold, and let $\mathcal{F}$ be the union of $\mathcal{O}(X, \mathbb{C}\backslash\{0,1\})$ and the set of constant maps from X to $\mathbb{C}$. Then $\mathcal{F}$ is compact with respect to τ_p and τ_0, and these topologies coincide on $\mathcal{F}$.*

The proof of Theorem 8.4.2 rely on Proposition 8.4.1, the classical Montel theorem, and general contraction properties of holomorphic mappings.

Theorem 8.4.2 is a corollary of the following two lemmas. The proof of the first will use Proposition 8.4.1. We shall denote the open unit disk in $\mathbb{C}$ by D.

Lemma 8.4.3. *The set $\mathcal{F}$ is closed in the product space $\mathbb{C}^X$.*

Proof. Suppose the net (f_α) in $\mathcal{F}$ converges to f in $\mathbb{C}^X$. We must show that $f \in \mathcal{F}$. Choose x_0 in X. We may assume that X contains the open unit ball B in the model Banach space E and that x_0 is the point 0 in B. It will be convenient to set $\Omega := \widehat{\mathbb{C}}\backslash\{0, 1, \infty\}$.

First, suppose $f(0) \in \widehat{\mathbb{C}}\backslash\Omega$. Let u in E be a unit vector. For each α, set $g_\alpha(z) := f_\alpha(zu)$, z in D. As $f_\alpha \to f$, the function $g(z) := f(zu)$ is the pointwise limit of the net (g_α). By the classical Montel theorem, the convergence is locally uniform (with respect to the chordal metric) on some subnet (g_{α_i}), and the limit function g is constant in D because $g(0) = f(0) = 0, 1$ or ∞. Therefore $f(zu) = f(0)$ for all z in D. By varying u, we obtain $f(x) = f(0)$ for all x in B. Now suppose $f(0) \in \Omega$. Since $f_\alpha(0) \to f(0)$, we may assume that $f_\alpha(0) \in \Omega$ for all α. As $f_\alpha \in \mathcal{F}, f_\alpha(X) \subset \Omega$ for all α. Let $\pi : D \to \Omega$ be a holomorphic covering map such that $\pi(0) = f(0)$. Choose holomorphic functions $g_\alpha : B \to D$ so that $f_\alpha = \pi \circ g_\alpha$ in B and $g_\alpha(0) \to 0$.

By Proposition 8.4.1, some subnet g_{α_i} converges uniformly on compact subsets of B to a holomorphic function g. As $g(0) = 0$ and $|g(x)| \leq 1$ for all x in B, the maximum principle (see [117, Chapter III, Theorem 3.18.3]) implies that $g(B) \subset D$. Therefore $f = \pi \circ g$ and is holomorphic in B.

As the point x_0 was arbitrary, f is a holomorphic mapping of X into $\mathbb{C}$. As X is connected and the inverse images of 0,1, and ∞ are open in $X, f \in \mathcal{F}$. □

The proof of the next lemma will use some properties of the Kobayashi metrics ρ_B and ρ_Ω of the ball B and domain Ω. These metrics satisfy

$$\rho_\Omega(h(x), h(y)) \leq \rho_B(x, y) \tag{8.1}$$

if $x \in B, y \in B$, and $h : B \to \Omega$ is holomorphic. In addition

$$\rho_B(0, x) = \operatorname{arctanh} \|x\| \quad \text{for all } x \text{ in } B. \tag{8.2}$$

Also, the Kobayashi metric ρ_D of the open unit disk D is the Poincaré metric, scaled so that $\rho_D(0, z) = \operatorname{arctanh}|z|$ for all z in D, and ρ_Ω is the quotient metric on Ω induced by the holomorphic covering map $\pi : D \to \Omega$, by [56, Chapter 5, Lemma 5.7]. All closed ρ_Ω-bounded sets are compact, as any such set is contained in $\pi(\{z \in D : |z| \leq r\})$ for some $r < 1$.

Lemma 8.4.4. *The map* $P(f,x)=f(x)$ *from* $\mathcal{F}\times X$ *to* $\mathbb{C}$ *is continuous when* $\mathcal{F}$ *is given the topology* τ_p *of pointwise convergence.*

Proof. Let $(f_\alpha, x_\alpha)\to(f,x)$ in $\mathcal{F}\times X$. Let U be a non-trivial closed disk in $\mathbb{C}$, centered at $f(x)$, such that $U\backslash\{f(x)\}$ is contained in Ω. We may assume that $f_\alpha(x)\in U$ for all α. We shall also assume that X contains the open unit ball B, that x is the point 0 in B, and that $x_\alpha\in B$ for all α.

Suppose $P(f,0)=f(0)=0,1$, or ∞. As $f_\alpha(0)\in U$, either $f_\alpha(0)=f(0)$ or $f_\alpha(0)\in\Omega$. In the first case, $f_\alpha(x_\alpha)=f(0)$. In the second case, $f_\alpha(X)\subset\Omega$, so f_α maps B into Ω. Therefore

$$\rho_\Omega\left(f_\alpha\left(x_\alpha\right), f_\alpha(0)\right)\le\rho_B\left(x_\alpha,0\right)=\operatorname{arctanh}\|x_\alpha\|\to 0$$

so $\rho_\Omega\left(f_\alpha\left(x_\alpha\right),\zeta\right)\to\infty$ for any given point ζ of Ω. Hence $f_\alpha\left(x_\alpha\right)$ approaches $f(0)$ along with $f_\alpha(0)$. This proves the continuity of P at any point where $P(f,x)\notin\Omega$. If $P(f,0)=f(0)\in\Omega$, then $U\subset\Omega$, so $f_\alpha(0)\in\Omega$ for all α, and f_α maps B into Ω. By (8.1) and (8.2),

$$\begin{aligned}\rho_\Omega\left(P\left(f_\alpha,x_\alpha\right),P(f,0)\right)&=\rho_\Omega\left(f_\alpha\left(x_\alpha\right),f(0)\right)\\&\le\rho_\Omega\left(f_\alpha\left(x_\alpha\right),f_\alpha(0)\right)+\rho_\Omega\left(f_\alpha(0),f(0)\right)\\&\le\rho_B\left(x_\alpha,0\right)+\rho_\Omega\left(f_\alpha(0),f(0)\right)\to 0,\end{aligned}$$

so P is also continuous at all points of $P^{-1}(\Omega)$. □

Proof of the theorem.. As the product space $\mathbb{C}^X$ is compact, Lemma 8.4.3 implies that $\mathcal{F}$ is compact with respect to the topology τ_p. The compact-open topology τ_0 is always stronger than τ_p. Conversely, by Kelley [126, Chapter 7, Theorem 5], Lemma 8.4.4 implies that τ_p is stronger than τ_0 on $\mathcal{F}$. □

9

Concluding Remarks

The theory of normal families has been an important tool in complex function theory for well over 100 years. More recently the theory of normal functions has assumed a significant role. The two theories interact productively.

The purpose of this book has been to explain these two sets of ideas, both in the classical one-variable setting and also in the context of several complex variables and Banach spaces. In addition, we treat normal families and normal function on complex spaces and manifolds.

We hope that this is a new and useful contribution to the literature, and we look forward to feedback from our readers.

DOI: 10.1201/9781032669861-9

Bibliography

[1] Marco Abate. "A characterization of hyperbolic manifolds". In: *Proc. Amer. Math. Soc.* 117.3 (1993), pp. 789–793. ISSN: 0002-9939. DOI: 10.2307/2159145.

[2] Marco Abate. "The Lindelöf principle and the angular derivative in strongly convex domains". In: *J. Analyse Math.* 54 (1990), pp. 189–228. ISSN: 0021-7670. DOI: 10.1007/BF02796148.

[3] Marco Abate and Roberto Tauraso. "The Lindelöf principle and angular derivatives in convex domains of finite type". In: *J. Aust. Math. Soc.* 73.2 (2002), pp. 221–250. ISSN: 1446-7887. DOI: 10.1017/S1446788700008818.

[4] Lars V. Ahlfors. *Complex analysis.* Third International Series in Pure and Applied Mathematics. McGraw-Hill Book Co., New York, 1978, pp. xi+331. ISBN: 0-07-000657-1.

[5] G. Aladro. "Localization of the Kobayashi distance". English. In: *Jour. Math. Anal. Appl.* 181 (1994), pp. 200–204.

[6] Gerardo Aladro and Steven G. Krantz. "A criterion for normality in $\mathbf{C}^n$". In: *J. Math. Anal. Appl.* 161.1 (1991), pp. 1–8. ISSN: 0022-247X. DOI: 10.1016/0022-247X(91)90356-5.

[7] Daniel S. Alexander. *A history of complex dynamics.* Aspects of Mathematics, E24. Friedr. Vieweg & Sohn, Braunschweig, 1994, pp. viii+165. ISBN: 3-528-06520-6. DOI: 10.1007/978-3-663-09197-4.

[8] H. Alexander. "Volumes of images of varieties in projective space and in Grassmannians". In: *Trans. Amer. Math. Soc.* 189 (1974), pp. 237–249. ISSN: 0002-9947. DOI: 10.2307/1996857.

[9] F. J. Almgren Jr. "Mass continuous cochains are differential forms". In: *Proc. Amer. Math. Soc.* 16 (1965), pp. 1291–1294. ISSN: 0002-9939. DOI: 10.2307/2035917.

[10] J. M. Anderson, J. Clunie, and Ch. Pommerenke. "On Bloch functions and normal functions". In: *J. Reine Angew. Math.* 270 (1974), pp. 12–37. ISSN: 0075-4102.

[11] Cesare Arzela. "Note on series of analytic functions". In: *Ann. of Math. (2)* 5.2 (1904), pp. 51–63. ISSN: 0003-486X. DOI: 10.2307/2007164.

[12] G. Ascoli. "The limit curve of a given family of curves." Italian. In: *Rom. Acc. L. Mem. (3)* 18 (1884), pp. 521–586.

[13] Volker Aurich. "Bounded holomorphic embeddings of the unit disk into Banach spaces". In: *Manuscripta Math.* 45.1 (1983), pp. 61–67. ISSN: 0025-2611. DOI: 10.1007/BF01168580.

[14] F. Bagemihl and W. Seidel. "Sequential and continuous limits of meromorphic functions". In: *Ann. Acad. Sci. Fenn. Ser. A I No.* 280 (1960), p. 17.

[15] D. Bargmann et al. "Families of meromorphic functions avoiding continuous functions". In: *J. Anal. Math.* 79 (1999), pp. 379–387. ISSN: 0021-7670. DOI: 10.1007/BF02788248.

[16] Jorge Alberto Barroso, Mário C. Matos, and Leopoldo Nachbin. "On bounded sets of holomorphic mappings". In: *Proceedings on Infinite Dimensional Holomorphy (Internat. Conf., Univ. Kentucky, Lexington, Ky., 1973).* Lecture Notes in Math., Vol. 364. Springer, Berlin, 1974, pp. 123–134.

[17] Jorge Alberto Barroso, Mário C. Matos, and Leopoldo Nachbin. "On holomorphy versus linearity in classifying locally convex spaces". In: *Infinite dimensional holomorphy and applications (Proc. Internat. Sympos., Univ. Estadual de Campinas, São Paulo, 1975).* North-Holland Math. Studies, Vol. 12; Notas de Mat., No. 54. North-Holland, Amsterdam, 1977, pp. 31–74.

[18] Jorge Alberto Barroso and Leopoldo Nachbin. "A direct sum is holomorphically bornological with the topology induced by a Cartesian product". In: *Portugal. Math.* 40.2 (1981), 252–256 (1985). ISSN: 0032-5155.

[19] Theodore J. Barth. "Families of holomorphic maps into Riemann surfaces". In: *Trans. Amer. Math. Soc.* 207 (1975), pp. 175–187. ISSN: 0002-9947. DOI: 10.2307/1997173.

[20] A. F. Beardon and D. Minda. "Normal families: a geometric perspective". In: *Comput. Methods Funct. Theory* 14.2-3 (2014), pp. 331–355. ISSN: 1617-9447. DOI: 10.1007/s40315-014-0054-2.

[21] Alan F. Beardon and David Minda. "On the pointwise limit of complex analytic functions". In: *Amer. Math. Monthly* 110.4 (2003), pp. 289–297. ISSN: 0002-9890. DOI: 10.2307/3647878.

[22] E. Bedford and S. I. Pinchuk. "Domains in $\mathbb{C}^2$ with noncompact groups of holomorphic automorphisms". Russian. In: *Mat. Sb., Nov. Ser.* 135.2 (1988), pp. 147–157.

[23] Steven R. Bell and Raghavan Narasimhan. "Proper holomorphic mappings of complex spaces". In: *Several complex variables, VI.* Vol. 69. Encyclopaedia Math. Sci. Springer, Berlin, 1990, pp. 1–38.

[24] Stefan Bergman. *The Kernel Function and Conformal Mapping.* Mathematical Surveys, No. 5. American Mathematical Society, New York, N. Y., 1950, pp. vii+161.

[25] Stefan Bergmann. "Zur Theorie von pseudokonformen Abbildungen". German. In: *Rec. Math. Moscou, n. Ser.* 1 (1936), pp. 79–96.

[26] Walter Bergweiler. "Bloch's principle". In: *Comput. Methods Funct. Theory* 6.1 (2006), pp. 77–108. ISSN: 1617-9447. DOI: 10.1007/BF03321119.

[27] Walter Bergweiler. "Rescaling principles in function theory". In: *Analysis and its applications (Chennai, 2000)*. Allied Publ., New Delhi, 2001, pp. 11–29. URL: https://analysis.math.uni-kiel.de/bergweiler/papers/56.pdf.

[28] A. S. Besicovitch. "On sufficient conditions for a function to be analytic, and on behaviour of analytic functions in the neighbourhood of non-isolated singular points". In: *Proc. London Math. Soc. (2)* 32.1 (1931), pp. 1–9. ISSN: 0024-6115. DOI: 10.1112/plms/s2-32.1.1.

[29] Zbigniew Błock. "Regularity of the fundamental solution for the Monge-Ampère operator". In: *Progress in partial differential equations, Vol. 1 (Pont-à-Mousson, 1997)*. Vol. 383. Pitman Res. Notes Math. Ser. Longman, Harlow, 1998, pp. 40–45.

[30] Douglas Bridges et al. "Picard's theorem". In: *Trans. Amer. Math. Soc.* 269.2 (1982), pp. 513–520. ISSN: 0002-9947. DOI: 10.2307/1998462.

[31] Robert B. Burckel. *Classical analysis in the complex plane*. Cornerstones. Springer, New York, 2021, pp. xxix+1123. ISBN: 978-1-0716-1963-6; 978-1-0716-1965-0. DOI: 10.1007/978-1-0716-1965-0.

[32] J. Byun. "Geometry of automorphism group orbits and Levi Geometry". Ph.D. Thesis. Pohang 790-784 Korea: Pohang University of Science and Technology, June 2002.

[33] Douglas M. Campbell and Gene Wickes. *Characterizations of normal meromorphic functions*. English. Complex analysis, Proc. Colloq., Joensuu/Finland 1978, Lect. Notes Math. 747, 55-72 (1979). 1979.

[34] L. A. Campbell, A. Howard, and T. Ochiai. "Moving holomorphic disks off analytic subsets". In: *Proc. Amer. Math. Soc.* 60 (1976), 106–108 (1977). ISSN: 0002-9939. DOI: 10.1090/s0002-9939-1976-0425186-0.

[35] L. Andrew Campbell and Roy H. Ogawa. "On preserving the Kobayashi pseudodistance". In: *Nagoya Math. J.* 57 (1975), pp. 37–47. ISSN: 0027-7630. URL: http://projecteuclid.org/euclid.nmj/1118795359.

[36] C. Caratheodory. "Book Review: *Theory of Functions of a Complex Variable, Vol. I* by Constantin Caratheodory". In: *Math. Mag.* 28.2 (1954), p. 122. ISSN: 0025-570X. URL: http://www.jstor.org/stable/3029375?origin=pubexport.

[37] C. Caratheodory. *Theory of functions of a complex variable. Vol. 2.* Chelsea Publishing Co., New York, 1954, p. 220.

[38] C. Carathéodory. “Stetige Konvergenz und normale Familien von Funktionen”. In: *Math. Ann.* 101.1 (1929), pp. 515–533. ISSN: 0025-5831. DOI: 10.1007/BF01454857.

[39] C. Carathéodory. “Sur le théorème général de M. *Picard.*” French. In: *C. R. Acad. Sci., Paris* 154 (1912), pp. 1690–1693. ISSN: 0001-4036.

[40] C. Carathéodory. “Über das Schwarzsche Lemma bei analytischen Funktionen von zwei komplexen Veränderlichen”. In: *Math. Ann.* 97.1 (1927), pp. 76–98. ISSN: 0025-5831. DOI: 10.1007/BF01447861.

[41] Elie Cartan. “Sur les domaines bornés homogènes de l’espace den variables complexes”. In: *Abh. Math. Sem. Univ. Hamburg* 11.1 (1935), pp. 116–162. ISSN: 0025-5858. DOI: 10.1007/BF02940719.

[42] David W. Catlin. “Estimates of invariant metrics on pseudoconvex domains of dimension two”. In: *Math. Z.* 200.3 (1989), pp. 429–466. ISSN: 0025-5874. DOI: 10.1007/BF01215657.

[43] Soo Bong Chae. *Holomorphy and calculus in normed spaces.* Vol. 92. Monographs and Textbooks in Pure and Applied Mathematics. Marcel Dekker, Inc., New York, 1985, pp. xii+421. ISBN: 0-8247-7231-8.

[44] E. M. Chirka. *Complex analytic sets.* Vol. 46. Mathematics and its Applications (Soviet Series). Kluwer Academic Publishers Group, Dordrecht, 1989, pp. xx+372. ISBN: 0-7923-0234-6. DOI: 10.1007/978-94-009-2366-9.

[45] E. M. Chirka. “Harnack inequalities, Kobayashi distances, and holomorphic motions”. In: *Tr. Mat. Inst. Steklova* 279.Analiticheskie i Geometricheskie Voprosy Kompleksnogo Analiza (2012), pp. 206–218. ISSN: 0371-9685. DOI: 10.1134/s0081543812080135.

[46] Chi Tai Chuang. *Normal families of meromorphic functions.* World Scientific Publishing Co., Inc., River Edge, NJ, 1993, pp. xii+473. ISBN: 981-02-1257-7. DOI: 10.1142/1904.

[47] Joseph A. Cima and Steven G. Krantz. “The Lindelöf principle and normal functions of several complex variables”. In: *Duke Math. J.* 50.1 (1983), pp. 303–328. ISSN: 0012-7094. DOI: 10.1215/S0012-7094-83-05014-7.

[48] E. M. Čirka. “The Lindelöf and Fatou theorems in $\mathbf{C}^n$”. In: *Mat. Sb. (N.S.)* 92(134) (1973), pp. 622–644, 648.

[49] John B. Conway. *Functions of one complex variable.* Vol. 11. Graduate Texts in Mathematics. Springer-Verlag, New York-Heidelberg, 1973, pp. xi+313.

[50] John P. D’Angelo. *Several complex variables and the geometry of real hypersurfaces.* Studies in Advanced Mathematics. CRC Press, Boca Raton, FL, 1993, pp. xiv+272. ISBN: 0-8493-8272-6.

[51] Kenneth R. Davidson. "Pointwise limits of analytic functions". In: *Amer. Math. Monthly* 90.6 (1983), pp. 391–394. ISSN: 0002-9890. DOI: 10.2307/2975578.

[52] Klas Diederich. "Das Randverhalten der Bergmanschen Kernfunktion und Metrik in streng pseudo-konvexen Gebieten". In: *Math. Ann.* 187 (1970), pp. 9–36. ISSN: 0025-5831. DOI: 10.1007/BF01368157.

[53] Klas Diederich and John Erik Fornæss. "Comparison of the Bergman and the Kobayashi metric". In: *Math. Ann.* 254.3 (1980), pp. 257–262. ISSN: 0025-5831. DOI: 10.1007/BF01457999.

[54] Jean Dieudonné. "Montel, Paul". In: *Dictionary of Scientific Biography: Suppliment II, Aleksandr Nikolaevich Lebedev-Fritz Zwicky*. Ed. by C.C. Gillispie and Frederic L. Holmes (Editor in chief) American Council of Learned Societies. first. Vol. 18. Dictionary of Scientific Biography. Charles Scribner's Sons Macmillan Library Reference USA Simon & Schuster Macmillan New York, 1990, pp. 649–650. ISBN: 0-684-19178-4.

[55] Seán Dineen. *Complex analysis on infinite-dimensional spaces*. Springer Monographs in Mathematics. Springer-Verlag London, Ltd., London, 1999, pp. xvi+543. ISBN: 1-85233-158-5. DOI: 10.1007/978-1-4471-0869-6.

[56] Seán Dineen. *The Schwarz lemma*. Oxford Mathematical Monographs. The Clarendon Press, Oxford University Press, New York, 1989, pp. x+248. ISBN: 0-19-853571-6.

[57] Adrien Douady. "Le problème des modules pour les sous-espaces analytiques compacts d'un espace analytique donné". In: *Ann. Inst. Fourier (Grenoble)* 16.fasc. 1 (1966), pp. 1–95. ISSN: 0373-0956. URL: http://www.numdam.org/item?id=AIF_1966__16_1_1_0.

[58] P. V. Dovbush. "Boundary behavior of normal holomorphic functions of several complex variables". English. In: *Sov. Math., Dokl.* 25 (1982), pp. 267–270. ISSN: 0197-6788.

[59] P. V. Dovbush. "Boundary behavior of normal holomorphic functions of several complex variables". In: *Dokl. Akad. Nauk SSSR* 263.1 (1982), pp. 14–17. ISSN: 0002-3264.

[60] P. V. Dovbush. "Lindelöf's theorem in $\mathbb{C}^n$". In: *Vestnik Moskov. Univ. Ser. I Mat. Mekh.* 6 (1981), pp. 33–36, 116. ISSN: 0201-7385.

[61] P. V. Dovbush. "Normal functions of many complex variables". English. In: *Mosc. Univ. Math. Bull.* 36.1 (1981), pp. 44–48. ISSN: 0027-1322.

[62] P. V. Dovbush. "Normal functions of several complex variables". In: *Vestnik Moskov. Univ. Ser. I Mat. Mekh.* 1 (1981), pp. 38–42, 107. ISSN: 0201-7385.

[63] P. V. Dovbush. “On a normality criterion of Mandelbrojt”. In: *Complex Var. Elliptic Equ.* 59.10 (2014), pp. 1388–1394. ISSN: 1747-6933. DOI: 10.1080/17476933.2013.854342.

[64] P. V. Dovbush. “On a normality criterion of W. Schwick”. In: *J. Geom. Anal.* 31.5 (2021), pp. 5355–5358. ISSN: 1050-6926. DOI: 10.1007/s12220-020-00481-4.

[65] P. V. Dovbush. “On Bloch and normal functions on complex Banach manifolds”. English. In: *Further progress in analysis. Proceedings of the 6th international ISAAC congress, Ankara, Turkey, August 13–18, 2007.* Hackensack, NJ: World Scientific, 2009, pp. 122–131. ISBN: 978-981-283-732-5.

[66] P. V. Dovbush. “On normal families in C^n”. In: *Complex Var. Elliptic Equ.* 67.1 (2022), pp. 1–8. ISSN: 1747-6933. DOI: 10.1080/17476933.2020.1797703.

[67] P. V. Dovbush. “On the Lindelöf theorem in $\mathbb{C}^n$”. Russian. In: *Ukr. Mat. Zh.* 40.6 (1988), pp. 796–799. ISSN: 0041-6053.

[68] P. V. Dovbush. “Zalcman’s lemma in C^n”. In: *Complex Var. Elliptic Equ.* 65.5 (2020), pp. 796–800. ISSN: 1747-6933. DOI: 10.1080/17476933.2019.1627529.

[69] Clifford J. Earle. “A Montel theorem for holomorphic functions on infinite dimensional spaces that omit the values 0 and 1”. In: *Comput. Methods Funct. Theory* 8.1-2 (2008), pp. 195–198. ISSN: 1617-9447. DOI: 10.1007/BF03321682.

[70] Clifford J. Earle et al. “Schwarz’s lemma and the Kobayashi and Carathéodory pseudometrics on complex Banach manifolds”. In: *Kleinian groups and hyperbolic 3-manifolds (Warwick, 2001).* Vol. 299. London Math. Soc. Lecture Note Ser. Cambridge Univ. Press, Cambridge, 2003, pp. 363–384. DOI: 10.1017/CBO9780511542817.017.

[71] M. A. Evgrafov. *Analytic functions.* Ed. by Bernard R. Gelbaum. W. B. Saunders Co., Philadelphia, Pa.-London, 1966, pp. x+336.

[72] Herbert Federer. *Geometric measure theory.* Die Grundlehren der mathematischen Wissenschaften, Band 153. Springer-Verlag New York, Inc., New York, 1969, pp. xiv+676.

[73] Herbert Federer and Wendell H. Fleming. “Normal and integral currents”. In: *Ann. Math. (2)* 72 (1960), pp. 458–520. ISSN: 0003-486X. DOI: 10.2307/1970227.

[74] Charles Fefferman. “The Bergman kernel and biholomorphic mappings of pseudoconvex domains”. In: *Invent. Math.* 26 (1974), pp. 1–65. ISSN: 0020-9910. DOI: 10.1007/BF01406845.

[75] J. E. Fornæss and E. Bedford. “A construction of peak functions on weakly pseudoconvex domains”. English. In: *Ann. Math.* 107 (1978), pp. 555–568.

[76] Tullio Franzoni and Edoardo Vesentini. *Holomorphic maps and invariant distances.* Vol. 69. Notas de Matemática [Mathematical Notes]. North-Holland Publishing Co., Amsterdam-New York, 1980, pp. viii+226. ISBN: 0-444-85436-3.

[77] B. L. Fridman. "An approximate Riemann mapping theorem in $\mathbb{C}^n$". In: *Math. Ann.* 275.1 (1986), pp. 49–55. ISSN: 0025-5831. DOI: 10.1007/BF01458582.

[78] W. H. J. Fuchs. *Topics in the theory of functions of one complex variable.* Van Nostrand Mathematical Studies, No. 12. D. Van Nostrand Co., Inc., Princeton, N.J.-Toronto, Ont.-London, 1967, pp. vi+193.

[79] Hirotaka Fujimoto. "On families of meromorphic maps into the complex projective space". In: *Nagoya Math. J.* 54 (1974), pp. 21–51. ISSN: 0027-7630. URL: http://projecteuclid.org/euclid.nmj/1118795097.

[80] B. A. Fuks. *Special chapters in the theory of analytic functions of several complex variables.* Translations of Mathematical Monographs, Vol. 14. American Mathematical Society, Providence, R.I., 1965, pp. vi+357.

[81] Ken-ichi Funahashi. "Normal holomorphic mappings and classical theorems of function theory". In: *Nagoya Math. J.* 94 (1984), pp. 89–104. ISSN: 0027-7630. DOI: 10.1017/S0027763000020857.

[82] V. I. Gavrilov. "On the distribution of values of functions meromorphic in the unit circle, which are not normal". In: *Mat. Sb. (N.S.)* 67 (109) (1965), pp. 408–427.

[83] F. W. Gehring. "Rings and quasiconformal mappings in space". In: *Trans. Amer. Math. Soc.* 103 (1962), pp. 353–393. ISSN: 0002-9947. DOI: 10.2307/1993834.

[84] Ian Graham. "Boundary behavior of the Carathéodory and Kobayashi metrics on strongly pseudoconvex domains in C^n with smooth boundary". In: *Trans. Amer. Math. Soc.* 207 (1975), pp. 219–240. ISSN: 0002-9947. DOI: 10.2307/1997175.

[85] Ian Graham. "Boundary behavior of the Carathéodory, Kobayashi, and Bergman metrics on strongly pseudoconvex domains in $\mathbf{C}^n$ with smooth boundary". In: *Bull. Amer. Math. Soc.* 79 (1973), pp. 749–751. ISSN: 0002-9904. DOI: 10.1090/S0002-9904-1973-13297-5.

[86] Jürgen Grahl and Shahar Nevo. "Spherical derivatives and normal families". In: *J. Anal. Math.* 117 (2012), pp. 119–128. ISSN: 0021-7670. DOI: 10.1007/s11854-012-0016-4.

[87] Hans Grauert and Helmut Reckziegel. "Hermitesche Metriken und normale Familien holomorpher Abbildungen". In: *Math. Z.* 89 (1965), pp. 108–125. ISSN: 0025-5874. DOI: 10.1007/BF01111588.

[88] Hans Grauert and Reinhold Remmert. *Coherent analytic sheaves.* Vol. 265. Grundlehren der mathematischen Wissenschaften [Fundamental Principles of Mathematical Sciences]. Springer-Verlag, Berlin, 1984, pp. xviii+249. ISBN: 3-540-13178-7. DOI: 10.1007/978-3-642-69582-7.

[89] Mark L. Green. "Holomorphic maps to complex tori". In: *Amer. J. Math.* 100.3 (1978), pp. 615–620. ISSN: 0002-9327. DOI: 10.2307/2373842.

[90] Mark L. Green. "The hyperbolicity of the complement of $2n+1$ hyperplanes in general position in P_n and related results". In: *Proc. Amer. Math. Soc.* 66.1 (1977), pp. 109–113. ISSN: 0002-9939. DOI: 10.2307/2041540.

[91] R. E. Greene and Steven G. Krantz. "Stability properties of the Bergman kernel and curvature properties of bounded domains". In: *Recent developments in several complex variables (Proc. Conf., Princeton Univ., Princeton, N. J., 1979).* Vol. 100. Ann. of Math. Stud. Princeton Univ. Press, Princeton, N.J., 1981, pp. 179–198.

[92] Robert E. Greene and Kang-Tae Kim. "The Riemann mapping theorem from Riemann's viewpoint". In: *Complex Anal. Synerg.* 3.1 (2017), Paper No. 1, 11. ISSN: 2524-7581. DOI: 10.1186/s40627-016-0009-7.

[93] Robert E. Greene, Kang-Tae Kim, and Steven G. Krantz. *The geometry of complex domains.* Vol. 291. Progress in Mathematics. Birkhäuser Boston, Ltd., Boston, MA, 2011, pp. xiv+303. ISBN: 978-0-8176-4139-9. DOI: 10.1007/978-0-8176-4622-6.

[94] Robert E. Greene and Steven G. Krantz. "Biholomorphic self-maps of domains". In: *Complex analysis, II (College Park, Md., 1985–86).* Vol. 1276. Lecture Notes in Math. Springer, Berlin, 1987, pp. 136–207. DOI: 10.1007/BFb0078959.

[95] Robert E. Greene and Steven G. Krantz. *Function theory of one complex variable.* Third. Vol. 40. Graduate Studies in Mathematics. American Mathematical Society, Providence, RI, 2006, pp. x+504. ISBN: 0-8218-3962-4. DOI: 10.1090/gsm/040.

[96] Lawrence Gruman and Christer O. Kiselman. "Le problème de Levi dans les espaces de Banach à base". In: *C. R. Acad. Sci. Paris Sér. A-B* 274 (1972), A1296–A1299. ISSN: 0151-0509.

[97] Robert C. Gunning. *Introduction to holomorphic functions of several variables. Vol. I.* The Wadsworth & Brooks/Cole Mathematics Series. Wadsworth & Brooks/Cole Advanced Books & Software, Pacific Grove, CA, 1990, pp. xx+203. ISBN: 0-534-13308-8.

[98] Robert C. Gunning and Hugo Rossi. *Analytic functions of several complex variables.* Prentice-Hall, Inc., Englewood Cliffs, NJ, 1965, pp. xiv+317.

[99] Kyong T. Hahn. "Asymptotic behavior of normal mappings of several complex variables". In: *Canad. J. Math.* 36.4 (1984), pp. 718–746. ISSN: 0008-414X. DOI: 10.4153/CJM-1984-041-9.

[100] Kyong T. Hahn. "Boundary behavior of normal and nonnormal holomorphic mappings". In: *Analysis and geometry 1987 (Taejŏn, 1987)*. Korea Inst. Tech., Taejŏn, 1987, pp. 123–147.

[101] Kyong T. Hahn. "Higher-dimensional generalizations of some classical theorems on normal meromorphic functions". In: *Complex Variables Theory Appl.* 6.2-4 (1986), pp. 109–121. ISSN: 0278-1077. DOI: 10.1080/17476938608814163.

[102] Kyong T. Hahn. "Holomorphic mappings of the hyperbolic space into the complex Euclidean space and the Bloch theorem". In: *Canadian J. Math.* 27 (1975), pp. 446–458. ISSN: 0008-414X. DOI: 10.4153/CJM-1975-053-0.

[103] Kyong T. Hahn. "Hyperbolicity of the complement of closed subsets in a compact Hermitian manifold". In: *Complex analysis and applications '87 (Varna, 1987)*. Publ. House Bulgar. Acad. Sci., Sofia, 1989, pp. 211–218.

[104] Kyong T. Hahn. "Inequality between the Bergman metric and Carathéodory differential metric". In: *Proc. Amer. Math. Soc.* 68.2 (1978), pp. 193–194. ISSN: 0002-9939. DOI: 10.2307/2041770.

[105] Kyong T. Hahn. "Nontangential limit theorems for normal mappings". In: *Pacific J. Math.* 135.1 (1988), pp. 57–64. ISSN: 0030-8730. URL: http://projecteuclid.org/euclid.pjm/1102688344.

[106] Kyong T. Hahn. "On completeness of the Bergman metric and its subordinate metric". In: *Proc. Nat. Acad. Sci. U.S.A.* 73.12 (1976), p. 4294. ISSN: 0027-8424. DOI: 10.1073/pnas.73.12.4294.

[107] Kyong T. Hahn. "Quantitative Bloch's theorem for certain classes of holomorphic mappings of the ball into $P_n(C)$". In: *J. Reine Angew. Math.* 283(284) (1976), pp. 99–109. ISSN: 0075-4102. DOI: 10.1515/crll.1976.283-284.99.

[108] Kyong T. Hahn and Kang T. Kim. "Hyperbolicity of a complex manifold and other equivalent properties". In: *Proc. Amer. Math. Soc.* 91.1 (1984), pp. 49–53. ISSN: 0002-9939. DOI: 10.2307/2045267.

[109] Lawrence A. Harris. "Fixed point theorems for infinite dimensional holomorphic functions". In: vol. 41. 1. 2004, pp. 175–192. DOI: 10.4134/JKMS.2004.41.1.175.

[110] Lawrence A. Harris. "Schwarz-Pick systems of pseudometrics for domains in normed linear spaces". In: *Advances in holomorphy (Proc. Sem. Univ. Fed. Rio de Janeiro, Rio de Janeiro, 1977)*. Vol. 34. North-Holland Math. Stud. North-Holland, Amsterdam-New York, 1979, pp. 345–406.

[111] W. K. Hayman. *Meromorphic functions.* Oxford Mathematical Monographs. Clarendon Press, Oxford, 1964, pp. xiv+191.

[112] W. K. Hayman. "Uniformly normal families". In: *Lectures on functions of a complex variable.* University of Michigan Press, Ann Arbor, Mich., 1955, pp. 199–212.

[113] Joachim A. Hempel. "Precise bounds in the theorems of Schottky and Picard". In: *J. London Math. Soc. (2)* 21.2 (1980), pp. 279–286. ISSN: 0024-6107. DOI: 10.1112/jlms/s2-21.2.279.

[114] G. M. Henkin. "An analytic polyhedron is not holomorphically equivalent to a strictly pseudoconvex domain". In: *Dokl. Akad. Nauk SSSR* 210 (1973), pp. 1026–1029. ISSN: 0002-3264.

[115] G. M. Henkin and E. M. Čirka. "Boundary properties of holomorphic functions of several complex variables". In: *Current problems in mathematics, Vol. 4 (Russian).* Akad. Nauk SSSR Vsesojuz. Inst. Naučn. i Tehn. Informacii, Moscow, 1975, 12–142. (errata insert).

[116] Einar Hille. *Analytic function theory. Vol. II.* Introductions to Higher Mathematics. Ginn and Company, Boston, Mass.-New York-Toronto, 1962, pp. xii+496.

[117] Einar Hille and Ralph S. Phillips. *Functional analysis and semi-groups.* American Mathematical Society Colloquium Publications, Vol. 31. American Mathematical Society, Providence, R.I., 1957, pp. xii+808.

[118] Jaakko Hyvönen and Juhani Riihentaus. "Removable singularities for holomorphic functions with locally finite Riesz mass". In: *J. London Math. Soc. (2)* 35.2 (1987), pp. 296–302. ISSN: 0024-6107. DOI: 10.1112/jlms/s2-35.2.296.

[119] Pentti Järvi. "An extension theorem for normal functions". In: *Proc. Amer. Math. Soc.* 103.4 (1988), pp. 1171–1174. ISSN: 0002-9939. DOI: 10.2307/2047105.

[120] J. A. Jenkins. "On explicit bounds in Schottky's theorem". In: *Canadian J. Math.* 7 (1955), pp. 76–82. ISSN: 0008-414X. DOI: 10.4153/CJM-1955-010-4.

[121] Robert Jentzsch. "Untersuchungen zur Theorie der Folgen analytischer Funktionen". In: *Acta Math.* 41.1 (1916), pp. 219–251. ISSN: 0001-5962. DOI: 10.1007/BF02422945.

[122] James E. Joseph and Myung H. Kwack. "Extension and convergence theorems for families of normal maps in several complex variables". In: *Proc. Amer. Math. Soc.* 125.6 (1997), pp. 1675–1684. ISSN: 0002-9939. DOI: 10.1090/S0002-9939-97-04117-8.

[123] James E. Joseph and Myung H. Kwack. "Hyperbolic imbedding and spaces of continuous extensions of holomorphic maps". In: *J. Geom. Anal.* 4.3 (1994), pp. 361–378. ISSN: 1050-6926. DOI: 10.1007/BF02921586.

[124] James E. Joseph and Myung H. Kwack. "Some classical theorems and families of normal maps in several complex variables". In: *Complex Variables Theory Appl.* 29.4 (1996), pp. 343–362. ISSN: 0278-1077. DOI: 10.1080/17476939608814902.

[125] Gaston Julia. "Sur les familles de fonctions analytiques de plusieurs variables". In: *Acta Math.* 47.1-2 (1926), pp. 53–115. ISSN: 0001-5962. DOI: 10.1007/BF02544108.

[126] John L. Kelley. *General topology.* D. Van Nostrand Co., Inc., Toronto-New York-London, 1955, pp. xiv+298.

[127] N. Kerzman. "Taut manifolds and domains of holomorphy in Cn". In: *Notices Am. Math. Soc.* 16 (1969), p. 675.

[128] Peter Kiernan. "Extensions of holomorphic maps". In: *Trans. Amer. Math. Soc.* 172 (1972), pp. 347–355. ISSN: 0002-9947. DOI: 10.2307/1996355.

[129] Peter Kiernan. "Hyperbolically imbedded spaces and the big Picard theorem". In: *Math. Ann.* 204 (1973), pp. 203–209. ISSN: 0025-5831. DOI: 10.1007/BF01351589.

[130] Kang-Tae Kim and Steven G. Krantz. "Characterization of the Hilbert ball by its automorphism group". In: *Trans. Amer. Math. Soc.* 354.7 (2002), pp. 2797–2818. ISSN: 0002-9947. DOI: 10.1090/S0002-9947-02-02895-7.

[131] Maciej Klimek. *Pluripotential theory.* Vol. 6. London Mathematical Society Monographs. New Series. The Clarendon Press, Oxford University Press, New York, 1991, pp. xiv+266. ISBN: 0-19-853568-6.

[132] Shoshichi Kobayashi. "Distance, holomorphic mappings and the Schwarz lemma". In: *J. Math. Soc. Japan* 19 (1967), pp. 481–485. ISSN: 0025-5645. DOI: 10.2969/jmsj/01940481.

[133] Shoshichi Kobayashi. "Geometry of bounded domains". In: *Trans. Amer. Math. Soc.* 92 (1959), pp. 267–290. ISSN: 0002-9947. DOI: 10.2307/1993156.

[134] Shoshichi Kobayashi. *Hyperbolic complex spaces.* Vol. 318. Grundlehren der mathematischen Wissenschaften [Fundamental Principles of Mathematical Sciences]. Springer-Verlag, Berlin, 1998, pp. xiv+471. ISBN: 3-540-63534-3. DOI: 10.1007/978-3-662-03582-5.

[135] Shoshichi Kobayashi. *Hyperbolic manifolds and holomorphic mappings.* Vol. 2. Pure and Applied Mathematics. Marcel Dekker, Inc., New York, 1970, pp. ix+148.

[136] Shoshichi Kobayashi. "Intrinsic metrics on complex manifolds". In: *Bull. Amer. Math. Soc.* 73 (1967), pp. 347–349. ISSN: 0002-9904. DOI: 10.1090/S0002-9904-1967-11745-2.

[137] Shoshichi Kobayashi. "Relative intrinsic distance and hyperbolic imbedding". In: 819. 1993, pp. 239–242.

[138] P. Koebe. "Über die Uniformisierung beliebiger analytischer Kurven. Dritte Mitteilung." German. In: *Nachr. Ges. Wiss. Göttingen, Math.-Phys. Kl.* 1908 (1908), pp. 337–358.

[139] A. Korányi. "Holomorphic and harmonic functions on bounded symmetric domains". In: *Geometry of Homogeneous Bounded Domains (C.I.M.E., 3 Ciclo, Urbino, 1967)*. Edizioni Cremonese, Rome, 1968, pp. 127–197.

[140] Adam Korányi. "Boundary behavior of Poisson integrals on symmetric spaces". In: *Trans. Amer. Math. Soc.* 140 (1969), pp. 393–409. ISSN: 0002-9947. DOI: 10.2307/1995145.

[141] Adam Korányi. "Harmonic functions on Hermitian hyperbolic space". In: *Trans. Amer. Math. Soc.* 135 (1969), pp. 507–516. ISSN: 0002-9947. DOI: 10.2307/1995029.

[142] Steven G. Krantz. *Complex analysis: the geometric viewpoint.* Second. Vol. 23. Carus Mathematical Monographs. Mathematical Association of America, Washington, DC, 2004, pp. xviii+219. ISBN: 0-88385-035-4. DOI: 10.5948/UPO9780883859681.

[143] Steven G. Krantz. *Function theory of several complex variables.* AMS Chelsea Publishing, Providence, RI, 2001, pp. xvi+564. ISBN: 0-8218-2724-3. DOI: 10.1090/chel/340.

[144] Steven G. Krantz. *Geometric analysis and function spaces.* Vol. 81. CBMS Regional Conference Series in Mathematics. Published for the Conference Board of the Mathematical Sciences, Washington, DC; by the American Mathematical Society, Providence, RI, 1993, pp. xii+202. ISBN: 0-8218-0734-X. DOI: 10.1090/cbms/081.

[145] Steven G. Krantz. *Geometric Analysis of the Bergman Kernel and Metric.* Graduate Texts in Mathematics, vol. 268. Birkhäuser Publishing, 2013, pp. xiv+292.

[146] Steven G. Krantz. *Real analysis and foundations.* Studies in Advanced Mathematics. CRC Press, Boca Raton, FL, 1991, pp. xiv+295. ISBN: 0-8493-7156-2.

[147] Steven G. Krantz. "The Lindelöf principle in several complex variables". In: *J. Math. Anal. Appl.* 326.2 (2007), pp. 1190–1198. ISSN: 0022-247X. DOI: 10.1016/j.jmaa.2006.03.059.

[148] Steven G. Krantz and Daowei Ma. "Bloch functions on strongly pseudoconvex domains". In: *Indiana Univ. Math. J.* 37.1 (1988), pp. 145–163. ISSN: 0022-2518. DOI: 10.1512/iumj.1988.37.37007.

[149] G. Kurepa. "Espaces pseudodistanciés." French. In: *C. R. Acad. Sci., Paris* 198 (1934), pp. 1563–1565. ISSN: 0001-4036.

[150] Myung H. Kwack. "Correction to: "Separately normal maps" [Complex Var. Elliptic Equ. **51** (2006), no. 1, 77–79; MR2201259]". In: *Complex Var. Elliptic Equ.* 52.8 (2007), pp. 755–756. ISSN: 1747-6933. DOI: 10.1080/17476930701303627.

[151] Myung H. Kwack. *Families of normal maps in several variables and classical theorems in complex analysis.* Vol. 33. Lecture Notes Series. Seoul National University, Research Institute of Mathematics, Global Analysis Research Center, Seoul, 1996, pp. iv+45.

[152] Myung H. Kwack. "Generalization of the big Picard theorem". In: *Ann. Math. (2)* 90 (1969), pp. 9–22. ISSN: 0003-486X. DOI: 10.2307/1970678.

[153] Myung H. Kwack. "Separately normal maps". In: *Complex Var. Elliptic Equ.* 51.1 (2006), pp. 77–79. ISSN: 1747-6933. DOI: 10.1080/02781070500369248.

[154] Wan Tzei Lai. "The exact value of Hayman's constant in Landau's theorem". In: *Sci. Sinica* 22.2 (1979), pp. 129–134. ISSN: 0582-236x.

[155] E. Landau. "Der Picard-Schottkysche Satz und die Blochsche Konstante." German. In: *Sitzungsber. Preuß. Akad. Wiss., Phys.-Math. Kl.* 1926 (1926), pp. 467–474.

[156] E. Landau. "On a generalization of Picard's theorem". German. In: *Berl. Ber.* 1904 (1904), pp. 1118–1133.

[157] Serge Lang. *Introduction to complex hyperbolic spaces.* Springer-Verlag, New York, 1987, pp. viii+271. ISBN: 0-387-96447-9. DOI: 10.1007/978-1-4757-1945-1.

[158] Serge Lang. *Introduction to differentiable manifolds.* Interscience Publishers (a division of John Wiley & Sons, Inc.), New York-London, 1962, pp. x+126.

[159] Serge Lang. *Real and functional analysis.* Third. Vol. 142. Graduate Texts in Mathematics. Springer-Verlag, New York, 1993, pp. xiv+580. ISBN: 0-387-94001-4. DOI: 10.1007/978-1-4612-0897-6.

[160] Peter Lappan. "A criterion for a meromorphic function to be normal". In: *Comment. Math. Helv.* 49 (1974), pp. 492–495. ISSN: 0010-2571. DOI: 10.1007/BF02566744.

[161] Peter Lappan. "A uniform approach to normal families". In: *Rev. Roumaine Math. Pures Appl.* 39.7 (1994), pp. 691–702. ISSN: 0035-3965.

[162] Peter Lappan. "Non-normal sums and products of unbounded normal functions". In: *Michigan Math. J.* 8 (1961), pp. 187–192. ISSN: 0026-2285. URL: http://projecteuclid.org/euclid.mmj/1028998570.

[163] Peter Lappan. "Sums of normal functions and Fatou points". In: *Michigan Math. J.* 10 (1963), pp. 221–224. ISSN: 0026-2285. URL: http://projecteuclid.org/euclid.mmj/1028998902.

[164] Olli Lehto and K. I. Virtanen. "Boundary behaviour and normal meromorphic functions". In: *Acta Math.* 97 (1957), pp. 47–65. ISSN: 0001-5962. DOI: 10.1007/BF02392392.

[165] Olli Lehto and K. I. Virtanen. "On the behaviour of meromorphic functions in the neighbourhood of an isolated singularity". In: *Ann. Acad. Sci. Fenn. Ser. A. I.* 1957.240 (1957), p. 9. ISSN: 0066-1953.

[166] Pierre Lelong. *Plurisubharmonic functions and positive differential forms. Translated by M.A.Dostal.* English. Notes on Mathematics and its Applications. New York-London-Paris: Gordon & Breach Science Publishers. X, 78 p. (1969). 1969.

[167] Pierre Lelong and Lawrence Gruman. *Entire functions of several complex variables.* Vol. 282. Grundlehren der mathematischen Wissenschaften [Fundamental Principles of Mathematical Sciences]. Springer-Verlag, Berlin, 1986, pp. xii+270. ISBN: 3-540-15296-2. DOI: 10.1007/978-3-642-70344-7.

[168] László Lempert. "Approximation de fonctions holomorphes d'un nombre infini de variables". In: *Ann. Inst. Fourier (Grenoble)* 49.4 (1999), pp. 1293–1304. ISSN: 0373-0956. URL: http://www.numdam.org/item?id=AIF_1999__49_4_1293_0.

[169] László Lempert. "Approximation of holomorphic functions of infinitely many variables. II". In: *Ann. Inst. Fourier (Grenoble)* 50.2 (2000), pp. 423–442. ISSN: 0373-0956. URL: http://www.numdam.org/item?id=AIF_2000__50_2_423_0.

[170] László Lempert. "La métrique de Kobayashi et la représentation des domaines sur la boule". In: *Bull. Soc. Math. France* 109.4 (1981), pp. 427–474. ISSN: 0037-9484. URL: http://www.numdam.org/item?id=BSMF_1981__109__427_0.

[171] László Lempert. "The Cauchy-Riemann equations in infinite dimensions". In: *Journées "Équations aux Dérivées Partielles" (Saint-Jean-de-Monts, 1998).* Univ. Nantes, Nantes, 1998, Exp. No. VIII, 8.

[172] László Lempert. "The Dolbeault complex in infinite dimensions. I". In: *J. Amer. Math. Soc.* 11.3 (1998), pp. 485–520. ISSN: 0894-0347. DOI: 10.1090/S0894-0347-98-00266-5.

[173] László Lempert. "The Dolbeault complex in infinite dimensions. II". In: *J. Amer. Math. Soc.* 12.3 (1999), pp. 775–793. ISSN: 0894-0347. DOI: 10.1090/S0894-0347-99-00296-9.

[174] E. Lindelöf. "Démonstration nouvelle d'un théorème fondamental sur les suites de fonctions monogènes". In: *Bull. Soc. Math. France* 41 (1913), pp. 171–178. ISSN: 0037-9484. URL: http://www.numdam.org/item?id=BSMF_1913__41__171_0.

[175] E. Lindelöf. *Sur un principe général de l'analyse et ses applications à la théorie de la représentation conforme.* French. Acta Soc. Fennicae 46, Nr. 4, 35 S. (1915). 1915.

[176] A. J. Lohwater and Ch. Pommerenke. "On normal meromorphic functions". In: *Ann. Acad. Sci. Fenn. Ser. A. I.* 550 (1973), p. 12. ISSN: 0066-1953. URL: https://www.acadsci.fi/mathematica/1973/no550pp01-12.pdf.

[177] K. H. Look. "Schwarz lemma in the theory of functions of several complex variables". In: *Sci. Record (N.S.)* 1.2 (1957), pp. 5–8. ISSN: 0559-1244.

[178] K. H. Look. "Schwarz lemma in the theory of functions of several complex variables". In: *Acta Math. Sinica* 7 (1957), pp. 370–420. ISSN: 0583-1431.

[179] Lisa Lorentzen. "General convergence in quasi-normal families". In: *Proc. Edinb. Math. Soc. (2)* 46.1 (2003), pp. 169–183. ISSN: 0013-0915. DOI: 10.1017/S001309150100027X.

[180] A. Lovater. "The boundary behavior of analytic functions". In: *Mathematical analysis, Vol. 10 (Russian).* 1973, 99–259 (errata insert). URL: https://www.mathnet.ru/php/archive.phtml?wshow=paper&jrnid=intm&paperid=24&option_lang=rus.

[181] G. R. MacLane. "Asymptotic values of holomorphic functions". In: *Rice Univ. Stud.* 49.1 (1963), p. 83. ISSN: 0035-4996.

[182] G. R. MacLane. "Holomorphic functions, of arbitrarily slow growth, without radial limits". In: *Michigan Math. J.* 9 (1962), pp. 21–24. ISSN: 0026-2285. URL: http://projecteuclid.org/euclid.mmj/1028998615.

[183] G. R. MacLane. "Meromorphic functions with small characteristic and no asymptotic values". In: *Michigan Math. J.* 8 (1961), pp. 177–185. ISSN: 0026-2285. URL: http://projecteuclid.org/euclid.mmj/1028998569.

[184] S. Mandelbrojt. "Sur les suites de fonctions holomorphes. Les suites correspondantes des fonctions dérivées. Fonctions entières." French. In: *J. Math. Pures Appl. (9)* 8 (1929), pp. 173–195. ISSN: 0021-7824. URL: http://portail.mathdoc.fr/JMPA/afficher_notice.php?id=JMPA_1929_9_8_A10_0.

[185] Donald E. Marshall. *Complex analysis.* Cambridge Mathematical Textbooks. Cambridge University Press, Cambridge, 2019, pp. xvi+272. ISBN: 978-1-107-13482-9.

[186] Gaven J. Martin and Volker Mayer. "Rigidity in holomorphic and quasiregular dynamics". In: *Trans. Amer. Math. Soc.* 355.11 (2003), pp. 4349–4363. ISSN: 0002-9947. DOI: 10.1090/S0002-9947-03-03160-X.

[187] F. Marty. “Recherches sur la répartition des valeurs d’une fonction méromorphe”. In: *Ann. Fac. Sci. Toulouse Sci. Math. Sci. Phys. (3)* 23 (1931), pp. 183–261. ISSN: 0996-0481. URL: http://www.numdam.org/item?id=AFST_1931_3_23__183_0.

[188] Pertti Mattila. *Geometry of sets and measures in Euclidean spaces.* Vol. 44. Cambridge Studies in Advanced Mathematics. Cambridge University Press, Cambridge, 1995, pp. xii+343. ISBN: 0-521-46576-1; 0-521-65595-1. DOI: 10.1017/CBO9780511623813.

[189] David Minda. “A heuristic principle for a nonessential isolated singularity”. In: *Proc. Amer. Math. Soc.* 93.3 (1985), pp. 443–447. ISSN: 0002-9939. DOI: 10.2307/2045610.

[190] David Minda and Glenn Schober. “Another elementary approach to the theorems of Landau, Montel, Picard and Schottky”. In: *Complex Variables Theory Appl.* 2.2 (1983), pp. 157–164. ISSN: 0278-1077. DOI: 10.1080/17476938308814039.

[191] Ruth Miniowitz. “Normal families of quasimeromorphic mappings”. In: *Proc. Amer. Math. Soc.* 84.1 (1982), pp. 35–43. ISSN: 0002-9939. DOI: 10.2307/2043804.

[192] B. S. Mitjagin and G. M. Henkin. “Linear problems of complex analysis”. In: *Uspehi Mat. Nauk* 26.4 (160) (1971), pp. 93–152. ISSN: 0042-1316.

[193] P. Montel. *Leçons sur les familles normales de fonctions analytiques et leurs applications. Recueillies et rédigées par J. Barbotte.* French. Paris: Gauthier-Villars. viii, 306 p. (1927). 1927.

[194] P. Montel. “On the sequences of analytic functions.” French. In: *C. R. Acad. Sci., Paris* 138 (1904), pp. 469–471. ISSN: 0001-4036.

[195] P. Montel. “Sur la représentation conforme.” French. In: *Journ. de Math. (7)* 3 (1917), pp. 1–54. URL: http://www.numdam.org/item/JMPA_1917_7_3__1_0/.

[196] P. Montel. “Sur les familles quasi-normales de fonctions analytiques”. In: *Bull. Soc. Math. France* 52 (1924), pp. 85–114. ISSN: 0037-9484. URL: http://www.numdam.org/item?id=BSMF_1924__52__85_1.

[197] P. Montel. “Sur les fonctions analytiques qui admettent deux valéurs exceptionnelles dans un domaine.” French. In: *C. R. Acad. Sci., Paris* 153 (1911), pp. 996–998. ISSN: 0001-4036.

[198] P. Montel. “Sur les suites infinies de fonctions.” French. In: *Ann. Sci. Éc. Norm. Supér. (3)* 24 (1907), pp. 233–334. ISSN: 0012-9593. DOI: 10.24033/asens.580.

[199] Paul Montel. *Leçons sur les récurrences et leurs applications.* Gauthier-Villars, Paris, 1957, pp. xi+268.

[200] Paul Montel. "Sur les familles de fonctions analytiques qui admettent des valeurs exceptionnelles dans un domaine". In: *Ann. Sci. École Norm. Sup. (3)* 29 (1912), pp. 487–535. ISSN: 0012-9593. URL: `http://www.numdam.org/item?id=ASENS_1912_3_29__487_0`.

[201] Paul Montel. "Sur les familles normales de fonctions analytiques". In: *Ann. Sci. École Norm. Sup. (3)* 33 (1916), pp. 223–302. ISSN: 0012-9593. URL: `http://www.numdam.org/item?id=ASENS_1916_3_33__223_0`.

[202] Jorge Mujica. *Complex analysis in Banach spaces*. Vol. 120. North-Holland Mathematics Studies. North-Holland Publishing Co., Amsterdam, 1986, pp. xii+434. ISBN: 0-444-87886-6.

[203] Jorge Mujica and Leopoldo Nachbin. "Linearization of holomorphic mappings on locally convex spaces". In: *J. Math. Pures Appl. (9)* 71.6 (1992), pp. 543–560. ISSN: 0021-7824.

[204] L. Nachbin. "Sur quelques aspects récents de l'holomorphie en dimension infinie". In: *Séminaire Goulaouic-Schwartz 1971–1972*. École Polytech., Centre de Math., Paris, 1972, Exp. No. 18, 9.

[205] Leopoldo Nachbin. "A glimpse at infinite dimensional holomorphy". In: *Proceedings on Infinite Dimensional Holomorphy (Internat. Conf., Univ. Kentucky, Lexington, Ky., 1973)*. Lecture Notes in Math., Vol. 364. Springer, Berlin, 1974, pp. 69–79.

[206] Leopoldo Nachbin. "Analogies entre l'holomorphie et la linéarité". In: *Séminaire Paul Krée, 4e année: 1977–1978. Équations aux dérivées partielles en dimension infinie*. Secrétariat Math., Paris, 1979, Exp. No. 1, 10.

[207] Leopoldo Nachbin. *Concerning holomorphy types for Banach spaces*. Memórias de Matemática da Universidade Federal do Rio de Janeiro, No. 3. [Mathematical Reports of the Federal University of Rio de Janeiro, No. 3]. Universidade Federal do Rio de Janeiro, Instituto de Matemática, Rio de Janeiro; Universidade Federal do Rio de Janeiro, Coordenação dos Programas de Pós-Graduação em Engenharia, Rio de Janeiro, 1971, pp. i+11.

[208] Leopoldo Nachbin. *Holomorphic functions, domains of holomorphy and local properties*. Vol. 1. North-Holland Mathematics Studies. North-Holland Publishing Co., Amsterdam-London; American Elsevier Publishing Co., Inc., New York, 1970, pp. vii+122.

[209] Leopoldo Nachbin. "Limites et perturbation des applications holomorphes". In: *Fonctions analytiques de plusieurs variables et analyse complexe (Colloq. Internat. CNRS, No. 208, Paris, 1972)*. "Agora Mathematica", No. 1. Gauthier-Villars, Paris, 1974, pp. 141–158.

[210] Leopoldo Nachbin. "On pure uniform holomorphy in spaces of holomorphic germs". In: *Results Math.* 8.2 (1985), pp. 117–122. ISSN: 0378-6218.

[211] Leopoldo Nachbin. “On vector-valued versus scalar-valued holomorphic continuation”. In: *Nederl. Akad. Wetensch. Proc. Ser. A 76=Indag. Math.* 35 (1973), pp. 352–354.

[212] Leopoldo Nachbin. “Recent developments in infinite dimensional holomorphy”. In: *Bull. Amer. Math. Soc.* 79 (1973), pp. 625–640. ISSN: 0002-9904. DOI: 10.1090/S0002-9904-1973-13244-6.

[213] Leopoldo Nachbin. “Some aspects and problems in holomorphy”. In: *Extracta Math.* 1.2 (1986), pp. 57–72. ISSN: 0213-8743.

[214] Leopoldo Nachbin. “Some holomorphically significant properties of locally convex spaces”. In: *Functional analysis (Proc. Brazilian Math. Soc. Sympos., Inst. Mat. Univ. Estad. Campinas, São Paulo, 1974).* Vol. 18. Lecture Notes in Pure and Appl. Math. Dekker, New York, 1976, pp. 251–277.

[215] Leopoldo Nachbin. “Some problems in the application of functional analysis to holomorphy”. In: *Advances in holomorphy (Proc. Sem. Univ. Fed. Rio de Janeiro, Rio de Janeiro, 1977).* Vol. 34. North-Holland Math. Stud. North-Holland, Amsterdam-New York, 1979, pp. 577–583.

[216] Leopoldo Nachbin. “Warum unendlichdimensionale Holomorphie?” In: *Jahrb. Überblicke Math, 1979.* Bibliographisches Inst., Mannheim, 1979, pp. 9–20.

[217] Leopoldo Nachbin. “Why holomorphy in infinite dimensions?” In: *Enseign. Math. (2)* 26.3-4 (1980), 257–269 (1981). ISSN: 0013-8584.

[218] Raghavan Narasimhan. *Several complex variables.* Chicago Lectures in Mathematics. The University of Chicago Press, Chicago, Ill.-London, 1971, pp. x+174.

[219] Rolf Nevanlinna. “On differentiable mappings”. In: *Analytic functions.* Princeton Univ. Press, Princeton, N.J., 1960, pp. 3–9.

[220] Toshio Nishino. *Function theory in several complex variables.* Vol. 193. Translations of Mathematical Monographs. American Mathematical Society, Providence, RI, 2001, pp. xiv+366. ISBN: 0-8218-0816-8. DOI: 10.1090/mmono/193.

[221] Toshio Nishino. “Sur une propriété des familles de fonctions analytiques de deux variables complexes”. In: *J. Math. Kyoto Univ.* 4 (1965), pp. 255–282. ISSN: 0023-608X. DOI: 10.1215/kjm/1250524660.

[222] Junjiro Noguchi. “Hyperbolic fibre spaces and Mordell’s conjecture over function fields”. In: *Publ. Res. Inst. Math. Sci.* 21.1 (1985), pp. 27–46. ISSN: 0034-5318. DOI: 10.2977/prims/1195179839.

[223] Junjiro Noguchi. “Moduli spaces of holomorphic mappings into hyperbolically imbedded complex spaces and locally symmetric spaces”. In: *Invent. Math.* 93.1 (1988), pp. 15–34. ISSN: 0020-9910. DOI: 10.1007/BF01393686.

[224] Junjiro Noguchi and Takushiro Ochiai. *Geometric function theory in several complex variables.* Vol. 80. Translations of Mathematical Monographs. American Mathematical Society, Providence, RI, 1990, pp. xii+283. ISBN: 0-8218-4533-0. DOI: 10.1090/mmono/080.

[225] Kiyoshi Noshiro. *Cluster sets.* Ergebnisse der Mathematik und ihrer Grenzgebiete, (N.F.), Heft 28. Springer-Verlag, Berlin-Göttingen-Heidelberg, 1960, pp. vii+135.

[226] Kiyoshi Noshiro. "Contributions to the theory of meromorphic functions in the unit-circle". English. In: *J. Fac. Sci., Hokkaido Univ., Ser. I* 7 (1939), pp. 149–159. ISSN: 0018-3482. DOI: DOI:10.14492/hokmj/1531123220.

[227] Takeo Ohsawa. *Analysis of several complex variables.* Vol. 211. Translations of Mathematical Monographs. American Mathematical Society, Providence, RI, 2002, pp. xviii+121. ISBN: 0-8218-2098-2. DOI: 10.1090/mmono/211.

[228] Kiyoshi Oka. "Sur les fonctions analytiques de plusieurs variables. X. Une mode nouvelle engendrant les domaines pseudo-convexes". In: *Jpn. J. Math.* 32 (1962), pp. 1–12. ISSN: 0075-3432. DOI: 10.4099/jjm1924.32.0_1.

[229] Kiyosi Oka. "Note sur les familles de fonctions analytiques multiformes etc". French. In: *J. Sci. Hiroshima Univ., Ser. A* 4 (1934), pp. 93–98. ISSN: 0386-3034. DOI: 10.32917/hmj/1558749763.

[230] W. F. Osgood. "Note on the functions defined by infinite series whose terms are analytic functions of a complex variable; with corresponding theorems for definite integrals". In: *Ann. of Math. (2)* 3.1-4 (1902), pp. 25–34. ISSN: 0003-486X. DOI: 10.2307/1967630.

[231] Xue Cheng Pang. "Bloch's principle and normal criterion". In: *Sci. China Ser. A* 32.7 (1989), pp. 782–791. ISSN: 1001-6511.

[232] E. Picard. "On single valued analytic functions in the neighbourhood of an essential singularity". French. In: *C. R. Acad. Sci., Paris* 88 (1879), pp. 745–747. ISSN: 0001-4036.

[233] S. I. Pincuk. "On proper holomorphic mappings of strictly pseudoconvex domains". English. In: *Sib. Math. J.* 15 (1975), pp. 644–649. ISSN: 0037-4466. DOI: 10.1007/BF00967440.

[234] S. I. Pinčuk. "Holomorphic inequivalence of certain classes of domains in C^n". In: *Mat. Sb. (N.S.)* 111(153).1 (1980), pp. 67–94, 159. ISSN: 0368-8666.

[235] S. I. Pinčuk. "Holomorphic mappings in C^n and the problem of holomorphic equivalence". In: *Current problems in mathematics. Fundamental directions, Vol. 9 (Russian).* Itogi Nauki i Tekhniki. Akad. Nauk SSSR, Vsesoyuz. Inst. Nauchn. i Tekhn. Inform., Moscow, 1986, pp. 195–223, 292.

[236] I. I. Pjateckii-Shapiro. "On a problem proposed by E. Cartan". In: *Dokl. Akad. Nauk SSSR* 124 (1959), pp. 272–273. ISSN: 0002-3264.

[237] H. Poincaré. "Les fonctions analytiques de deux variables et la représentation conforme." French. In: *Rend. Circ. Mat. Palermo* 23 (1907), pp. 185–220. ISSN: 0009-725X. DOI: 10.1007/BF03013518.

[238] E. A. Poletskii and B. V. Shabat. "Invariant metrics". In: *Current problems in mathematics. Fundamental directions, Vol. 9 (Russian).* Itogi Nauki i Tekhniki. Akad. Nauk SSSR, Vsesoyuz. Inst. Nauchn. i Tekhn. Inform., Moscow, 1986, pp. 73–125, 292.

[239] M. B. Porter. "Concerning series of analytic functions". In: *Ann. of Math. (2)* 6.4 (1905), pp. 190–192. ISSN: 0003-486X. DOI: 10.2307/2007247.

[240] M. B. Porter. "On functions defined by an infinite series of analytic functions of a complex variable". In: *Ann. of Math. (2)* 6.1 (1904), pp. 45–48. ISSN: 0003-486X. DOI: 10.2307/2007157.

[241] Wade C. Ramey. "Boundary behavior of bounded holomorphic functions along maximally complex submanifolds". In: *Amer. J. Math.* 106.4 (1984), pp. 975–999. ISSN: 0002-9327. DOI: 10.2307/2374329.

[242] R. Michael Range. "The Carathéodory metric and holomorphic maps on a class of weakly pseudoconvex domains". In: *Pacific J. Math.* 78.1 (1978), pp. 173–189. ISSN: 0030-8730. URL: http://projecteuclid.org/euclid.pjm/1102806309.

[243] Murali Rao et al. *Complex analysis.* Second. World Scientific Publishing Co. Pte. Ltd., Hackensack, NJ, 2015, pp. x+414. ISBN: 978-981-4579-59-9. DOI: 10.1142/9062.

[244] Hans-Jörg Reiffen. "Die Carathéodorysche Distanz und ihre zugehörige Differentialmetrik". In: *Math. Ann.* 161 (1965), pp. 315–324. ISSN: 0025-5831. DOI: 10.1007/BF01359970.

[245] Reinhold Remmert. *Theory of complex functions.* Vol. 122. Graduate Texts in Mathematics. Springer-Verlag, New York, 1991, pp. xx+453. ISBN: 0-387-97195-5. DOI: 10.1007/978-1-4612-0939-3.

[246] Juhani Riihentaus. "A nullset for normal functions in several variables". In: *Proc. Amer. Math. Soc.* 110.4 (1990), pp. 923–933. ISSN: 0002-9939. DOI: 10.2307/2047738.

[247] Juhani Riihentaus. "An extension theorem for meromorphic functions of several variables". In: *Ann. Acad. Sci. Fenn. Ser. A I Math.* 4.1 (1979), pp. 145–149. ISSN: 0066-1953. DOI: 10.5186/aasfm.1978-79.0422.

[248] Juhani Riihentaus. "Removable singularities for Bloch and normal functions". In: *Czechoslovak Math. J.* 43(118).4 (1993), pp. 723–741. ISSN: 0011-4642. DOI: 10.21136/CMJ.1993.128430.

[249] Antonio Ros. "The Gauss map of minimal surfaces". In: *Differential geometry, Valencia, 2001.* World Sci. Publ., River Edge, NJ, 2002, pp. 235–252.

[250] Jean-Pierre Rosay. "Sur une caractérisation de la boule parmi les domaines de $\mathbf{C}^n$ par son groupe d'automorphismes". In: *Ann. Inst. Fourier (Grenoble)* 29.4 (1979), pp. ix, 91–97. ISSN: 0373-0956. URL: http://www.numdam.org/item?id=AIF_1979__29_4_91_0.

[251] Wolfgang Rothstein and Klaus Kopfermann. *Funktionentheorie meherer komplexer Veränderlicher.* Bibliographisches Institut, Mannheim, 1982, p. 256. ISBN: 3-411-01623-X.

[252] H. L. Royden. "A criterion for the normality of a family of meromorphic functions". In: *Ann. Acad. Sci. Fenn. Ser. A I Math.* 10 (1985), pp. 499–500. ISSN: 0066-1953.

[253] H. L. Royden. "Hyperbolicity in complex analysis". In: *Ann. Acad. Sci. Fenn. Ser. A I Math.* 13.3 (1988), pp. 387–400. ISSN: 0066-1953. DOI: 10.5186/aasfm.1988.1329.

[254] H. L. Royden. "Remarks on the Kobayashi metric". In: *Several complex variables, II (Proc. Internat. Conf., Univ. Maryland, College Park, Md., 1970).* 1971, 125–137. Lecture Notes in Math., Vol. 185.

[255] H. L. Royden. "The extension of regular holomorphic maps". In: *Proc. Amer. Math. Soc.* 43 (1974), pp. 306–310. ISSN: 0002-9939. DOI: 10.2307/2038884.

[256] Lee A. Rubel. "Four counterexamples to Bloch's principle". In: *Proc. Amer. Math. Soc.* 98.2 (1986), pp. 257–260. ISSN: 0002-9939. DOI: 10.2307/2045694.

[257] Walter Rudin. *Function theory in the unit ball of* $\mathbb{C}^n$. Classics in Mathematics. Springer-Verlag, Berlin, 2008, pp. xiv+436. ISBN: 978-3-540-68272-1.

[258] Walter Rudin. *Principles of mathematical analysis.* Third. International Series in Pure and Applied Mathematics. McGraw-Hill Book Co., New York-Auckland-Düsseldorf, 1976, pp. x+342.

[259] Walter Rudin. *Real and complex analysis.* Third. McGraw-Hill Book Co., New York, 1987, pp. xiv+416. ISBN: 0-07-054234-1.

[260] Joel L. Schiff. *Normal families.* Universitext. Springer-Verlag, New York, 1993, xii+236. ISBN: 0-387-97967-0. DOI: 10.1007/978-1-4612-0907-2.

[261] Joel L. Schiff. *Topics in complex analysis.* Vol. 88. De Gruyter Studies in Mathematics. De Gruyter, Berlin, [2022] ©2022, pp. xvi+274. ISBN: 978-3-11-075769-9; 978-3-11-075782-8; 978-3-11-075787-3. DOI: 10.1515/9783110757828.

[262] F. Schottky. "On Picard's theorem and Borel's inequality". German. In: *Berl. Ber.* 1904 (1904), pp. 1244–1263.

[263] Johann Schröder. "Das Iterationsverfahren bei allgemeinerem Abstandsbegriff". In: *Math. Z.* 66 (1956), pp. 111–116. ISSN: 0025-5874. DOI: 10.1007/BF01186599.

[264] Wilhelm Schwick. "Normality criteria for families of meromorphic functions". In: *J. Analyse Math.* 52 (1989), pp. 241–289. ISSN: 0021-7670. DOI: 10.1007/BF02820480.

[265] Wilhelm Schwick. "On a normality criterion of H. L. Royden". In: *New Zealand J. Math.* 23.1 (1994), pp. 91–92. ISSN: 1171-6096.

[266] W. Seidel. "Holomorphic functions with spiral asymptotic paths". In: *Nagoya Math. J.* 14 (1959), pp. 159–171. ISSN: 0027-7630. URL: http://projecteuclid.org/euclid.nmj/1118800146.

[267] Wladimir Seidel. "On the cluster values of analytic functions". In: *Trans. Amer. Math. Soc.* 34.1 (1932), pp. 1–21. ISSN: 0002-9947. DOI: 10.2307/1989501.

[268] Bernard Shiffman. "On the removal of singularities of analytic sets". In: *Michigan Math. J.* 15 (1968), pp. 111–120. ISSN: 0026-2285. URL: http://projecteuclid.org/euclid.mmj/1028999912.

[269] Barry Simon. *Advanced complex analysis.* A Comprehensive Course in Analysis, Part 2B. American Mathematical Society, Providence, RI, 2015, pp. xvi+321. ISBN: 978-1-4704-1101-5. DOI: 10.1090/simon/002.2.

[270] E. M. Stein. *Boundary behavior of holomorphic functions of several complex variables.* Mathematical Notes, No. 11. Princeton University Press, Princeton, N.J.; University of Tokyo Press, Tokyo, 1972, pp. x+72.

[271] Elias M. Stein. *Singular integrals and differentiability properties of functions.* Princeton Mathematical Series, No. 30. Princeton University Press, Princeton, N.J., 1970, pp. xiv+290.

[272] Norbert Steinmetz. *Nevanlinna theory, normal families, and algebraic differential equations.* Universitext. Springer, Cham, 2017, pp. xviii+235. ISBN: 978-3-319-59799-7. DOI: 10.1007/978-3-319-59800-0.

[273] T.-J. Stieltjes. "Recherches sur les fractions continues". In: *Ann. Fac. Sci. Toulouse Sci. Math. Sci. Phys.* 8.4 (1894), J1–J122. ISSN: 0996-0481. URL: http://www.numdam.org/item?id=AFST_1894_1_8_4_J1_0.

[274] Paula Takatsuka. "Normal families of holomorphic functions on infinite dimensional spaces". In: *Port. Math. (N.S.)* 63.3 (2006), pp. 351–362. ISSN: 0032-5155.

[275] Do Duc Thai, Pham Nguyen Thu Trang, and Pham Dinh Huong. "Families of normal maps in several complex variables and hyperbolicity of complex spaces". In: *Complex Var. Theory Appl.* 48.6 (2003), pp. 469–482. ISSN: 0278-1077. DOI: 10.1080/0278107031000094963.

[276] Richard M. Timoney. "Bloch functions in several complex variables. I". In: *Bull. London Math. Soc.* 12.4 (1980), pp. 241–267. ISSN: 0024-6093. DOI: 10.1112/blms/12.4.241.

[277] Richard M. Timoney. "Bloch functions in several complex variables. II". In: *J. Reine Angew. Math.* 319 (1980), pp. 1–22. ISSN: 0075-4102. DOI: 10.1515/crll.1980.319.1.

[278] E. C. Titchmarsh. *The theory of functions*. Second. Oxford University Press, Oxford, 1939, pp. x+454.

[279] Jussi Väisälä. "On normal quasiconformal functions". In: *Ann. Acad. Sci. Fenn. Ser. A. I.* 266 (1959), p. 33. ISSN: 0066-1953.

[280] G. Valiron. "Sur les théorèmes de M. M. Bloch, Landau, Montel et Schottky." French. In: *C. R. Acad. Sci., Paris* 183 (1926), pp. 728–730. ISSN: 0001-4036.

[281] Sergio Venturini. "Pseudodistances and pseudometrics on real and complex manifolds". In: *Ann. Mat. Pura Appl. (4)* 154 (1989), pp. 385–402. ISSN: 0003-4622. DOI: 10.1007/BF01790358.

[282] G. Vitali. "On the series of analytic functions." Italian. In: *Annali di Mat. (3)* 10 (1904), pp. 65–82. ISSN: 0373-3114. DOI: 10.1007/BF02419376.

[283] G. Vitali. "On the series of analytic functions." Italian. In: *Torino Atti* 39 (1904), pp. 22–32.

[284] Chikara Watanabe. "On normality of a family of holomorphic functions". In: *Publ. Res. Inst. Math. Sci.* 9 (1973/74), pp. 631–648. ISSN: 0034-5318. DOI: 10.2977/prims/1195192446.

[285] Hassler Whitney. "Tangents to an analytic variety". In: *Ann. of Math. (2)* 81 (1965), pp. 496–549. ISSN: 0003-486X. DOI: 10.2307/1970400.

[286] B. Wong. "Characterization of the unit ball in $\mathbf{C}^n$ by its automorphism group". In: *Invent. Math.* 41.3 (1977), pp. 253–257. ISSN: 0020-9910. DOI: 10.1007/BF01403050.

[287] H. Wu. "Normal families of holomorphic mappings". In: *Acta Math.* 119 (1967), pp. 193–233. ISSN: 0001-5962. DOI: 10.1007/BF02392083.

[288] Kosaku Yosida. "On a class of meromorphic functions". English. In: *Proc. Phys.-Math. Soc. Japan, III. Ser.* 16 (1934), pp. 227–235. ISSN: 0370-1239. DOI: https://doi.org/10.11429/ppmsj1919.16.0_227.

[289] M. G. Zaidenberg. "Picard's theorems and hyperbolicity". In: *Sibirsk. Mat. Zh.* 24.6 (1983), pp. 44–55. ISSN: 0037-4474.

[290] M. G. Zaidenberg. "Schottky-Landau growth estimates for s-normal families of holomorphic mappings". In: *Math. Ann.* 293.1 (1992), pp. 123–141. ISSN: 0025-5831. DOI: 10.1007/BF01444708.

[291] Lawrence Zalcman. "A heuristic principle in complex function theory". In: *Amer. Math. Monthly* 82.8 (1975), pp. 813–817. ISSN: 0002-9890. DOI: 10.2307/2319796.

[292] Lawrence Zalcman. "Normal families: new perspectives". In: *Bull. Amer. Math. Soc. (N.S.)* 35.3 (1998), pp. 215–230. ISSN: 0273-0979. DOI: 10.1090/S0273-0979-98-00755-1.

Alphabetical Index